…UDIES IN THE HISTORY OF SCIENCE AND TECHNOLOGY

James Rodger Fleming (Colby College) and Roger D. Launius (National Air and Space Museum), Series Editors

This series presents original, high-quality, and accessible works at the cutting edge of scholarship within the history of science and technology. Books in the series aim to disseminate new knowledge and new perspectives about the history of science and technology, enhance and extend education, foster public understanding, and enrich cultural life. Collectively, these books will break down conventional lines of demarcation by incorporating historical perspectives into issues of current and ongoing concern, offering international and global perspectives on a variety of issues, and bridging the gap between historians and practicing scientists. In this way they not only advance scholarly conversation within and across traditional disciplines but also help define new areas of intellectual endeavor.

Published by Palgrave Macmillan:

Continental Defense in the Eisenhower Era: Nuclear Antiaircraft Arms and the Cold War
By Christopher J. Bright

Confronting the Climate: British Airs and the Making of Environmental Medicine
By Vladimir Janković

Globalizing Polar Science: Reconsidering the International Polar and Geophysical Years
Edited by Roger D. Launius, James Rodger Fleming, and David H. DeVorkin

Eugenics and the Nature-Nurture Debate in the Twentieth Century
By Aaron Gillette

John F. Kennedy and the Race to the Moon
By John M. Logsdon

A Vision of Modern Science: John Tyndall and the Role of the Scientist in Victorian Culture
By Ursula DeYoung

Searching for Sasquatch: Crackpots, Eggheads, and Cryptozoology
By Brian Regal

Inventing the American Astronaut
By Matthew H. Hersch

The Nuclear Age in Popular Media: A Transnational History
Edited by Dick van Lente

Exploring the Solar System: The History and Science of Planetary Exploration
Edited by Roger D. Launius

The Sociable Sciences: Darwin and His Contemporaries in Chile
By Patience A. Schell

The First Atomic Age: Scientists, Radiations, and the American Public, 1895–1945
By Matthew Lavine

NASA in the World: Fifty Years of International Collaboration in Space
By John Krige, Angelina Long Callahan, and Ashok Maharaj

Empire and Science in the Making: Dutch Colonial Scholarship in Comparative Global Perspective, 1760–1830
Edited by Peter Boomgaard

Anglo-American Connections in Japanese Chemistry: The Lab as Contact Zone
By Yoshiyuki Kikuchi

Eismitte in the Scientific Imagination: Knowledge and Politics at the Center of Greenland
By Janet Martin-Nielsen

Empire and Science in the Making

Dutch Colonial Scholarship in Comparative Global Perspective, 1760–1830

Edited by
Peter Boomgaard

palgrave
macmillan

EMPIRE AND SCIENCE IN THE MAKING

First published in 2013 by
PALGRAVE MACMILLAN®
in the United States—a division of St. Martin's Press LLC,
175 Fifth Avenue, New York, NY 10010.

Where this book is distributed in the UK, Europe and the rest of the world, this is by Palgrave Macmillan, a division of Macmillan Publishers Limited, registered in England, company number 785998, of Houndmills, Basingstoke, Hampshire RG21 6XS.

Palgrave Macmillan is the global academic imprint of the above companies and has companies and representatives throughout the world.

ISBN: 978–1–137–33401–5

Library of Congress Cataloging-in-Publication Data

Empire and science in the making : Dutch colonial scholarship in comparative global perspective, 1760–1830 / edited by Peter Boomgaard.
pages cm.—(Palgrave studies in the history of science and technology)
Includes bibliographical references.
ISBN 978–1–137–33401–5 (hardback : alkaline paper)
1. Science—Netherlands—Colonies—History—18th century. 2. Science—Netherlands—Colonies—History—19th century. 3. Learning and scholarship—Netherlands—Colonies—History—18th century. 4. Learning and scholarship—Netherlands—Colonies—History—19th century. 5. Imperialism—Social aspects—History—18th century. 6. Imperialism—Social aspects—History—19th century. 7. Netherlands—Colonies—Intellectual life. 8. Great Britain—Colonies—Intellectual life. 9. Spain—Colonies—Intellectual life. 10. Russia—Colonies—Intellectual life. I. Boomgaard, P., 1946–

Q127.N2E56 2013
509.171′249209033—dc23 2013035038

A catalogue record of the book is available from the British Library.

This book is printed on paper suitable for recycling and made from fully managed and sustained forest sources. Logging, pulping and manufacturing processes are expected to conform to the environmental regulations of the country of origin.

Design by Newgen Knowledge Works (P) Ltd., Chennai, India.

First edition: October 2013

10 9 8 7 6 5 4 3 2 1

This book is dedicated to my wife, Raquel Reyes, and my children, Pim, Gaïa, and Cosmo

Contents

Contributors

David Arnold is emeritus professor of History at the University of Warwick, UK, having previously taught at the School of Oriental and African Studies, London. A founder member of the Subaltern Studies group, he has written on the history of disease and medicine in colonial India, on crime and policing, environmental history, and the history of science. His published work include *Colonizing the Body: State Medicine and Epidemic Disease in Nineteenth-Century India* (1993), *Gandhi* (2001), and *Everyday Technology: Machines and the Making of India's Modernity*' (2013).

Klaas van Berkel is Rudolf Agricola professor of History at the University of Groningen, the Netherlands. He has published books on the history of science and the universities in the Netherlands and its colonies from 1500 to the present, including a two-volume history of the Royal Netherlands Academy of Arts and Sciences (2008–2011). His most recent book is *Isaac Beeckman on Matter and Motion: Mechanical Philosophy in the Making* (2013).

Marieke Bloembergen (PhD University of Amsterdam, 2001) is a cultural historian specialized in Dutch and Indonesian colonial and postcolonial history. She works as senior researcher at the Royal Netherlands Institute of Southeast Asian and Caribbean Studies in Leiden (KITLV). As a postdoctoral researcher at Utrecht University, she completed a monograph on policing and perceptions of security in the Dutch East Indies (published in 2009, Indonesian translation in 2011). Currently she is finishing, together with Martijn Eickhoff, a research project on archaeology and heritage formation in colonial and postcolonial Indonesia, in the context of the NWO-funded research project "Sites, Bodies and Stories."

Peter Boomgaard is senior researcher at KITLV, Leiden, professor (emeritus) of History, University of Amsterdam, and, during 2012/13, fellow at the Rachel Carson Center, Munich. Among his sole-authored books are *Frontiers of Fear: Tigers and People in the Malay World, 1600–1950* (2001), and *Southeast Asia: An Environmental History*

(2007). He has (co)edited many volumes, including *Globalization, Environmental Change, and Social History* (2010), and "Long-Term Changes in Land-Tenure Arrangements in Pre-Modern and Early-Modern Southeast Asia" (*Journal of the Economic and Social History of the Orient*, Vol. 54, 2011).

Martijn Eickhoff (PhD University of Amsterdam, 2003) is a cultural historian who specializes in the historical culture of dictatorial and colonial regimes, with a specific interest in archaeological practices. He works as senior researcher at NIOD—Institute for War, Holocaust, and Genocide Studies in Amsterdam and is lecturer at the Radboud University Nijmegen, the Netherlands. In the context of the NWO-funded "Sites, Bodies and Stories" research project and in collaboration with Marieke Bloembergen, he is currently conducting a project on colonial and postcolonial archaeology and heritage formation in Indonesia. Martijn Eickhoff has also published widely on scholarly research in times of war and regime change.

Siegfried Huigen is associate professor of Dutch Literature and Cultural History at the University of Stellenbosch, in South Africa, and university professor at the Uniwersytet Wrocławski, in Poland. His research interests are early modern travel writing and the history of colonial science and scholarship. He is the author of *De weg naar Monomotapa* (1996) and *Knowledge and Colonialism; Eighteenth-century Travellers in South Africa* (2007 and 2009), and (co)editor of several books on South African politics of memory and the history of colonial knowledge.

Gerry van Klinken is senior researcher at the KITLV and professor of Southeast Asian history at the University of Amsterdam. His recent sole-authored books include *The Making of Middle Indonesia* (in press) and *Communal violence and Democratization in Indonesia* (2007). He now coordinates the multidisciplinary research project "Elite Network Shifts" (computational network analysis on electronic news archives), and researches the history of citizenship practices in Indonesia. He also has a master's degree in geophysics, and has taught physics at universities in Malaysia and Indonesia for 10 years.

Michael Laffan (PhD Sydney University, 2001) is professor of History at Princeton University, where he teaches on Southeast Asian Islam and Indian Ocean connections. He is the author of *Islamic Nationhood and Colonial Indonesia* (2003) and, more recently, the *Makings of Indonesian Islam* (2011). His current project traces the movements and legacies of exiles ranging from Cairo and Colombo to Batavia (now Jakarta) and Cape Town.

Gert Oostindie is director of the KITLV/Royal Netherlands Institute of Southeast Asian and Caribbean Studies in Leiden and professor of Caribbean History at Leiden University. His principal areas of research have been the Caribbean and Dutch (post)colonial history in a comparative perspective. He has (co)published and edited over 25 books, including *Dutch Atlantic Connections, 1680–1815* (2014), *Curaçao in the Age of Revolutions, 1795–1800* (2011), *Postcolonial Netherlands. Sixty-five years of Forgetting, Commemorating, Silencing* (2011), and *Paradise Overseas: The Dutch Caribbean; Colonialism and Its Transatlantic Legacies* (2005).

Raquel A. G. Reyes is a research fellow in History at SOAS (London). She is the author of *Love, Passion and Patriotism: Sexuality and the Philippine Propaganda Movement, 1882–1892* (2008) and coeditor of *Sexual Diversity in Asia, c.600–1950* (2012). She writes on the history of science and medicine in Southeast Asia and is currently working on a project that traces historical interactions between science, religion, and sexual knowledge.

Peter Rietbergen (PhD Nijmegen, 1983) is professor of postmedieval cultural history at the Radboud University, Nijmegen, the Netherlands. Among his publications are *De eerste landvoogd: Pieter Both* (1987), *Japan verwoord: Nihon door Nederlandse ogen* (2003), and *Europa's India, ca 1800–2000* (2007). His *Geschiedenis van Nederland in Vogelvlucht* (1993–2013) has gone through six revised editions, as has its English version, *A Short History of the Netherlands* (1993–2013). His *Europe: A Cultural History* (1998; rev. edition 2006) was translated into Chinese, Korean and Polish. As Nicolaas Berg he wrote a historical novel about the Dutch EIC and Japan, *Dood op Deshima* (2001).

Acknowledgments

The history of this book started in 2008, the bicentennial of the Royal Netherlands Academy of Arts and Sciences (KNAW). They let it be known there was money available for workshops with participants from afar. During, before, and after the workshop, we were very ably assisted by one of the KNAW's personnel, Martine Wagenaar.

Based on the information received, we thought that another branch of the KNAW would take care of the publication too, but when I sent in the revised versions of the papers, we got the message, after a delay of half a year, that the KNAW had scrapped the series because of cutbacks.

Luckily, Palgrave Macmillan was willing to publish the book at short notice. I dealt mainly with Chris Chappell and Sarah Whalen, who helped me with great patience to turn a manuscript into a book.

In the last stage of getting the book together, before it could go into production, I was a fellow at the Rachel Carson Center for Environment and Society, in Munich, and the center, in the person of Maeve Storey was willing to help me with the correction of the notes and bibliography.

Introduction

From the Mundane to the Sublime: Science, Empire, and the Enlightenment (1760s–1820s)

Peter Boomgaard

Among historians of science, the eighteenth century has often been neglected in favor of the two adjacent epochs that cast such giant shadows—the seventeenth-century (First) Scientific Revolution, associated with names such as Descartes, Locke, and Newton, and the nineteenth-century Industrial Revolution.[1] Regarding the era in between, it has been said that "until recently, general histories of science have tended toward an impoverished estimation of this period."[2]

In contrast, among philosophers and historians of ideas, the "long" eighteenth century, called the Age of Reason or the Enlightenment, was always quite popular.[3] Recently, the period has been "discovered" by historians of science, a development that has contributed to the pleasure of writing this chapter, the more so as a complete volte-face appears to have been made, as witness the following quote, in which the eighteenth century is regarded as "the period that, if any, deserves to be called the 'Age of Science.'"[4]

This volume deals mainly with the tail end of this age—the period roughly between the 1760s and the 1820s. It has been dubbed the Second Scientific Revolution, or that of Romantic science, and recently, the Age of Wonder.[5] It was a period of intense rivalry between the imperial powers Great Britain, France, Spain, and the Netherlands, including large-scale wars in Europe and elsewhere.

It was also a time of voyages of discovery—over land or across the oceans—after a long period of near absence of such undertakings. The period from the 1760s to the 1820s has been called the second great age of European exploration, after the first one that lasted from the fifteenth to the mid-seventeenth century. The second age started with Cook and Bougainville and ended with Darwin.[6] It appears, therefore, that the First Scientific Revolution and the First Great Age of Discoveries coincided more or less, as did the second one of both.

It is, of course, also the period of the French Revolution, linked to the Enlightenment in various respects, although it spelled the end of the French Academy of Sciences, and cost great scientists such as Lavoisier and Condorcet their head.[7]

This collection does not pretend to give an overview of the science and empire nexus during the period under consideration.[8] The book deals mainly with overseas colonies or trading posts of the Dutch; for comparative reasons, chapters are included on areas forming part of the British and the Spanish Empires. These were arguably the two largest Empires with overseas, often tropical or subtropical possessions. The British Empire, which has inspired a huge historiography regarding scholarly activities during this period, was on its way up, while the Spanish Empire, the colonial scholarship of which has recently attracted more attention from (English-speaking) historians, was doing poorly. The emphasis, however, will be on the Dutch Empire, as publications in English on Dutch colonial scholarship during this period are rare.

Creation and Circulation of Knowledge

What was known about the European overseas settlements and colonies around these years? What was known, in the Netherlands and in other parts of the Western world, about climate, geology, geography, botany, zoology, and diseases of tropical and subtropical regions (and regions in the temperate zone as well) of America, Africa, and Asia? What was known about their languages, history, religion, arts, and people?

What was the relationship between European and indigenous scholars in these settlements, and between science in the metropole and indigenous science? What was the relationship between the Dutch scholars and their English, French, and Spanish competitors, particularly in periods of war?

What was done with all this information back home? Were there attempts to keep it secret, or did scholars attempt to integrate it as soon as possible in the relevant sciences and humanities in general, in order to arrive at new theories? Were there attempts, predating those of Alexander von Humboldt, to draw up research programs, in which scholars were sent out to the overseas areas with long and detailed questionnaires?

Were the scholars in the overseas settlements sufficiently aware of the state of play of their respective disciplines back home? Did governments play a role in all this? These are some of the questions that have been looked at, not only regarding the Dutch, but also for the Spanish and the British Empires, while one chapter deals exclusively with indigenous Javanese scholarship.

This type of research can be thought of as having been inspired by the actor-network "theory" (ANT) to which the name of Bruno Latour is connected, a theory that maps relations between people, things, and concepts, together shaping a single—transient—network. It appears likely that such networks in the field of "colonial" science existed during the period under consideration, examples of which are presented in this volume.[9]

Having formulated these questions, a closer look at the learned world of the eighteenth century is in order, the background to the stories told in this volume.

Observation and Experiments

During the eighteenth century, scientists increasingly distanced themselves from the authority of Classical Antiquity and Scholasticism. It was now regarded as unimportant how many teeth Aristotle had said a horse had—one just went outside, looked in the horse's mouth and counted. Direct, straightforward empirical observation was more and more acknowledged as the basis of all science.[10]

Direct observation was strongly connected to counting, weighing, and measuring, which brings us to the enthusiasm of Enlightenment scholars for mathematics, statistics, and quantifiability in general.[11] This has led some historians of science to argue that during the Enlightenment, the process of quantification was the key to scientific change. Exact sciences without mathematics became unthinkable, and even in the emerging social sciences, counting and measuring became far from unusual. The Englishman Francis Bacon (1561–1626) is often cited as the origin of the strong mathematical streak of the Enlightenment.

In order to quantify observations, tools and auxiliary instruments were often indispensable. One needs measuring scales, large balances, thermometers, clocks, watches, chronometers, barometers, air pumps, model steam engines, windmills, static electrical equipment, etcetera. Even more basic, before one can count or measure one has to be able to see the objects one wants to study properly, hence the need for microscope and telescope. Much of this was to be found in a dedicated space, the laboratory, a term, at least in England, dating from the seventeenth century. In general, therefore, the eighteenth century was also a period of invention/innovation and improvement of scientific instruments.

As instruments became more fine-tuned, the wish for precise measurements became stronger. At the same time, the need for standardization of units of measurements came to be felt increasingly, although much of this did not become a reality prior to the nineteenth or even the twentieth century.[12] It is an interesting feature of the history of Southeast Asia, that some standardization, for instance in units of weight, had already taken place prior to 1800, but that these units were Chinese, not European. Both standardization and the production of precision instruments would play an important role in trade and in the industrialization process.

Useful, Practical, Linked to Technology, and in the Service of God

There was a general inclination during the Age of Enlightenment to emphasize the usefulness of knowledge and science, even of social sciences and the humanities. Science was useful in two ways, according to most eighteenth-century scholars: it could be used for the common good, improving the human condition, and it was a way to worship God, and to learn his law, which was also the law of nature. Being interested in science was, therefore, a morally sanctioned, prestigious activity with a high status, and, as we will see presently, "doing" science—going to public lectures, watching experiments and dissections of human or animal corpses, being a member of a debating club or a learned society—was very popular with the upper and middle classes, and not only with those whom we can regard as professional scholars.[13]

Enlightenment science, therefore, was often mundane, pragmatic, practical, and very much "applied" science, dealing as it did with shipbuilding, maps, agriculture, manure, water quality, and health. It was also practical in the sense that, as we have seen, there was a strong emphasis on the invention, adoption, and improvement of instruments, of which increasingly the highest precision was expected.

Here, in other words, science shades imperceptibly into technology, which brings us to the relationship between the two, a hotly debated issue. Was the American economist W.W. Rostow right in stating that the self-sustained economic growth in Europe was a product of the marriage between science and technology in the eighteenth century? Or did that marriage not take place until much later? And is the absence of such a fusion between science and technology outside eighteenth-century Europe the reason that, according to many historians, the West did so much better than the rest during the eighteenth century and thereafter? Whatever the answer to those questions, eighteenth-century European "savants" were certainly very much interested in technology, as witness that famous monument of the Enlightenment in France, the *Encyclopédie* (1751–1772). In describing artisanal production processes in great detail, it finally carried out a wish that Francis Bacon had voiced over a century earlier.[14]

Studying nature—reading "the Book of Nature," as contemporary scientists used to say—was for most eighteenth-century European scholars, who wanted to understand God's creation, as important as, or perhaps even more important than, reading that other book, the Holy Scripture.[15] Atheism or Materialism was rare among Enlightenment scholars in most countries, with the possible exception of France, where the Enlightenment was characterized by a strong anticlerical streak. Generally speaking there was not much of a contradiction between Christianity, particularly Protestantism, and most aspects of eighteenth-century science. Our image of the relationship between science and religion in Europe prior to the nineteenth century is probably too much influenced by Catholic dealings with the ideas of Copernicus and Galileo on the one hand, and the writings of Voltaire and La Mettrie on the other. Most European scholars regarded themselves as good Christians and were proud of it. It has even been argued that Protestantism, with its emphasis on reading (the Bible), stimulated print culture, and therefore science.[16]

Collecting Naturalia and Artificialia

To observe natural phenomena directly, the objects in question had to be collected, ordered, identified, classified, compared, and displayed.[17] Europe, during the early modern period, knew several kinds of collections. Characteristic for this era was the curiosity cabinet, which typically brought together (parts of dead) plants and animals, minerals, fossils, and artifacts, including ethnographic objects and "scientific" instruments, such as telescopes and globes, while old coins and other antiquities were also popular.[18] If we look at plants alone, we find *herbaria*, that is, collections of dried plants pasted on paper and bound together in books, and herbals or illustrated catalogues of plants arranged in some kind of order; they, too, were often to be found in curiosity cabinets. And finally there were the botanical gardens, often connected to a university.

With Francis Bacon's notion in mind that knowledge itself was a virtue (while he also held that knowledge was power!), many European "*virtuosi*" from the Renaissance onward collected natural and artificial objects for their cabinet of curiosities or *Wunderkammer*, also called *Kunstkammer*.[19] No longer limited to kings and aristocrats from the sixteenth century onward, cabinets of curiosities were henceforth also in the possession of middle-class professionals like medical men and pharmacists, who collected mainly, but by no means exclusively, plants ("simples") for medicinal purposes.

In the sixteenth century, bringing natural and artificial objects together in a curiosity cabinet was not necessarily done with scientific purposes in mind. The collectors were often non-naturalists—lawyers, merchants, and clergymen—who collected out of curiosity and because large, well-ordered collections brought fame and prestige, and a stream of well-placed and famous, including royal, visitors. Aesthetic considerations played a role as well, and many a painter, like Rembrandt, had a small cabinet of curiosities at his disposal. Finally, a cabinet of curiosities was probably a good investment, and entire cabinets, or parts of it, were bought and sold all the time. Thus, curiosity cabinets fitted right in among the middle and upper strata of a society characterized by Mercantile Capitalism.

Collecting for Science

However, by the late seventeenth and early eighteenth centuries, cabinets put together by professionals—professors of natural history, physicians, and pharmacists—and for scientific reasons became the norm; collecting out of curiosity alone was increasingly criticized.[20] The collections became larger, as rich collectors bought up the possessions of the less well-to-do, and during the eighteenth century some of these cabinets would form the basis of large public museum collections. This happened for instance to that of the British physician and collector Sir Hans Sloane (1660–1753), a typical representative of the Enlightenment, participant in a large network of scholars that included Newton, Locke, and Linné [Linnaeus]; around Sloane's collection the British Museum was founded, opening its doors to the public in 1759.

Something similar happened in the case of the Amsterdam apothecary (pharmacist) Albertus Seba (1665–1736), one of the most important collectors of the eighteenth century, owner of a famous collection of *naturalia*, to whom Tsar Peter the Great of Russia paid a visit in 1717, buying the collection for a very large sum of money. Seba immediately started a new collection, which was the one inspected by Linnaeus, who visited him in 1735. The first collection of Seba, together with that of another Dutchman, Frederik Ruysch, became the basis of the *Kunstkammer* of the Russian Academy of Sciences.[21]

European herbal and physic (or botanical) gardens date from the mid-sixteenth century, the earliest ones being found in Italy. They were usually linked to universities or royal courts. The *Hortus Botanicus* of Leiden, founded in 1587, is regarded as the earliest scientific garden in Europe. Its first director (*praefectus horti*), Carolus Clusius (1526–1609), a seasoned traveler and experienced networker, was well acquainted with the newest literature (Da Orta, Acosta, Monardes) about recently discovered and described "exotic" (medicinal) plants from Asia and the Americas, and planted seeds he received from across the globe, being possibly the first one to grow tulips in the Netherlands.[22]

The equally famous botanical garden of Amsterdam was founded later, in 1638. As the Dutch Republic during the seventeenth century was the hub of international long-distance trade, seeds, bulbs, and cuttings from the Americas, Africa, and Asia arrived here in increasing numbers and found their way to the Leiden and Amsterdam botanical gardens.[23] So many "new" plants were found in the areas outside Europe, that the idea that the botanical gardens should contain specimens of all species had to be given up during the eighteenth century.

At the end of the eighteenth century, when it was no longer Amsterdam but London where much of the international trade flow was directed, the Royal Botanic Gardens at Kew (now London) became the most important "imperial" botanical garden.[24] The rich landowner and amateur botanist Sir Joseph Banks (1743–1820) oversaw Kew from 1772 until his death. Sloane and Banks, who stand at either end of the Age of Enlightenment, would both become president of the Royal Society—on which more presently—and were both naturalists. The British Museum was, in a sense, the creation of Sloane, Kew Gardens that of Banks—from dead plants to live ones. How much the interest in botany had grown between 1500 and 1800 can be illustrated with the following quotation:

> By the end of the eighteenth century, Europe possessed some sixteen hundred botanical gardens connecting scientific enterprise,

> plant acclimatization, plant transfers, and experimentation around the world.[25]

The astonishing growth in size and/or numbers of cabinets of curiosity, *herbaria,* herbals, and botanical gardens was largely due to European—mainly, but not exclusively, overseas—expansion. Around 1500, at the start of this expansion, Europe was familiar with the proto-herbals of Antiquity—the plant lists of Theophrastus and Dioscorides, who knew 500 and 600 species, respectively. Shortly after 1600, when many specimens had been brought back to Europe from the Americas and from Asia, Gaspar Bauhin classified and described 6,000 plants in 12 volumes. By 1700 the number of species known to European natural historians had increased even more—John Ray listed 17,000 species in his *Historia Plantarum Generalis* (1686–1704). That was the situation when Linnaeus, who later claimed to have added 8,000 new items to the corpus, came to the Netherlands, introducing his own classification system, a system that was able to accommodate the continuous flood of new species from outside Europe, which no longer fitted in the old one. Soon, he would send out his pupils across the globe to gather, describe, and name more specimens.[26]

Partly influenced by the activities and writings of Linnaeus, collecting specimens from nature became almost a craze or fad during the period discussed here.[27] As Mary Louise Pratt, who coined the phrases "contact zone" and "imperial gaze," has it:

> Travel and travel writing would never be the same again. In the second half of the eighteenth century, whether or not an expedition was primarily scientific, or the traveller a scientist, natural history played a part in it. Specimen gathering, the building up of collections, the naming of new species, the recognition of known ones, became standard themes in travel and travel books.[28]

Publication of natural history works accelerated during the eighteenth century, particularly from the 1770s to the 1790s. Illustrating books on natural history had become a specialized art in itself, involving illustrious artists such as Maria Sibylla Merian, who made beautiful colored drawings (watercolors) of Suriname's nature.[29]

Writing about the heyday of systematic collecting and classifying of specimens by Europeans in non-European areas from ca. 1750, Pratt remarked upon the link between, or at least the coincidence of, European expansion and its scientific features with (relatively) new forms of social discipline inside and outside Europe:

> Academic scholarship on the Enlightenment, resolutely Eurocentered, has often neglected Europe's aggressive colonial and imperial ventures as models, inspirations, and testing grounds for modes of social discipline which, imported back into Europe in the eighteenth century, were adapted to construct the bourgeois order. The systematization of nature coincides with the height of the slave trade, the plantation system, colonial genocide in North America and South Africa."[30]

Here, therefore, as we saw in the case of Foucault, the darker side of the Empire-Science-Enlightenment nexus rears its ugly head.

Science and Empire

Until the 1960s and even 70s, when the large majority of former colonies had become independent or was on the road to independence, most books and essays on the history of science, technology, and empire would in principle argue that the introduction of Western science and technology in much of America, Africa, and Asia had been one of the beneficial effects of imperialism/colonialism. Colonialism, in other words, had made possible the penetration of Western science, and so had been its handmaiden.[31]

However, from the 1970s onward, the politico-scholarly mood changed, and many studies were published that argued that Western science and technology, in turn, had also been the handmaidens of Empire, and in the 1970s, 80s, and 90s that was not meant as a compliment. In fact, it was now often argued, science served colonial expansion and empire building, and had made these processes possible; colonial scientific enterprise was characterized by political and economic motivations (in addition to being motivated by scholarly curiosity).[32]

It was also argued that Western science did not start from scratch in the regions that had become colonies, because there had been locally developed strands of science and technology that could be built upon, the presence of which led to hybridization.[33] It was also argued, particularly regarding India and China, that elements of what later would be regarded as Western science and technology, had their origins in Asia anyway!

As science and technology often had developed independently in those regions that would become colonies of European states or trading companies, the old "diffusionist" model (of which Basalla was the classical exponent)—science and technology spreading from "the West" to the colonies, which were passive recipients—became increasingly untenable. In fact, it was now argued by various scholars that scientific knowledge reaching the West from Asia, the Americas, and Africa, changed Western science considerably, as the new knowledge had to be incorporated in the "old" Western knowledge systems, which was impossible without major change.[34]

While not so long ago (Western) science was regarded as a monolithic entity, recently the social character and cultural plurality of science has been emphasized, and hence its local character.[35]

Finally, it has also been argued recently—not surprisingly, in view of the developments just mentioned—that the Enlightenment should not be seen as Europe's gift to the world:

> [It] cannot be understood as the sovereign and autonomous accomplishment of European intellectuals alone.... Enlightenment ideas need to be understood as a response to cross-border interaction and global integration.[36]

In much of this literature, the question that was hardly asked was: How much did Westerners know about local, indigenous science and technology in the Americas, Africa, and Asia? The main exception to this rule was Edward Said, who, however, in his study entitled *Orientalism* (1978), argued that much of Western science regarding "the Orient" had got it all wrong.[37]

During the last two decades, interest in what was known in Europe about the areas that had been or were about to be colonized increased considerably, particularly in the field of "natural history."[38] This trend has been further developed in the present volume.

The Growth of Empires and the Increasing Flow of Information

At the end of the fifteenth century, three things occurred that would change European science beyond recognition—the arrival of the Portuguese in India by ship, the "discovery" of America, and the invention and spread of movable type printing in Europe.[39] While around 1500 most Europeans hardly had any knowledge of what happened more than a day's journey away, by 1800, knowledge in Europe of the non-European world was, at least among the upper and middle classes, fairly widespread. Ever since the sixteenth century, a stream of—often illustrated—publications about the "new" lands had flooded the European markets, which were reprinted and reedited constantly. Of course that knowledge was often superficial, inaccurate, and stereotyped, but there was a growing awareness of and interest in the existence of other societies, cultures, religions, histories, and *naturalia* in faraway places. Moreover, the quality of the information usually improved considerably between 1500 and 1800, as European visitors and scholars learned American, African, and Asian languages, and as maps were improved continuously, partly because indigenous geographic information was gradually included and travelers increasingly entered uncharted territories, partly because better techniques became available and were applied.[40]

The effects of this vastly increased and qualitatively improved knowledge on European scholars and society at large were mixed.[41] Many people from all walks of life were interested in the exotic "other," in mysterious regions and people, in strange morals and customs. Although this appears to indicate that people were mostly interested in those aspects in which non-Europeans differed from Europeans, many writers were of the opinion that all people in principle had the same potentialities. The differences that could be observed were, in this view, largely caused by environmental factors, like differences in climate.

However, confronted with more information on "primitive" societies, for instance, in the Pacific and Africa, many Europeans at the same time regarded the accomplishments of their own societies as far superior to anything America, Africa, or Asia had to offer. It became customary to think of civilizations as stages, of which Europe with its advanced science, technology, and economy represented the highest one, and Africans and other "tribal" peoples the lowest stage.[42] According to historian Christopher Bayly, these notions of inequality increased in the British Empire during this period:

> The spirit which animated British relations with subject races had changed quite sharply. Whatever the promptings of Enlightenment philosophy, the English had always considered Asians and Africans inferior, whether because of climate, despotic government or ignorance of Christian virtue. But these attitudes had rarely been codified or rigid and they had had only limited effect on the political hierarchies of colonial territories which perforce had included indigenous rulers and merchants. Between 1780 and 1820, however, Asians, Eurasians, Africans and even non-British and non-Protestant Europeans were widely excluded from positions of authority in government, while steps were taken to decontaminate the springs of British executive power from the influence of "native corruption."[43]

There were lively discussions about the relationship between humans and apes, of which the various species were not clearly distinguished until late in the eighteenth century.[44] Skin color was also often discussed, with special interest in albino Africans and albino Papuans. Some Enlightenment scholars thought that humans were an undivided species, and saw skin color as a feature shaped by environment and climate, but others thought more in terms of "race," although what we now call racist categories are more a feature of the (late) nineteenth century.

On the other hand, people were also impressed by several non-European—particularly Asian—ancient civilizations, and a number of European (mainly French) thinkers argued that Europeans could learn from Persian, Indian, and, perhaps particularly, Chinese history.[45]

As knowledge about groups of people outside Europe increased, the inclination to regard certain inequalities as unacceptable increased as well. The existence of slavery and the slave trade were hotly debated in many European countries—if all people were in principle equal, no one should be the slave of someone else. If all people were equal, a situation in which one nation lorded it over another one could not be accepted either, and we see how a number of Enlightenment intellectuals, such as Alexander von Humboldt (1769–1859) turned against colonialism. It was also increasingly seen as unacceptable that so many people outside Europe were non-Christians, and therefore, in the eyes of many Europeans, doomed. At the end of the eighteenth century, therefore, missionary activity, directed by newly established missionary societies, was intensifying.[46]

Finally, the second half of the eighteenth century also set the stage for the Romantic Era, which was characterized by enthusiasm and admiration for "the sublime" (mountains, tropical nature). Here, again, Humboldt can be called upon to testify, overawed as he was by the grandiose nature of Central and South America. Another Romantic notion, this time in the field of anthropology, was the view of (for instance) the—recently "discovered"—Pacific islanders as "noble savages," much closer to nature and therefore happier than European intellectuals.[47]

The effect of all this new information on individual sciences was, of course, also considerable and quite varied. In fact, around 1800, we see the birth of the modern way of classifying disciplines, and terms such as natural philosophy, experimental philosophy, natural history, moral sciences, "antiquarianism," etcetera, were on their way out, making room for terms such as botany, zoology, ethnography, geography, archaeology, and statistics.[48] Bayly sums up the culmination of knowledge—much of which was linked to "Empire"—as follows:

> [T]he massive accumulation and categorization of data about natural phenomena, had begun in the early modern period and speeded up in the eighteenth century. Observations of the heavens, the mapping of geological strata, the beginnings of archaeology in southern Italy and the Egyptian desert, the listing of the variety of species as a consequence of Pacific expeditions...: all these developments reflected the inquisitive burrowings of Enlightenment scholars and travelers.... Goethe and Alexander von Humboldt...stood at the center of a huge web of taxonomical speculation, creating analytical tools for subjects as widely separated today as geology, botany, and perceptual psychology. In the field of linguistic analysis, the East India Company judge William Jones and the later German Sanskrit scholar Franz Bopp were similar great synthesizers.... The Pacific was drawn into the context of Asia by linguists and early anthropologists.

Bayly concludes with the following line:

> Governments had come to believe that they had a role in discovering new facts and felt that they would gain both honor and resources from doing so.[49]

Voyages of Discovery

Governments had, indeed, played a role regarding the funding or subsidizing of the voyages of discovery that were typical for this period, which has been called the second great age of European exploration.[50] Names associated with it are those of Byron, Wallis and Carteret, Cook, Banks, Bougainville, La Pérouse, Malaspina, Vancouver, d'Entrecasteaux, Flinders, and, finally, Darwin.[51]

This aspect of Science and Empire during this period is the most well known among nonspecialist scholars, particularly as regards the figures of Cook and Darwin, whose adventures and accomplishments have been publicized widely. But even specialists are fascinated by the sheer bulk of data that these expeditions produced, the beautiful publications, and the many treasures from these journeys that came to form the basis of European museum collections. Alan Frost even states that "these expeditions form one of the bases of our present science."[52]

These voyages of discovery have two features in common—they were undertaken in the framework of empire, and they had explicit scientific goals. Undertaking such voyages had become easier, as better ships had become available, better food during the journey, and new instruments to measure longitude at sea. Voyages in the Pacific have received most attention from specialists and the general public alike. The Romantic appeal of the first European contacts with island "paradises" such as Tahiti and Hawai'i surely has something to do with this.

Much less known, it seems, are the voyages, over land, undertaken on the initiative of the Russian Academy of Sciences, St. Petersburg, in the largely uncharted territories of European and Asian Russia, during the eighteenth century.[53] Most of these voyages were led and organized by German scholars, but in the interest of and financed by the Russian state, whose imperial expansion they served. Thus the Germans, who *had* no Empire of their own—but, in a formal sense, *were* an Empire themselves, though in reality more a loosely structured federation—did take part in imperial scholarly endeavors, something that most historians (of science and empire) do not appear to be aware of.

All these expeditions produced masses of data, published in multivolume, beautifully illustrated books, which were then often translated into other languages. Preliminary results were often published in one or more of the many learned journals that by the late eighteenth century were produced by the various—mainly, but not exclusively—European academies and learned societies, to which we now turn.

The Learned or Scientific Societies and Academies

Scientific societies and academies were another feature of the Scientific Revolution(s) and the Enlightenment.[54] As was the case with the botanical gardens, their origins can be traced back to Italy (e.g., Accademia dei Lincei, Rome, of which Galileo was a member), in the early seventeenth century. As time went by, their character became more official, and they increasingly performed public, civic functions. The best-known representatives of these "transformed" societies are the Royal Society of London for the Promotion of Natural Knowledge, created by royal charter in 1662, and the Académie Royale des Sciences, in Paris, founded in 1666.[55] Here, scholars—both those who held university positions and those who did not—would meet, discuss new findings, conduct experiments, and, of course, socialize. The eighteenth-century learned societies and academies incorporated a large number of disciplines, including some rather practical and technical ones (cartography, naval sciences), in addition to the arts. The more well known of these scientific societies—often called academies—were chartered, financed, or at least subsidized by the crown, their members were given remunerations ("pensions"), and money was available for instruments and equipment for experiments to be conducted in their laboratories. In return, such academies often carried out certain tasks for the state. On occasion, they sponsored or at least subsidized voyages of discovery. A very important function of the societies was that they published learned journals, such as the *Philosophical Transactions* of the Royal Society and the *Mémoires* of the Académie Royale. They also had libraries, sometimes quite extensive ones. Scientific societies and academies occasionally had links to observatories and botanical gardens.

The Dutch Republic had its fair share of learned societies, but not being a centralized monarchy, it did not have a central organization such as the one in Paris and London, the Royal Prussian Academy in Berlin (1700)[56] or the Academia Scientiarum Imperialis in St. Petersburg (1724), and the ones in Stockholm (1739) and Copenhagen (1742).[57]

The learned societies in the Republic were founded by the burghers of a province or a city. Many such societies were established in the seventeenth and eighteenth centuries, often short-lived, small affairs, very specialized or very broad. To give just one (early) example of such a learned society, the Collegium Mechanicum could be mentioned, founded in Rotterdam in 1620 by the well-known physicist, engineer, and meteorologist Isaac Beeckman (1588–1637), pupil of Simon Stevin, and teacher and friend of Descartes.[58]

The oldest and best known of the larger societies, and one that is still going strong, is the Hollandsche Maatschappij der Wetenschappen [Holland Society of Arts and Sciences], founded in 1752.[59] Its motto was "in God's honor, and to the benefit of the Fatherland (*Deo et Patriae*)." Over the next 25 years, similar societies were established in the provinces of Zeeland and Utrecht, and in the cities of Rotterdam, Amsterdam, and Haarlem. All these societies acquired patents from their province and placed themselves under the patronage of the Stadholder, Willem V. They all offered prizes for essays on relevant topics, often of an applied nature, regarding typical Dutch preoccupations with water (polders, dikes, windmills, sluices, siltation), and they published *Verhandelingen* [*Transactions*]. The members were well-to-do citizens interested in the natural sciences, but this was often combined with an interest in the humanities (poetry, history). The societies hired lecturers in the exact sciences, and organized sessions, during which experiments were carried out, often with—sometimes quite expensive—state-of-the-art instruments, such as an electrification machine.

During the eighteenth century, learned societies were also formed in the areas of European expansion. Examples are, in the Western hemisphere, the American Philosophical Society in Philadelphia (1768), the Academia Científica do Rio de Janeiro (1772), and the Société Royale des Sciences et Arts in Haiti (1784); and in Asia the Bataviaasch Genootschap van Kunsten en Wetenschappen [Batavia Society of Arts and Sciences] (1778), and the Asiatic Society in Calcutta (1784).[60]

These European societies and academies were in a sense complementary to the universities. Universities were teaching in Latin, and the lectures and demonstrations organized in and by the learned societies were given in the vernacular, which attracted partly a different public. It appears likely that the universities were geared toward more traditional topics, while the learned societies and the academies often presented cutting-edge experiences—people back from voyages of discovery, new experiments with electricity, etcetera. All these societies and academies had their journals, where new information could be published relatively easily and cheaply. In various countries, individual universities were not doing well during (part of) the eighteenth century, and there the learned societies were picking up the slack. Sometimes there were links between a university and an academy, as was the case in Bologna, St. Petersburg, and Göttingen.

The learned societies partly attracted a new public, including people who did not have university training. It is customary to call such persons amateurs, as opposed to university-employed scholars. This term would no doubt fit many of the members, to whom the meetings were purely a social event and a highbrow type of entertainment. However, in the case of wealthy and often aristocratic members of these academies and societies, like, for instance, Christiaan Huygens (1629–1695), fellow of the London Royal Society and of the French Academy of Sciences, and one of the leading physicists, astronomers, and mathematicians of the period, the term " amateur" would be a misnomer. Huygens, and many others with him, must be regarded as a non-university-linked gentleman-scholar, who participated in the network of learned societies and academies, and who often contributed more to the development of the sciences and the humanities than many a university-linked scholar.

By the second half of the eighteenth century, these societies constituted a network. They exchanged their journals on a regular basis with institutions toward which they felt an affinity, and they nominated members of such societies as honorary members of their own club. Famous—or just very rich and sociable—scholars, therefore, could be members of a whole range of learned societies.

Concluding Remarks

Between ca. 1500 and ca. 1800, the way knowledge was collected, described, depicted, classified, created, and circulated would be radically altered. In this collection of essays, the authors look mainly at the period 1760s–1820s, sometimes dubbed as the Second Scientific Revolution, or that of Romantic Science. It is also the era during which the French Revolution took place, while it was, at the same time, the period that saw the resumption of the voyages of discovery, from Cook to Darwin.

It is the tail end of the period of the cabinets of curiosity, which became more "scientific" and specialized as time went by. Many learned or scientific societies and academies became more "scholarly" too, partly complementing, partly challenging the universities. Most of the academies were supported or even entirely funded and founded by the state (monarch), and, they in turn, partly or entirely financed voyages of discovery.

Most of the scholars who were active during this period were members of one or usually more of these societies, gave lectures there, conducted and witnessed experiments, made use of their libraries, and published the data they had collected, their findings, and their speculations in the *Transactions* of the various societies and academies. Some of these scholars were "amateurs" in the sense that neither were they employed by a university, nor did they make a living as a teacher or researcher, but the term amateur does not really do them justice, gentleman-scholar possibly being a more appropriate word.

Thus, in the field of "colonial" science, networks were created of things, scholars (and "amateurs"), and concepts. It is shown in the various chapters that, even though the imperial powers were often at war during the period under consideration, scholars of various nationalities communicated freely as parts of these networks, although some forms of knowledge were probably kept secret.

It was an era of invention and improvement of scientific instruments, emphasizing the link between science and technology, a link that has always been strong in the Netherlands, where windmills, shipbuilding, navigation, the production of canon and guns, and the invention or improvement of microscopes, telescopes, and pendulum clocks speak of the industrial and mercantile side of scientific development.

Such instruments played a role in observing and describing things that are either too small, or too far away to be properly seen by the naked eye. They were instrumental in more accurate registration, quantification, measuring, and counting, leading not only to increased "governmentality," but also to more accurate maps, and safer sea travel.

Knowledge in the Age of Enlightenment, as a rule, had to be useful and practical, and was supposed to be employed for the improvement of the human condition. Much of the knowledge, therefore, generated, discussed, published, and circulated during the period under consideration was rather mundane, and tended to be applied science. However, as Romanticism became fashionable in this era, there was much interest in the sublime from the late eighteenth century onward as well.

Generally speaking, there was often no conflict between Western religion and science, and many scholars were fairly orthodox Catholics or Protestants, who saw no contrast between missionary activities in Asia and Africa on the one hand, and the spread of Western science and technology on the other. Neither was there a contrast between the sublime and the divine!

It was finally, also a period that saw the birth of new disciplines, such as statistics, geography, geology, archaeology, and ethnography.

Lessons Learned

What did the Dutch know about their overseas territories around 1800? How was this knowledge created and circulated? Those are the kind of questions we attempt to answer in this book. The findings are summarized in this section.

The period around 1800 was not the finest hour of the Netherlands and their overseas territories, and the period from the 1760s to the early years of the nineteenth century is not known for its scholarly excellence. Economically speaking, neither the Netherlands nor most of its overseas establishments were doing well. Lack of money usually implies lack of patronage for the arts and sciences, which did not bode well for the creation and circulation of knowledge during this epoch.

This expectation is largely borne out by the findings of the essays in this collection, but it does not mean that the period was totally barren.

The first point to be made about the creation of knowledge regarding the Dutch overseas territories is that during the decades prior to, say, 1815, there were no expensive expeditions or other large-scale projects undertaken by the Netherlands or their representatives abroad, no doubt because there was no central authority with large funds at its disposal, such as a monarch, or an institution such as a royal academy, which could have acted as a patron of the arts and sciences. This was a serious drawback in comparison with the British, the French, the Spanish, and even the Russians, who all participated in the wave of scientific expeditions.

However, up to the 1790s, and partly even thereafter, the networks of VOC and WIC were still in working order, which facilitated the circulation of knowledge, although not always to the degree the scholars concerned might have wished. Nevertheless, the presence of these networks should not be underestimated. In the case of Japan, it was the only channel through which knowledge about the country reached the West (and through which knowledge about Europe reached Japan). In fact, the Dutch trade routes all over the world—but particularly those of VOC and WIC—can be seen as conduits of knowledge.[61]

Networks of scholars were not a novel phenomenon around the 1800s. There had been a network of naturalists in Renaissance Europe, in which the first species from "non-Western" territories were discussed and given a place in the existing systems. This network did not have a center, in contrast to that (or those) dating from the seventeenth and early eighteenth centuries, which, thanks to the VOC and WIC, were centered on the Dutch Republic.[62] During the period studied here, Sweden became an important node in the natural history network (botany/zoology), solely due to the activities of Linné and his pupils, who, in fact, carried out a research program predating that of Humboldt by almost a century. London became the center of most if not all networks, as it had become the economic center of the Western world. However, the Netherlands, with places such as Leiden and Amsterdam, with their universities, libraries, and botanical gardens, had not lost its nodal place in various networks. Circulation of knowledge has a cumulative character.

In the period under discussion, the old networks became extended. Here, I only mention the learned societies that were established, in addition to the journals of these societies and the royal academies that were founded in Europe.

Alas, circulation of knowledge through old, established networks might also have a repetitive character, and often would be no longer valid, as shown in some of the essays. Thus, the popular nineteenth and even twentieth-century Western notion that Asian and other non-Western societies were basically unchanging may have originated. Knowledge being circulated through encyclopedic publications, a feature of the Enlightenment, does run the risk of spreading dated information.[63]

Some of the information was not only outdated, but had been flawed from the very beginning. Of course it is well known that nonsensical information about overseas regions was reaching Europe through the usual networks, although ridiculous stories—like the one about birds of paradise that were supposed to have no feet—often got corrected as time went by. But yes, some of the information that was created and circulated was pseudo-knowledge and could lead to pseudo-science. The social-history-of-truth discussion shows, however, that the problem of credibility was not confined to information from overseas.[64]

How far was misinformation created and circulated on purpose (and was, therefore, disinformation), in order to mislead people from competing empires or enterprises? It is sometimes difficult to believe that scholars or bureaucrats did not know better than the information that was published would suggest.

This edited volume presents various cases of information that never reached its destination. Of course we all know—but may never have regarded this information in an epistemological light—that many European ships did not return from abroad because they sunk owing to bad weather, exploding gunpowder, being shot to pieces in a sea battle, or being captured by pirates. This led not only to a loss of people and commodities, but also of knowledge, as we know from many scholars who lost their manuscripts, naturalia, curiosa, etcetera in a shipwreck. In addition to this, information, in whatever shape (manuscripts, objects), could also reach Spain, or Sweden, or the Netherlands, but not be published or exhibited. People could have fallen out of favor with the high and mighty, the amounts of money needed to print a voluminous manuscript could be prohibitive for the funding agency, or it might have wanted to keep the information in question secret, and would not allow printing until many decades later. In some cases, publication did not occur until recently, or not at all. I think this waste of scholarly energy and resources has been underestimated so far, as most scholars were only aware of isolated instances. In this volume, mention is made of the ill-fated collections and writings of Malaspina, Rolander, and Titsingh, and the reasons they remained in obscurity for two centuries. However, keeping knowledge in manuscript form, and therefore not printing it, or at least not distributing it if printed, may have occurred so frequently that it should be regarded as unexceptional. At present there is lack of data to come to a definitive conclusion on this point.[65]

Much of the creation of knowledge in overseas territories was not undertaken by scholars of the colonizing polity alone. In the contributions on Spanish and Dutch scholarly enterprise, there are many examples of foreign nationals involved in "colonial" research. Not only were they evidently welcome in territories that often had been closed to foreigners (examples from Indonesia, the Philippines, and Japan), but there are also various examples of how they could dispose of the information and objects they had collected as they saw fit, taking it with them to their country of origin and publishing it in foreign journals. Here, apparently, the rivalry between the empires was not of overriding importance. However, in the essay about the British Empire this feature is not present, and one wonders whether this is significant. Had the British more secrets to keep than the Spaniards or the Dutch?

The journals of the newly established learned societies, remarkably often contain (translations of) articles copied from journals of other learned societies, established in other countries. It is also shown that scholar-members of one society were often members of foreign societies as well. Thus, information from a journal published by a Dutch learned society in the Netherlands, or by a journal published by a Dutch learned society established in what is now Indonesia, could easily find its way in a journal published by a British, French, German, or even Russian learned ociety. We appear to be dealing here with a truly international, recently established scholarly network, in which "colonial" (overseas) knowledge was circulating freely, reaching learned audiences from places other than the mother country soon after having been created.

Knowledge created in the overseas establishments, and circulated through the usual (and new) channels, was often very similar to—or even the same as—what had been circulated in the sixteenth or seventeenth centuries, and frequently also structured in the same way. New information was created as well, and circulated in addition to the older knowledge, and sometimes it was differently structured. The latter was the case when new disciplines came into being after, say, 1750. This happened, for instance, with ethnography/anthropology, geography, and statistics.

Statistics, a notion and discipline that came up in the second half of the eighteenth century, originally had a broader meaning than it does today, although figures were certainly important, more important than they used to be. Various essays in this volume mention "statistical" surveys and publications, new instruments of the colonial state.

Examples are also presented of publications in the fields of ethnography and geography, of which one instance at least (of an ethnographic study) can be regarded as "modern" or "enlightened," in the sense that it attempts a systematic comparison, while others were more old fashioned and much less systematic. Comparing peoples explicitly was an Enlightenment feature, in principle based on the notion of equality of all people. On the other hand, comparisons also fueled the notion of Western superiority, suggesting the need for guidance by the colonizing Westerners (although there were also discussions regarding the question of whether civilizing savages would make them happier). As witnessed in the practice of slavery, the notion of human equality was in most areas more theoretical than practical.

There is little doubt that the volume of scholarly information reaching the Netherlands (or for that matter other Western countries) increased, be it in printed or manuscript form, as objects, even in the shape of humans from overseas territories, were transported to the West as curiosities. For instance, there was a constant stream of dried plants and/or drawings of plants from overseas territories to Europe. There was a clear division of labor: the collecting was done in the overseas areas, but the more "theoretical" elaboration occurred in Europe, where the specimens ended up in museums and botanical gardens. This division even applied to individuals at different points in their careers—they went as young scholars to the overseas territories, where they took part in the creation of knowledge by collecting new species. They then returned to the Netherlands, where at a more advanced age they became professors or directors of botanical gardens, responsible for the more theoretical aspects of botany.

"New" science often depended on the availability of new instruments, also a typical Enlightenment feature. As new instruments usually implied that greater precision was possible, greater accuracy became more and more the standard that scholars were expected to adhere to. This applied across the board, from astronomy to geography (e.g., mapmaking), and throughout the period discussed here. Nevertheless, as we have seen, it is not difficult to find examples of outdated and inaccurate data in the essays presented here.

The new instruments were often expensive, which is one of the reasons why measuring and surveying was often done by the military. They had the funding, the manpower, and the training, and, moreover, during this period they were often present in the overseas territories anyway. In one case, it was one military officer Colin Mackenzie who was first charged with gathering information in newly acquired territories in India, while two decades later he was doing the same thing in Java during the so-called British Interregnum, the period of Raffles. In sum, much knowledge about the various empires was created and circulated by military physicians and military engineers.

The military presence was much stronger in the British Empire than it was in the Dutch territories, which no doubt partly explains the difference in scholarly production between the two.

The strong involvement of the military might have contributed to the fact that much of the new knowledge was practical knowledge, "applied science" rather than "pure science." As Europeans became increasingly involved with the "improvement" of not only agriculture, livestock keeping, forestry, etcetera, on European-run large-scale plantations, but also indigenous subsistence production, the need for such mundane knowledge increased. The way climate influenced primary production became increasingly a matter of European research as well. In Enlightenment science, utility was strongly emphasized.

And thus, finally, indigenous people and their knowledge came into view, although they had never been entirely absent. They had been the ones who informed the seventeenth-century botanists such as Bontius and Rumphius, who acknowledged their contributions, but rarely mentioned names. For information on agriculture, etcetera, the role of indigenous informants was at least as important, if not more so, but European researchers collecting this type of information were even less inclined than their predecessors to tell us much about them, something that would not really change until much later. Again, for most Enlightenment researchers, equality was more a theoretical than a practical notion. In some areas, such as Java, the indigenous notion of what constituted (evidence-based) scientific or scholarly knowledge was quite different from the European concept.

Looking at the image the contributions to this volume conjure up, it is clear that the Dutch Empire, precisely during this period going through a difficult patch, was no longer the center of the scholarly networks that it used to be, and lacked the central institutions that financed large-scale "colonial" research in other empires. There were no parallels to the scientific voyages launched by the British, the French, or even the Spaniards, whose home country was not doing well either. Dutch scholarly production regarding the overseas territories was unimpressive.

However, new knowledge from overseas territories—in addition to old knowledge being repeated—was reaching the mother country through old and new networks. New disciplines (sciences) spread to the Netherlands without much of a delay, and provided old and new data with novel structures. And yet, prior to, say, 1815, as far as the Netherlands and its Empire was concerned, these were sciences in the making, and Dutch scholarship had to reinvent itself. It was probably not a coincidence that the new and differently structured knowledge about, for example, Indonesia, was published by scholars such as Marsden, Raffles, and Crawfurd, all citizens of the British Empire, of which the metropole was the new center of scholarly networks.

From the 1820s, the Netherlands had a more centralized structure than before, both as a polity, and as regards to the scholarly world. Conditions for participating more fully in the creation and circulation of knowledge regarding overseas (colonial) territories were now better, and during the later part of the nineteenth century, Dutch "colonial" scholarship came (again) into its own.

Notes

1. H. Floris Cohen, *The Scientific Revolution: A Historiographical Inquiry* (Chicago: The University of Chicago Press, 1994); Lisa Jardine, *Ingenious Pursuits: Building the Scientific Revolution* (London: Little, Brown and Company, 1999).
2. William Clark, Jan Golinski, and Simon Schaffer, "Introduction," in *The Sciences in Enlightened Europe*, ed. William Clark, Jan Golinski, and Simon Schaffer (Chicago/London: The University of Chicago Press, 1999), 15.
3. Peter Jack Gay, *The Enlightenment: An Interpretation* (New York: Knopf, 1966–69); Jonathan Israel, *A Revolution of the Mind: Radical Enlightenment and the Intellectual Origins of Modern Democracy* (Princeton/Oxford: Princeton University Press, 2010); Jonathan Israel, *Democratic Enlightenment: Philosophy, Revolution, and Human Rights 1750–1790* (Oxford: Oxford University Press, 2011).
4. Clark, Golinski, and Schaffer, "Introduction," 16.
5. Andrew Cunningham and Nicholas Jardine, eds., *Romanticism and the Sciences* (Cambridge, etc.: Cambridge University Press, 1990); Stefano Poggi and Maurizio Bossi, eds., *Romanticism in Science: Science in Europe, 1790–1840* (Dordrecht, etc.: Kluwer Academic Publishers, 1994); Richard Holmes, *The Age of Wonder: How the Romantic Generation Discovered the Beauty and Terror of Science* (London: Harper Press, 2008), xv–xxi. Others used the term Second Scientific Revolution for the late nineteenth century.
6. Daniel A. Baugh, "Seapower and Science: The Motives for Pacific Exploration," in *Scientific Aspects of European Expansion*, ed. William K. Storey (Aldershot: Variorum, 1996); Alan Frost, "Science

for Political Purposes: European Explorations of the Pacific Ocean, 1764–1806," in *Scientific Aspects of European Expansion*, ed. William K. Storey (Aldershot: Variorum, 1996); Holmes, *The Age of Wonder*, xvi.

7. Thomas L. Hankins, *Science and the Enlightenment* (Cambridge, etc.: Cambridge University Press, 1985), 173.
8. For Science and Empire Studies as a recently constituted academic community, see Kapil Raj, *Relocating Modern Science: Circulation and the Construction of Knowledge in South Asia and Europe, 1650–1900* (Houndmills/New York: Palgrave Macmillan, 2007), 3–4.
9. Bruno Latour, *Science in Action: How to Follow Scientists and Engineers through Society* (Cambridge, MA: Harvard University Press, 1987); Bruno Latour, *Reassembling the Social: An Introduction to Actor-Network Theory* (Oxford: Oxford University Press, 2005); c.f. also David Philip Miller and Peter Hanns Reill, eds., *Visions of Empire: Voyages, Botany, and Representations of Nature* (Cambridge: Cambridge University Press, 1996); Julie Berger Hochstrasser, "The Butterfly Effect: Embodied Cognition and Perceptual Knowledge in Maria Sibylla Merian's *Metamorphosis insectorum Surinamensium*," in *The Dutch Trading Companies as Knowledge Networks*, ed. Siegfried Huigen, Jan L. de Jong, and Elmer Kolfin (Leiden/Boston: Brill, 2010).
10. Much of the information presented here is taken from Roy Porter, "The Enlightenment in England," in *The Enlightenment in National Context*, ed. Roy Porter and Mikuláš Teich (Cambridge, etc.: Cambridge University Press, 1981); Simon Schama, "The Enlightenment in the Netherlands," in *The Enlightenment in National Context*, ed. Roy Porter and Mikuláš Teich (Cambridge, etc.: Cambridge University Press, 1981); Hankins, *Science and the Enlightenment*; Steven Shapin and Simon Schaffer, *Leviathan and the Air-Pump: Hobbes, Boyle, and the Experimental Life* (Princeton: Princeton University Press, 1985); Clark, Golinski and Schaffer, "Introduction"; Lissa Roberts, "Going Dutch: Situating Science in the Dutch Enlightenment," in *The Sciences in Enlightened Europe*, ed. William Clark, Jan Golinski, and Simon Schaffer. (Chicago/London: The University of Chicago Press, 1999); Kim Sloan, "'Aimed at Universality and Belonging to the Nation': The Enlightenment and the British Museum," in *Enlightenment: Discovering the World in the Eighteenth Century*, ed. Kim Sloan (London: The British Museum Press, 2003); Benedikt Stuchtey, *Science across the European Empires, 1800–1950* (Oxford: Oxford University Press, 2005); Jan Golinski, *British Weather and the Climate of Enlightenment* (Chicago/London: The University of Chicago Press, 2007).
11. This element was stressed by Foucault, who saw a connection between the formation of knowledge, which constituted power (which then, in turn, produced more knowledge), and the techniques of punishment

and discipline (penology and biopolitics) in Enlightenment Europe, notions related to the Foucauldian term "governmentality." Thus, Foucault emphasized the dark side of the Enlightenment (Clark, Golinski and Schaffer, "Introduction," 21).

12. See, for instance, Joseph O'Connell, "The Creation of Universality by the Circulation of Particulars," *Social Studies of Science* 23, 1 (1993).
13. Much recent research regarding the history of eighteenth-century science emphasizes nonelite participation, informal networks, and knowledge production at nonacademic sites, like coffee houses, pubs, and breweries (e.g., Raj, *Relocating Modern Science,* 10).
14. On this discussion, see Hankins, *Science and the Enlightenment*; Jan Luiten van Zanden, "The Great Convergence from a West-European Perspective: Some Thoughts and Hypotheses," *Itinerario: European Journal of Overseas History* 24, 3/4 (2000); Lissa Roberts and Simon Schaffer, "Preface," in *The Mindful Hand: Inquiry and Invention from the Late Renaissance to Early Industrialisation*, ed. Lissa Roberts, Simon Schaffer, and Peter Dear (Amsterdam: Koninklijke Nederlandse Akademie van Wetenschappen, 2007); Joel Mokyr, *The Enlightened Economy: An Economic History of Britain 1700–1850* (New Haven/London: Yale University Press, 2009); Lissa Roberts, "Situating Science in Global History: Local Exchanges and Networks of Circulation," *Itinerario: International Journal on the History of European Expansion and Global Interaction* 33, 1 (2009): 9–27.
15. The term physico-theology has been coined for this science-religion nexus.
16. Porter, "The Enlightenment in England"; Schama, "The Enlightenment in the Netherlands"; Hankins, *Science and the Enlightenment*; Richard Drayton, *Nature's Government: Science, Imperial Britain, and the "Improvement" of the World* (New Haven/London: Yale University Press, 2000), 7, 21; Ferngren, Gary B., ed., *Science and Religion: A Historical Introduction* (Baltimore/London: The Johns Hopkins University Press, 2002); Klaas van Berkel and Arjo Vanderjagt, eds., *The Book of Nature in Early Modern and Modern History* (Leuven, etc.: Peeters, 2006); Eric Jorink, *Reading the Book of Nature in the Dutch Golden Age, 1575–1715* (Leiden: Brill, 2010).
17. This section is largely based on Paula Findlen, "Courting Nature," in *Cultures of Natural History*, ed. N. Jardine, J. A. Secord, and E. C. Spary (Cambridge: Cambridge University Press, 1996); Katie Whitaker, "The Culture of Curiosity," in *Cultures of Natural History*, ed. N. Jardine, J. A. Secord, and E. C. Spary (Cambridge: Cambridge University Press, 1996); Drayton, *Nature's Government*; Robert Huxley, "Natural History Collectors and Their Collections: 'Simpling Macaronis' and Instruments of Empire," in *Enlightenment: Discovering the World in the Eighteenth Century*, ed. Kim Sloan (London: The British Museum Press, 2003); Martin Rudwick,

"Picturing Nature in the Age of Enlightenment," *Proceedings of the American Philosophical Society* 149, 3 (2005); Claudia Swan, "Collecting *Naturalia* in the Shadow of Early Modern Dutch Trade," in *Colonial Botany: Science, Commerce, and Politics in the Early Modern World*, ed. Londa Schiebinger and Claudia Swan (Philadelphia: University of Pennsylvania Press, 2005).

18. On cabinets of curiosity, see Joy Kenseth, ed., *The Age of the Marvelous* (Dartmouth, NH: Hood Museum of Art, 1991); E. Bergvelt and R. Kistemaker, eds., *De Wereld binnen Handbereik: Nederlandse Kunst- en Rariteitenverzamelingen, 1585–1735* (Zwolle/ Amsterdam: Waanders/Amsterdams Historisch Museum, 1992); Huxley, "Natural History Collectors"; Jorink, *Reading the Book of Nature*, 257–345.
19. *Virtuosi*, sometimes called *curiosi*, should be translated as scholars or savants.
20. This section is largely based on Drayton, *Nature's Government*; Lucile H. Brockway, *Science and Colonial Expansion: The Role of the British Royal Botanic Gardens* (New Haven/London: Yale University Press, 2000); Geoffrey C. Gunn, *First Globalization: The Eurasian Exchange, 1500–1800* (Lanham, etc.: Rowman and Littlefield, 2003), 67–77; Sloan, "Aimed at Universality."
21. Jozien J. Driessen-van het Reve, *De Kunstkamera van Peter de Grote; De Hollandse Inbreng Gereconstrueerd uit de Brieven van Albert Seba en Johann Daniel Schumacher uit de Jaren 1711–1752* (Hilversum: Verloren, 2006).
22. Jorge Cañizares-Esguerra, *Nature, Empire, and Nation: Explorations of the History of Science in the Iberian World* (Stanford: Stanford University Press, 2006), 14–45; Harold J. Cook, *Matters of Exchange: Commerce, Medicine, and Science in the Dutch Golden Age* (New Haven/London: Yale University Press, 2007), 82–132, 191–207; Florike Egmond, *The World of Carolus Clusius: Natural History in the Making, 1550–1610* (London: Pickering and Chatto, 2010).
23. On the botanical gardens of Leiden and Amsterdam, see H. Veendorp and L. G. M. Baas Becking, *Hortus Academicus Lugduno Batavus 1587–1937* (Leiden: Rijksherbarium/ Hortus Botanicus, 1990); D. O. Wijnands, E. J. A. Zevenhuizen, and J. Heninger, *Een Sieraad voor de Stad: De Amsterdamse Hortus Botanicus 1638–1993* (Amsterdam: Amsterdam University Press, 1994); Cook, *Matters of Exchange*.
24. On Kew Gardens, see Ray Desmond, *Kew: The History of the Royal Botanic Gardens* (London: The Harvill Press, 1995).
25. Londa Schiebinger and Claudia Swan, eds., *Colonial Botany: Science, Commerce, and Politics in the Early Modern World* (Philadelphia: University of Pennsylvania Press, 2005), 13.
26. Mary Louise Pratt, *Imperial Eyes: Travel Writing and Transculturation* (London/New York: Routledge, 1992), 31; Drayton, *Nature's*

Government, 12–18; Desmond, *Kew: The History*; Lisbet Koerner, *Linnaeus: Nature and Nation* (Cambridge MA: Harvard University Press, 1999); Anna Pavord, *The Naming of Names: The Search for Order in the World of Plants* (London: Bloomsbury Publishing, 2005); Therese O'Malley and Amy R. W. Meyers, eds., *The Art of Natural History: Illustrated Treatises and Botanical Paintings, 1400–1850* (Washington: National Gallery of Art, 2008).

27. On collecting *naturalia* during this period, see also Boomgaard, ch. 5, and Reyes, this volume.
28. Pratt, *Imperial Eyes,* 27. On the visualization of tropical—often colonial—nature and culture, see also Felix Driver and Luciana Martins, eds., *Tropical Visions in an Age of Empire* (Chicago/London: The University of Chicago Press, 2005).
29. Ray Desmond, *Great Natural History Books and Their Creators* (London: The British Library, 2003), 17–18; O'Malley and Meyers, *The Art of Natural History*; Ella Reitsma, *Maria Sibylla Merian & Daughters: Women of Art and Science* (Zwolle: Waanders Pubishers, 2008). On Suriname, see Hochstrasser, "The Butterfly Effect", and Oostindie, this volume.
30. Pratt, *Imperial Eyes,* 36.
31. George Basalla, "The Spread of Western Science," *Science* 156 (1967).
32. David Arnold, *Science, Technology and Medicine in Colonial India* (Cambridge: Cambridge University Press, 2000), 14–15; Roy MacLeod, "Introduction," in *Nature and Empire: Science and the Colonial Enterprise,* ed. Roy MacLeod (Ithaca, NY: Dept. of Science and Technology Studies, Cornell University, 2000), 11; Hans-Jürgen Lüsebrink, "Von der Faszination zur Wissenssystematisierung: die koloniale Welt im Diskurs der europäischen Aufklärung," in *Das Europa der Aufklärung und die aussereuropäische koloniale Welt,* ed. Hans-Jürgen Lüsebrink (Göttingen: Wallstein Verlag, 2006), 16;Daniela Bleichmar et al., "Preface," in *Science in the Spanish and Portuguese Empires, 1500–1800,* ed. Daniela Bleichmar et al. (Stanford: Stanford University Press, 2009), xxi–xxii.
33. Robert DeKosky and Douglas Allchin, eds., An *Introduction to the History of Science in Non-Western Traditions* (Seattle: History of Science Society, 2008); the many volumes of Joseph Needham's *Science and Civilization in China*, and, on India, Deepak Kumar, *Science and Empire: Essays in Indian Context* (Delhi: Anamika Prakashan, 1991); Arnold, *Science, Technology*; Sheldon Pollock, ed., *Forms of Knowledge in Early Modern Asia: Explorations in the Intellectual History of India and Tibet, 1500–1800* (Durham/London: Duke University Press, 2011). However, regarding Java, see Van Klinken, this volume.
34. MacLeod, "Introduction."
35. Arnold, *Science, Technology,* 1; MacLeod, "Introduction," 10.

36. Sebastian Conrad, "Enlightenment in Global History: A Historiographical Critique," *The American Historical Review* 117, 4 (2012): 1001.
37. Edward W. Said, *Orientalism* (New York: Pantheon Books, 1978); Pratt, *Imperial Eyes*; Frost, "Science for Political Purposes"; Michael Adas, "A Field Matures: Technology, Science, and Western Colonialism," *Technology and Culture* 38, 2 (1997); John Gascoigne, *Science in the Service of Empire: Joseph Banks, the British State and the Uses of Science in the Age of Revolution* (Cambridge: Cambridge University Press, 1998).
38. Brockway, *Science and Colonial Expansion*; Schiebinger and Swan, *Colonial Botany*; Cook, *Matters of Exchange*; Raj, *Relocating Modern Science.*
39. This section is mainly based on Baugh, "Seapower and Science"; Frost, "Science for Political Purposes"; Drayton, *Nature's Government*; Gunn, *First Globalization*; Schiebinger and Swan, *Colonial Botany*; Stuchtey, *Science across the European Empires*; Cook, *Matters of Exchange.*
40. On maps and languages, see William K. Storey, ed., *Scientific Aspects of European Expansion* (Aldershot: Variorum, 1996), 161–246; Gunn, *First Globalization*, 113–44, 223–48.
41. For this topic in general, see Siep Stuurman, *Global Equality and Inequality in Enlightenment Thought* (Utrecht: Werkgroep 18e Eeuw, 2010); on the Netherlands ca. 1800, see Angelie Sens, *"Mensaap, Heiden, Slaaf": Nederlandse Visies op de Wereld rond 1800* (The Hague: Sdu Uitgevers, 2001).
42. C. A. Bayly, *The Birth of the Modern World 1780–1914: Global Connections and Comparisons* (Malden, MA/Oxford: Blackwell Publishing, 2004), 106–19; Emma C. Spary, "Self Preservation: French Travels between *Cuisine* and *Industrie*," in *The Brokered World: Go-Betweens and Global Intelligence 1770–1820*, ed. Simon Schaffer et al. (Sagamore Beach: Science History Publications, 2009), 357.
43. C. A. Bayly, *Imperial Meridian: The British Empire and the World 1780–1830* (London/New York: Longman, 1989), 7.
44. Robert Wokler, "Anthropology and Conjectural History in the Enlightenment," in *Inventing Human Science: Eighteenth-Century Domains*, ed. Christopher Fox, Roy Porter, and Robert Wokler. (Berkeley, etc.: University of California Press, 1995); Sens, *Mensaap, Heiden, Slaaf.*
45. Gunn, *First Globalization*, 145–68.
46. Bayly, *Imperial Meridian*, 133–92; Sens, *Mensaap, Heiden, Slaaf*, 63–128; Aaron Sachs, *The Humboldt Current: Nineteenth-Century Exploration and the Roots of American Environmentalism* (London: Viking, 2006), 13.
47. On the "noble savages" of the Pacific, see Bernard Smith, *European Vision and the South Pacific* (New Haven: Yale University Press,

1960); on Humboldt and Romanticism, see Sachs, *The Humboldt Current*, 41–72.

48. On the development from antiquarianism to archaeology, see Kim Sloan, ed., *Enlightenment: Discovering the World in the Eighteenth Century* (London: The British Museum Press, 2003), 166–221; see also Bloembergen and Eickhoff, this volume.
49. Bayly, *The Birth of the Modern World*, 313–14.
50. Much of the following section is taken from Baugh, "Seapower and Science"; Frost, "Science for Political Purposes"; Gascoigne, *Science in the Service of Empire*; Sloan, *Enlightenment: Discovering the World*, 224–75; Nicholas Thomas, *Discoveries: The Voyages of Captain Cook* (London: Penguin, 2003); Holmes, *The Age of Wonder*.
51. For more on voyages of discovery, see essays by Boomgaard, Huigen, and Reyes, this volume.
52. Frost, "Science for Political Purposes," 67.
53. Edouard I. Kolchinsky, "The Role of Eighteenth Century Expeditions in the Development of Natural History," *Proceedings of the California Academy of Sciences* 55, supplement 2, no. 8 (2004); Han F. Vermeulen, "Early History of Ethnography and Ethnology in the German Enlightenment: Anthropological Discourse in Europe and Asia, 1710–1808" (PhD diss., Leiden University, 2008); David Moon, "The Russian Academy of Sciences Expeditions to the Steppes in the Late-Eighteenth Century" (paper presented at the World Congress of Environmental History, Copenhagen, 2009).
54. On the learned societies in general, see Daniel-Odon Hurel and Gérard Laudin, eds., *Académies et Sociétés Savantes en Europe (1650–1800)* (Paris: Honoré Champion, 2000); Arjan van Dixhoorn, "Epilogue," in *The Reach of the Republic of Letters: Literary and Learned Societies in Late Medieval and Early Modern Europe*, ed. Arjan van Dixhoorn and Susie Speakman Sutch. (Leiden: Brill, 2008); for the Netherlands, see J. J. Kloek and W. W. Mijnhardt, *1800–Blauwdrukken voor een Samenleving* (The Hague: SDU, 2001); and Van Berkel, this volume; on French learned societies, see Huigen, this volume.
55. On the Royal Society, see Bill Bryson, ed., *Seeing Further: The Story of Science & the Royal Society* (London: Harper Press, 2010).
56. Its original name (1700) was Kurfürstlich Brandenburgische Sozietät der Wissenschaften.
57. In the Netherlands such a centralized institution would be founded under King Louis Napoleon, in 1808: the Koninklijk Instituut van Wetenschappen, Letterkunde en Schoone Kunsten, predecessor of the still existing Koninklijke Nederlandse Akademie van Wetenschappen [Royal Netherlands Academy of Arts and Sciences] (Klaas van Berkel, *De Stem van de Wetenschap: Geschiedenis van de Koninklijke Nederlandse Akademie van Wetenschappen. Deel 1: 1808–1914* [Amsterdam: Bert Bakker, 2008], 53).

58. On Beeckman, see Klaas van Berkel, "Isaac Beeckman (1588–1637) en de Mechanisering van het Wereldbeeld" [Isaac Beeckman (1588–1637) and the Mechanization of the World Picture] (PhD diss., University of Utrecht, 1983).
59. The term "Holland" refers here to the province of Holland.
60. For more on the Bataviaasch Genootschap, see Boomgaard, chapter 5; for India under the British, see Kumar, *Science and Empire;* Arnold, *Science, Technology;* and Arnold, this volume.
61. See also Huigen, de Jong, and Kolfin, *The Dutch Trading Companies.*
62. See Peter Boomgaard, "Dutch Medicine in Asia, 1600–1900," in *Warm Climates and Western Medicine: The Emergence of Tropical Medicine, 1500–1900*, ed. David Arnold (Amsterdam/Atlanta: Rodopi, 1996), 48; Cook, *Matters of Exchange*; Hochstrasser, "The Butterfly Effect," 62–63 (footnotes).
63. On eighteenth-century encyclopedic works, see Hans-Jürgen Lüsebrink, ed., *Das Europa der Aufklärung und die aussereuropäische koloniale Welt* (Göttingen: Wallstein Verlag, 2006).
64. Steven Shapin, *A Social History of Truth: Civility and Science in Seventeenth-Century England* (Chicago/London: University of Chicago Press, 1994).
65. Nicolaas Witsen (1641–1717), one of the directors of the Amsterdam Chamber of the VOC and a patron of the arts and sciences, had scholarly information, acquired through VOC channels, printed for himself, without publishing it (Bruno Naarden, "Witsen's Study of Inner Eurasia," in *The Dutch Trading Companies as Knowledge Networks,* ed. Siegfried Huigen, Jan L. de Jong, and Elmer Kolfin. (Leiden/Boston: Brill, 2010)).

Bibliography

Adas, Michael. "A Field Matures: Technology, Science, and Western Colonialism." *Technology and Culture* 38, 2 (1997): 478–87.

Arnold, David. *Science, Technology and Medicine in Colonial India.* The New Cambridge History of India, Series No. 3, Vol. 5. Cambridge: Cambridge University Press, 2000.

Basalla, George. "The Spread of Western Science." *Science* 156 (1967): 611–22.

Baugh, Daniel A. "Seapower and Science: The Motives for Pacific Exploration." In *Scientific Aspects of European Expansion*, edited by William K. Storey, 85–139. Aldershot: Variorum, 1996.

Bayly, C. A. *Imperial Meridian: The British Empire and the World 1780–1830.* London/New York: Longman, 1989.

———. *The Birth of the Modern World 1780–1914: Global Connections and Comparisons.* Malden, MA/Oxford: Blackwell Publishing, 2004.

Bergvelt, E. and R. Kistemaker, eds. *De Wereld binnen Handbereik: Nederlandse Kunst- en Rariteitenverzamelingen, 1585–1735.* 2 Vols. Zwolle/Amsterdam: Waanders/Amsterdams Historisch Museum, 1992.

Berkel, Klaas van. "Isaac Beeckman (1588–1637) en de Mechanisering van het Wereldbeeld [Isaac Beeckman (1588–1637) and the Mechanization of the World Picture]. PhD diss., University of Utrecht, 1983.

———. *De Stem van de Wetenschap: Geschiedenis van de Koninklijke Nederlandse Akademie van Wetenschappen. Deel 1: 1808–1914.* Amsterdam: Bert Bakker, 2008.

Berkel, Klaas van, and Arjo Vanderjagt, eds. *The Book of Nature in Early Modern and Modern History.* Leuven, etc.: Peeters, 2006.

Bleichmar, Daniela, Paula de Vos, Kristin Huffine, and Kevin Sheehan. "Preface." In *Science in the Spanish and Portuguese Empires, 1500–1800*, edited by Daniela Bleichmar, Paula de Vos, Kristin Huffine, and Kevin Sheehan, xix–xxii. Stanford: Stanford University Press, 2009.

Boomgaard, Peter. "Dutch Medicine in Asia, 1600–1900." In *Warm Climates and Western Medicine: The Emergence of Tropical Medicine, 1500–1900*, edited by David Arnold, 42–64. Amsterdam/Atlanta: Rodopi, 1996.

Brockway, Lucile H. *Science and Colonial Expansion: The Role of the British Royal Botanic Gardens.* New Haven/London: Yale University Press, 2000. First published 1979 by Academic Press, Inc.

Bryson, Bill, ed. *Seeing Further: The Story of Science & the Royal Society.* London: Harper Press, 2010.

Cañizares-Esguerra, Jorge. *Nature, Empire, and Nation: Explorations of the History of Science in the Iberian World.* Stanford: Stanford University Press, 2006.

Clark, William, Jan Golinski, and Simon Schaffer. "Introduction." In *The Sciences in Enlightened Europe*, edited by William Clark, Jan Golinski, and Simon Schaffer, 3–31. Chicago/London: The University of Chicago Press, 1999.

Cohen, H. Floris. *The Scientific Revolution: A Historiographical Inquiry.* Chicago: The University of Chicago Press, 1994.

Conrad, Sebastian. "Enlightenment in Global History: A Historiographical Critique." *The American Historical Review* 117, 4 (2012): 998–1027.

Cook, Harold J. *Matters of Exchange: Commerce, Medicine, and Science in the Dutch Golden Age.* New Haven/London: Yale University Press, 2007.

Cunningham, Andrew, and Nicholas Jardine, eds. *Romanticism and the Sciences.* Cambridge, etc.: Cambridge University Press, 1990.

DeKosky, Robert, and Douglas Allchin, eds. *An Introduction to the History of Science in Non-Western Traditions.* Seattle, WA: History of Science Society, 2008.

Desmond, Ray. *Kew: The History of the Royal Botanic Gardens.* London: The Harvill Press, 1995.

———. *Great Natural History Books and Their Creators.* London: The British Library, 2003.

Dixhoorn, Arjan van. "Epilogue." In *The Reach of the Republic of Letters: Literary and Learned Societies in Late Medieval and Early Modern Europe*, edited by Arjan van Dixhoorn and Susie Speakman Sutch, 423–61. 2 Vols. Leiden: Brill, 2008.

Drayton, Richard. *Nature's Government: Science, Imperial Britain, and the "Improvement" of the World.* New Haven/London: Yale University Press, 2000.

Driessen-van het Reve, Jozien J. *De Kunstkamera van Peter de Grote: De Hollandse Inbreng Gereconstrueerd uit de Brieven van Albert Seba en Johann Daniel Schumacher uit de Jaren 1711–1752.* Hilversum: Verloren, 2006.

Driver, Felix, and Luciana Martins, eds. *Tropical Visions in an Age of Empire.* Chicago/London: The University of Chicago Press, 2005.

Egmond, Florike. *The World of Carolus Clusius: Natural History in the Making, 1550–1610.* London: Pickering and Chatto, 2010.

Ferngren, Gary B., ed. *Science and Religion: A Historical Introduction.* Baltimore/London: The Johns Hopkins University Press, 2002.

Findlen, Paula. "Courting Nature." In *Cultures of Natural History*, edited by N. Jardine, J. A. Secord, and E. C. Spary, 57–74. Cambridge: Cambridge University Press, 1996.

Frost, Alan. "Science for Political Purposes: European Explorations of the Pacific Ocean, 1764–1806." In *Scientific Aspects of European Expansion*, edited by William K. Storey, 67–84. Aldershot: Variorum, 1996.

Gascoigne, John. *Science in the Service of Empire: Joseph Banks, the British State and the Uses of Science in the Age of Revolution.* Cambridge: Cambridge University Press, 1998.

Gay, Peter Jack. *The Enlightenment: An Interpretation.* 2 Vols. New York: Knopf, 1966–69.

Golinski, Jan. *British Weather and the Climate of Enlightenment.* Chicago/London: The University of Chicago Press, 2007.

Gunn, Geoffrey C. *First Globalization: The Eurasian Exchange, 1500–1800.* Lanham, etc.: Rowman and Littlefield, 2003.

Hankins, Thomas L. *Science and the Enlightenment.* Cambridge, etc.: Cambridge University Press, 1985.

Hochstrasser, Julie Berger. "The Butterfly Effect: Embodied Cognition and Perceptual Knowledge in Maria Sibylla Merian's *Metamorphosis insectorum Surinamensium.*" In *The Dutch Trading Companies as Knowledge Networks*, edited by Siegfried Huigen, Jan L. de Jong, and Elmer Kolfin, 59–101. Leiden/Boston: Brill, 2010.

Holmes, Richard. *The Age of Wonder: How the Romantic Generation Discovered the Beauty and Terror of Science.* London: Harper Press, 2008.

Huigen, Siegfried, Jan L. de Jong, and Elmer Kolfin, eds. *The Dutch Trading Companies as Knowledge Networks.* Leiden/Boston: Brill, 2010.

Hurel, Daniel-Odon, and Gérard Laudin, eds. *Académies et Sociétés Savantes en Europe (1650–1800).* Paris: Honoré Champion, 2000.

Huxley, Robert. "Natural History Collectors and Their Collections: 'Simpling Macaronis' and Instruments of Empire." In *Enlightenment: Discovering the World in the Eighteenth Century*, edited by Kim Sloan, 80–91. London: The British Museum Press, 2003.

Israel, Jonathan. *A Revolution of the Mind: Radical Enlightenment and the Intellectual Origins of Modern Democracy*. Princeton/Oxford: Princeton University Press, 2010.

———. *Democratic Enlightenment: Philosophy, Revolution, and Human Rights 1750–1790*. Oxford: Oxford University Press, 2011.

Jardine, Lisa. *Ingenious Pursuits: Building the Scientific Revolution*. London: Little, Brown and Company, 1999.

Jorink, Eric. *Reading the Book of Nature in the Dutch Golden Age, 1575–1715*. Leiden: Brill, 2010.

Kenseth, Joy, ed. *The Age of the Marvelous*. Dartmouth, NH: Hood Museum of Art, 1991.

Kloek, J. J., and W. W. Mijnhardt. *1800—Blauwdrukken voor een Samenleving*. The Hague: SDU, 2001.

Koerner, Lisbet. *Linnaeus: Nature and Nation*. Cambridge, MA: Harvard University Press, 1999.

Kolchinsky, Edouard I. "The Role of Eighteenth Century Expeditions in the Development of Natural History." *Proceedings of the California Academy of Sciences* 55, supplement 2, no. 8 (2004): 106–16.

Kumar, Deepak. *Science and Empire: Essays in Indian Context*. Delhi: Anamika Prakashan, 1991.

Latour, Bruno. *Science in Action: How to Follow Scientists and Engineers through Society*. Cambridge, MA: Harvard University Press, 1987.

———. *Reassembling the Social: An Introduction to Actor-Network Theory*. Oxford: Oxford University Press, 2005.

Lüsebrink, Hans-Jürgen, ed. *Das Europa der Aufklärung und die aussereuropäische koloniale Welt*. Göttingen: Wallstein Verlag, 2006.

———. "Von der Faszination zur Wissenssystematisierung: die koloniale Welt im Diskurs der europäischen Aufklärung." In *Das Europa der Aufklärung und die aussereuropäische koloniale Welt*, edited by Hans-Jürgen Lüsebrink, 9–18. Göttingen: Wallstein Verlag, 2006.

MacLeod, Roy. "Introduction." In *Nature and Empire: Science and the Colonial Enterprise*, edited by Roy MacLeod, 1–13. Osiris, 2nd series, 15. Ithaca, NY: Dept. of Science and Technology Studies, Cornell University, 2000.

Miller, David Philip, and Peter Hanns Reill, eds. *Visions of Empire: Voyages, Botany, and Representations of Nature*. Cambridge: Cambridge University Press, 1996.

Mokyr, Joel. *The Enlightened Economy: An Economic History of Britain 1700–1850*. New Haven/London: Yale University Press, 2009.

Moon, David. "The Russian Academy of Sciences Expeditions to the Steppes in the Late-Eighteenth Century." Paper presented at the World Congress of Environmental History, Copenhagen, 2009.

Naarden, Bruno. "Witsen's Study of Inner Eurasia." In *The Dutch Trading Companies as Knowledge Networks*, edited by Siegfried Huigen, Jan L. de Jong, and Elmer Kolfin, 211–39. Leiden/Boston: Brill, 2010.

O'Connell, Joseph. "The Creation of Universality by the Circulation of Particulars." *Social Studies of Science* 23, 1 (1993): 129–73.

O'Malley, Therese, and Amy R. W. Meyers, eds. *The Art of Natural History: Illustrated Treatises and Botanical Paintings, 1400–1850.* Washington: National Gallery of Art, 2008.

Pavord, Anna. *The Naming of Names: The Search for Order in the World of Plants.* London: Bloomsbury Publishing, 2005.

Poggi, Stefano, and Maurizio Bossi, eds. *Romanticism in Science: Science in Europe, 1790–1840.* Dordrecht, etc: Kluwer Academic Publishers, 1994.

Pollock, Sheldon, ed. *Forms of Knowledge in Early Modern Asia: Explorations in the Intellectual History of India and Tibet, 1500–1800.* Durham/London: Duke University Press, 2011.

Porter, Roy. "The Enlightenment in England." In *The Enlightenment in National Context*, edited by Roy Porter and Mikuláš Teich, 1–18. Cambridge, etc.: Cambridge University Press, 1981.

Pratt, Mary Louise. *Imperial Eyes: Travel Writing and Transculturation.* London/New York: Routledge, 1992.

Raj, Kapil. *Relocating Modern Science: Circulation and the Construction of Knowledge in South Asia and Europe, 1650–1900.* Houndmills/New York: Palgrave Macmillan, 2007.

Reitsma, Ella. *Maria Sibylla Merian & Daughters: Women of Art and Science.* Zwolle: Waanders Pubishers, 2008.

Roberts, Lissa. "Going Dutch: Situating Science in the Dutch Enlightenment." In *The Sciences in Enlightened Europe*, edited by William Clark, Jan Golinski, and Simon Schaffer, 350–88. Chicago/London: The University of Chicago Press, 1999.

———. "Situating Science in Global History: Local Exchanges and Networks of Circulation." *Itinerario: International Journal on the History of European Expansion and Global Interaction* 33, 1 (2009): 9–27.

Roberts, Lissa, and Simon Schaffer. "Preface." In *The Mindful Hand: Inquiry and Invention from the Late Renaissance to Early Industrialisation*, edited by Lissa Roberts, Simon Schaffer, and Peter Dear, xiii–xxvii. Amsterdam: Koninklijke Nederlandse Akademie van Wetenschappen, 2007.

Rudwick, Martin. "Picturing Nature in the Age of Enlightenment." *Proceedings of the American Philosophical Society* 149, 3 (2005): 279–303.

Sachs, Aaron. *The Humboldt Current: Nineteenth-Century Exploration and the Roots of American Environmentalism.* London: Viking, 2006.

Said, Edward W. *Orientalism.* New York: Pantheon Books, 1978.

Schama, Simon. "The Enlightenment in the Netherlands." In *The Enlightenment in National Context*, edited by Roy Porter and Mikuláš Teich, 54–71. Cambridge, etc.: Cambridge University Press, 1981.

Schiebinger, Londa, and Claudia Swan, eds. *Colonial Botany: Science, Commerce, and Politics in the Early Modern World.* Philadelphia: University of Pennsylvania Press, 2005.

Sens, Angelie. *"Mensaap, Heiden, Slaaf": Nederlandse Visies op de Wereld rond 1800.* The Hague: Sdu Uitgevers, 2001.
Shapin, Steven. *A Social History of Truth: Civility and Science in Seventeenth-Century England.* Chicago/London: University of Chicago Press, 1994.
Shapin, Steven, and Simon Schaffer. *Leviathan and the Air-Pump: Hobbes, Boyle, and the Experimental Life.* Princeton: Princeton University Press, 1985.
Sloan, Kim. "'Aimed at Universality and Belonging to the Nation': The Enlightenment and the British Museum." In *Enlightenment: Discovering the World in the Eighteenth Century*, edited by Kim Sloan, 12–25. London: The British Museum Press, 2003.
———, ed. *Enlightenment: Discovering the World in the Eighteenth Century.* London: The British Museum Press, 2003.
Smith, Bernard. *European Vision and the South Pacific.* New Haven: Yale University Press, 1960.
Spary, Emma C. "Self Preservation: French Travels between *Cuisine* and *Industrie*." In *The Brokered World: Go-Betweens and Global Intelligence 1770–1820*, edited by Simon Schaffer, Lissa Roberts, Kapil Raj, and James Delbourgo, 355–86. Sagamore Beach: Science History Publications, 2009.
Storey, William K., ed. *Scientific Aspects of European Expansion.* Aldershot: Variorum, 1996.
Stuchtey, Benedikt. *Science across the European Empires, 1800–1950.* Oxford: Oxford University Press, 2005.
Stuurman, Siep. *Global Equality and Inequality in Enlightenment Thought.* Burgerhartlezing, Werkgroep 18e Eeuw, No. 3. Utrecht: Werkgroep 18e Eeuw, 2010.
Swan, Claudia. "Collecting *Naturalia* in the Shadow of Early Modern Dutch Trade." In *Colonial Botany: Science, Commerce, and Politics in the Early Modern World*, edited by Londa Schiebinger and Claudia Swan, 223–36. Philadelphia: University of Pennsylvania Press, 2005.
Thomas, Nicholas. *Discoveries: The Voyages of Captain Cook.* London: Penguin, 2003.
Veendorp, H., and L. G. M. Baas Becking. *Hortus Academicus Lugduno Batavus 1587–1937.* 1938. Reprint, Leiden: Rijksherbarium/Hortus Botanicus, 1990.
Vermeulen, Han F. "Early History of Ethnography and Ethnology in the German Enlightenment: Anthropological Discourse in Europe and Asia, 1710–1808." PhD diss., Leiden University, 2008.
Whitaker, Katie. "The Culture of Curiosity." In *Cultures of Natural History*, edited by N. Jardine, J. A. Secord, and E. C. Spary, 75–90. Cambridge: Cambridge University Press, 1996.
Wijnands, D. O., E. J. A. Zevenhuizen, and J. Heninger. *Een Sieraad voor de Stad; De Amsterdamse Hortus Botanicus 1638–1993.* Amsterdam: Amsterdam University Press, 1994.

Wokler, Robert. "Anthropology and Conjectural History in the Enlightenment." In *Inventing Human Science: Eighteenth-Century Domains*, edited by Christopher Fox, Roy Porter, and Robert Wokler, 31–52. Berkeley, etc.: University of California Press, 1995.

Zanden, Jan Luiten van. "The Great Convergence from a West-European Perspective: Some Thoughts and Hypotheses." *Itinerario: European Journal of Overseas History* 24, 3/4 (2000): 9–28.

Chapter 1

Science and the Colonial War-State: British India, 1790–1820

David Arnold

Despite the recent growth in studies of the history of science, medicine, technology, and environment relating to the non-European world, most accounts of the rise of the British Empire in India in the late eighteenth and early nineteenth centuries contain few, if any, references to these new disciplinary approaches.[1] There is extensive discussion about the changing nature of the English East India Company as it morphed from being an essentially commercial enterprise to an extensive territorial power, and there has been wide-ranging consideration of the changing patterns of trade and political relations between Britain and the local "country powers." But there is scant acknowledgment of the role of science in this seminal episode in the making of modern empire—whether as a means by which British power was materially enhanced relative to that of Indian and European adversaries or as a means by which the British attained a new confidence in the beneficial nature and transformative effects of their rule. Part of the purpose of this chapter is, then, to suggest ways in which the history of science (broadly understood) might be foregrounded in relation to this moment of imperial arrival and to evaluate both its empirical and ideological role in the process of empire building in India.

The second aim of this discussion is to further address, through the science of the late eighteenth and early nineteenth centuries, the nature and function of "colonial knowledge" in this transitional period.[2] Scholars have debated the extent to which such a body of "knowledge" (if, indeed, it existed and amounted to more than mere fragments of "information") reflected a close and reciprocal engagement by the British with Indian knowledge systems as well as with scientific circles in metropolitan society, the ways in which it explicitly served imperial needs or occupied a more detached and independent position, and the nature of the human agencies and the scholarly and administrative means employed in its pursuit. But the question remains as to how the scientific knowledge of early colonial India was assembled and disseminated, and how closely this can be aligned with British military needs and strategic imperatives.

Imperial Wars and the Colonial State

Between the mid-eighteenth and early nineteenth centuries, the British, acting through their East India Company, were involved in a series of prolonged diplomatic and military confrontations with an array of Indian states, "country powers" that were themselves formidable military regimes. Among the principal conflicts were, first, those with the southern state of Mysore until the death of Tipu Sultan and the fall of his capital, Seringapatam, in 1799, which made possible British control over the southern Deccan; second, the wars with the Marathas from 1780 until their conclusive military and political defeat in 1817–18, which opened the way to British domination of western and central India; and, third, the Anglo-Gurkha war of 1814–16 in which the British overpowered the Nepalese and gained dominance over a large swathe of the sub-Himayalan region. Aside from these local wars and the territorial acquisitions that resulted from them, the British were also locked in conflict with rival European powers—the Dutch, until ousted from Ceylon and several small enclaves on the Indian mainland in the 1790s, the French, who, despite their defeat in the Seven Years' War, continued to be seen as dangerous opponents and potential allies for Mysore and other hostile states, and the Danes, whose Indian settlements, including Serampore in Bengal, were annexed during the Napoleonic Wars.

In order to sustain these lengthy and hard-fought struggles, the British developed a large and powerful military organization in the subcontinent. The Company's own army (divided between the three presidencies or provinces of Bengal, Madras, and Bombay) consisted of Indian (*sepoy*) regiments supplemented by a smaller number of European battalions. The Indian Army was further supported since the late eighteenth century by detachments of the British Army. At the height of the Napoleonic Wars, this "formidable war machine" could count on a fighting strength of more than a quarter of a million men. Beginning with around 25,000 soldiers in 1761, by 1810 the Company army had soared to 189,000 men.[3] The combined strength of Company and royal troops peaked in 1826 at the time of the first

Anglo-Burmese war with a total of 292,000 men. This, as C. A. Bayly has observed, put at the Company's disposal, "one of the biggest European-style armies in the world" at the time. But it was also costly to maintain: expenditure on the army in India was put at £8 million in 1813, rising to £13 million by 1826.[4] The colonial war-state did not entirely disappear in the 1820s. Military conflict continued in Burma in the 1820s and again in the early 1850s and in Punjab until the annexation of the province in 1849, and the widespread mutiny and rebellion of 1857–58 generated fresh military priorities and strategic preoccupations. But the heavily military nature of the company administration can be said to have diminished with the end of the Maratha wars in 1818, and thereafter colonial governance assumed a more civilian character. Despite the mutiny episode, British control over India grew more secure as the era of Anglo-French conflict receded, and scientific activity became correspondingly less attached to military needs, if not to imperial aggrandizement.

But in the short term, in the late eighteenth and early nineteenth centuries, one consequence of a rapidly expanding military presence in India was the need to utilize vast amounts of income derived from the taxation of trade and from land revenues to pay for the upkeep of the army. Funding the military became, and long remained, one of the principal items of state expenditure in British India. A further consequence of having a large army, whose prestige and authority had grown immensely as a result of its successes, and of the heavy reliance placed on army officers to provide the European personnel for an administration that had greatly expanded in its size and responsibilities, was the rise of what Bayly terms "the growing military character of the Company."[5] Encouraged by the patriotic fervor and imperialist spirit generated by decades of war against Indian and European rivals, this new militarism was most evident during the final struggle against Mysore and at a time when the perceived threat of French intervention was at its height. The Company's militaristic ethos was particularly prominent during the governor-generalship of the unabashedly expansionist governor-general Lord Wellesley between 1798 and 1805.[6] Wellesley saw in the envy of the French and the "unqualified hostility" of Tipu Sultan the foremost danger to Britain's eastern empire, a threat that could only be overcome by the necessary "expense and hazard of war."[7]

Wellesley was a dynamic figure as well as a divisive one. He assumed personal direction of the exceptionally aggressive phase of British expansionism directed against Mysore and the Marathas. The momentum his policies generated lasted well beyond his recall in 1805 and despite the growing alarm of the Court of Directors, the Company's managing body in London, at the escalating costs of the Indian administration, its mounting debts, and spiraling political entanglements. But, as seemed to befit his august office, Wellesley was also pleased to present himself as a patron of science, and his many scientific interests and objectives provide an insight into the nature and operations of what can conveniently be called the "Company science" of the time.

Wellesley looked first of all to the military establishment of doctors and engineers for a kind of scientific accountancy. In seeking to justify his war against Tipu Sultan in 1799 and rebut widespread criticism, Wellesley stated that his ultimate aim was not just to protect existing British possessions and commercial interests in India but also to save Mysore from the baleful consequences of tyrannical rule—"to improve its cultivation, to extend its commerce, and to secure the welfare of its inhabitants."[8] To carry this grand ambition into effect, Wellesley called for surveys of the territories that war and diplomacy had brought under British rule. In February 1800, immediately after the fall of Seringapatam, he despatched the surgeon-naturalist Francis Buchanan on a one-man mission to Mysore to investigate the current state of agriculture, trade, and manufacturing. On the one hand, this was a propaganda exercise intended to demonstrate how oppressive Tipu's rule had been and how inimical to material progress, and, on the other, it was designed to show how, aided by science, the province might in time become peaceful and prosperous. Buchanan was more than equal to the task. His report, while affecting an air of scientific detachment, was imbued with the spirit of capitalist "improvement," which informed so much British agrarian policy in India in the late eighteenth and early nineteenth centuries, and replete with observations on the primitive and neglected state of agriculture and manufacturing.[9] Here and elsewhere, under Wellesley, science was used to justify imperial purpose (as well as to serve the governor-general's personal ambitions), to critique Indian poverty, ignorance, and misrule, and to sketch out prospects for a more prosperous and orderly society.[10]

In fact, Buchanan's Mysore mission was not without precedent or parallel. Indeed, what Bernard S. Cohn termed the "survey modality" was one of the principal means by which the company deployed science in its service.[11] Following the earlier acquisition of territory from Mysore and Hyderabad in 1792, Lord Cornwallis, Wellesley's predecessor, had sent a Madras Army engineer, Colin Mackenzie, to carry out "the first attempt to methodise and embody the geography of the Dekkan." With the defeat of Tipu seven years later, Mackenzie's task—already vast—was expanded to include surveying and mapping this entire "terra incognita," and reporting on its history, mythology, languages and antiquities, its geology, agriculture, and manufactures. Work was still incomplete when Mackenzie left the post in 1808.[12] Similarly, shortly after Wellesley's departure, in 1807 Buchanan was commissioned by the Government of India to conduct a detailed survey of the Company's possessions in eastern India, again encompassing an immense range of objects of enquiry, an undertaking that kept him occupied until 1814.[13]

What was the "science" men like Mackenzie and Buchanan were charged with pursuing? Much of what passed for science in the minds of the governors-general of the period was, firstly, the carrying out of state-related inventories of natural and human resources and the potential of the Company territories for trade and revenue. Secondly, it signified the assembling of vast amounts of miscellaneous information, drawn from observation, oral and written sources, indigenous as well as European, and covering almost every imaginable subject from natural history, meteorology, diets and health, through agriculture and manufacturing, to religious beliefs, regional languages, and vernacular literature. It was assumed that these multidisciplinary enquiries would help the rulers to map out the resources of their new territories and gain an informed understanding of their new subjects. In short, what we now think of as science was part of a largely undifferentiated accumulation of miscellaneous, but hopefully "useful," knowledge. In fact, until the 1830s the term "science" was rarely used. Surgeon-botanists like Buchanan or William Roxburgh, his predecessor as superintendent of the botanic gardens at Calcutta, were far more likely to identify themselves with "natural history."[14] However, the colonial orientation of these early surveys, and the imperial impetus behind them, was consistently evident. Intermixed with the scientific observations and statistical data was a great deal of subjective commentary from an essentially European perspective on the aesthetic properties of the landscape, the superstitious nature of Indian beliefs and customs, and the compelling need for agrarian improvement.[15]

Further evidence of Wellesley's personal belief in the utility of science comes from the curriculum he drew up for Fort William College in Calcutta, an institution he founded in 1800 to train young civil servants newly arrived from Britain. The program for instruction included natural history, botany, chemistry, and astronomy, though less, one senses, as practical subjects than as fitting educational accomplishments for a new and superior breed of imperial administrators.[16]

But Wellesley was drawn to the decorative as well as the didactic. At his country residence at Barrackpore he assembled a menagerie of rare birds and animals, charging Buchanan and others with collecting exotic species for it and paying an Indian artist to depict prized species. Wellesley tried to persuade the Court of Directors in London that this was a branch of the study of natural history worthy of their "magnificence and liberality," but the directors scorned it as one of the governor-general's many extravagances and the menagerie did not long outlast Wellesley's recall.[17]

India's Scientific Establishment

The rise of the colonial war-state had far-reaching implications for the organization as well as the patronage of science in India. The rapid expansion of the army, combined with soaring levels of sickness and mortality among Indian and European troops, led to the creation of a military medical establishment. The early history of the medical services of Bengal, Madras, and Bombay has been traced back to the mid-eighteenth century. By 1785 the Court of Directors had sanctioned a peacetime establishment of 234 doctors, allowing one European "surgeon" (doctor) and one Indian assistant for each regiment of the Company army. By 1824 the number of European surgeons and assistant surgeons had risen to 630: 50 more were added in the mid-1820s to meet the emergency of the First Burma War.[18]

The primary duty of European surgeons was to attend to the medical needs of the army and, especially, in times of war, this remained their paramount responsibility, but the presence in India of so many university-trained doctors also created a substantial body of scientifically minded physicians. Whether from personal interest or at the behest of the Company, the surgeons turned their attention to a range of scientific concerns—from botany, geology, and zoology to astronomy, meteorology, and ethnography—thereby largely obviating the need for outsiders, unattached to the Company payroll, to carry out its scientific tasks. Indeed, partly for reasons of security, very few non-Company scientists were permitted by the Court of Directors to visit India: even the illustrious Alexander von Humboldt was denied entry, perhaps for fear he would be as critical of British rule in India as he had been of Spanish rule—and slavery—in the Americas.[19] Besides practicing medicine, Company surgeons also commented on the nature and treatment (by both Western and Indian therapies) of diseases common among the soldiers in their charge or among the population at large. The pioneering study of dysentery among

European troops in India by the army physician George Ballingall, published in 1818, epitomizes this close observational and therapeutic engagement with the diseases of India as seen from a European, military perspective.[20]

Medicine was a scientific discipline and profession of exceptional significance in the early colonial context. The compelling needs of the army gave medicine an urgency unmatched by most other domains of scientific enquiry and made necessary the creation of one of the largest scientific establishments to be found anywhere in the colonial world at the time. But medicine was also closely connected with other scientific fields. In keeping with the education and training surgeons had received in Britain, the search for, and use of, medicinal drugs in India called for an intimate knowledge of plants and of their toxic as well as therapeutic properties.[21] Given the perceived influence of climate, vegetation, and soil on the etiology of disease, physicians were pioneers of medical topography, just as their observations of society and culture—and of "race"—placed them among the first ethnographers. Medical training encouraged close observation and careful recording: the use of postmortems exemplified this observational empiricism and presented a cultural contrast, to which the British attached great importance, with Indians' apparent neglect of anatomy and their repugnance at the practice of dissection.[22] Through the classificatory legacy of Linnaean science, practitioners of medicine and botany were predisposed to see "systems" in the domain of nature and in human suffering, to believe in the need for rationality and order in a world that often otherwise appeared alien and chaotic.

But Company surgeons were not simply state employees. Indeed, they were often frustrated by the Company's reluctance to support their scientific pursuits or patronize through publication subsidies the costly fruits of their scientific labor. Among the personal incentives to scientific enquiry were isolation and boredom. As one naturalist put it, the objective must be "to rescue . . . those hours of leisure from indolent neglect," and to use the "fertile advantages" of the situation in which they found themselves to give "no less of pleasure to the individual, than eventually to the public benefit."[23] Perhaps more than any other science, botany was imbued with sentiment, especially since many of the temperate plants found in the Himalayan foothills and other upland areas were redolent of Britain's own flora and so a source of nostalgic recollections of home and childhood. Despite the practical value attached to ascertaining local names and indigenous uses, the study of plants often created a greater sense of affinity with Europe than with India.[24]

Surgeons and others, whose presence in India was required by the military, medical, and administrative needs of the Company, often found the time for science, apart from having the requisite educational background. Their interests were further stimulated by the creation of the Asiatic Society in Bengal, founded by Sir William Jones in 1784, followed by similar societies in Bombay (1804) and Madras (1812), which collectively set out to investigate "Oriental Arts, Sciences and Literatures."[25] Significantly, these bodies were overwhelmingly, if not quite exclusively by the 1830s, European in their membership as well as in their intellectual orientation. Although they published their proceedings, these worthy tomes tended to circulate largely within India itself, among the European elite, and, to the frustration of many contributors, were little known in Europe. Keeping up subscriptions among a mobile, transient, and widely dispersed European population (as army officers and civil servants were transferred from one posting to another, retired, or returned home on furlough) and maintaining publication schedules were formidable tasks. *Asiatick Researches*, the journal of the Asiatic Society in Bengal and one of the principal outlets for the publication of scientific articles in India, fell several years behind and some articles were out-of-date by the time the issues containing them appeared.

Bodies like the Asiatic Society of Bengal tended in practice to be more literary and antiquarian than scientific in their outlook and to show a general preference for observation over theory.[26] In 1808 the society set up a committee "to promote the knowledge of Natural History, Philosophy, [and] Medicine." The "Physical Class," as this came to be called, soon lapsed however, despite being temporarily revived in 1818. The aim of "all physical science," it was said, was to "acquire an accurate knowledge of facts by observation and experiment, and to apply those facts to a synthetical explanation of particular phenomena."[27] But in the age before Humboldt (and in a country that, to its regret, lacked a scientist of his stature and vision) the desire to achieve "synthetical explanations" was largely lost sight of while the goal of acquiring "accurate knowledge of facts by observation and experiment" came closer to being realized.[28]

The Court of Directors took some pride in the scientific activities of its servants—whether because they seemed conducive to its material benefit or because their work gave the prestige of science to an organization often identified in Britain with unprincipled expansionism, with its unpopular commercial monopoly (until this was lost, apart from the China trade, in 1813), and with corruption, cruelty, and famine. The botanic gardens established at Calcutta in 1786 and that founded in 1817 at Saharanpur in northern India were leading scientific institutions—bases from which plant-hunting expeditions were despatched, within whose walls scientifically ordered botanic gardens were laid out on Linnaean principles, where herbaria were assembled and preserved for reference, and where major studies in plant taxonomy and volumes of illustrated flora were produced.[29] Beyond these two main sets of institutional sites—the scientific societies and the botanic gardens—much of the science of the period was itinerant and ad hoc. Surgeon-naturalists seized the opportunities created by military campaigns and administrative duties, supplemented by occasional roving commissions from the government, to travel across the subcontinent and into adjacent areas like Nepal, Burma, and the Malaya peninsula.[30] War against the Dutch extended this scientific reconnaissance into Ceylon in the 1790s, and then, with the invasion of Java in 1811, further afield into Indonesia. As in India itself, this geographically extended investigation ranged widely over history, ethnography, archaeology, botany, zoology, and geology. It sometimes, too, led to the acquisition and translation of indigenous sources (texts, inscriptions, oral traditions), but little use appears to have been made by the British of earlier Dutch investigations.[31]

Although among the most active, surgeons were not the only individuals involved in science. Army officers of both the Company and the Crown were also scientific travelers, surveyors, and recorders of scientific data of various kinds.[32] It was not exceptional for an army officer, Lt.-Col. Robert Kyd, to be behind the founding of the Calcutta botanic garden in 1786.[33] From 1799 onwards, William Lambton, a British infantry officer, carried out the detailed and meticulous trigonometric surveys that began to map and measure the subcontinent and which helped to establish Britain's sense of scientific ownership over India.[34]

Imperial Ideology and Practical Science

The practitioners of science in British India were, in general, ideologically as well as professionally committed, if not to empire in the abstract, then at least to Britain's self-appointed "civilizing mission." This was evident both with respect to the local struggle against Mysore and the Marathas and the near-global contestation against the French and Dutch. Many of the scientific figures of the period made framing references to these struggles in their correspondence and in their published books and articles—sometimes in order to explain the importance of their work or the difficulties they encountered, but at other times because the ethos of imperial warfare and global competition was so pervasive as to made such references appear natural and apposite.

Thus, William Lambton, the founder of India's trigonometric survey, noted in an article published in 1808 (but written several years earlier) that it was the owing to the "success of the British arms during the late glorious campaign" against Tipu Sultan that the British had acquired control of the route across southern India from one side of the peninsula to the other, thereby making possible an uninterrupted survey from Malabar on the Arabian Sea to the Coromandel coast on the Bay of Bengal.[35] A decade later, he compared British survey work in India with the "grand operations" carried out by the French in Europe in measuring the distance from Dunkirk to Barcelona. This and others feats of modern surveying in Europe helped establish "a question of great importance in physical astronomy," that is, the shape and dimensions of the earth, but Lambton could not resist remarking patriotically on the work of his own survey in India and on the superior opportunities for surveying that the possession of vast swathes of India gave the British.[36] A third article, dated 1815 but not published until 1820, described how, in carrying the trigonometric survey northward from southern into central India, Lambton's survey parties had had to negotiate the politically difficult country still dominated by the Nizam of Hyderabad and the warring Marathas. But, in his view, their hazardous operations were fully justified because they enabled the accurate mapping of an area of strategic importance to the British.[37] Lambton was not alone in his imperial awareness. The surgeon-naturalist Benjamin Heyne, a former Moravian missionary who served with Mackenzie on the Mysore survey, similarly identified closely with British military success against the French and the country powers and with the establishment of Company rule in India. His "historical and statistical tracts," written between1800 and 1811, covered diverse topics from indigenous medicine and agricultural implements to copper mining and iron smelting. His writings are replete with references to the benefits of British administration and the failings of "native" society and government. Indeed, so highly did Heyne rate Company rule that he feared that the threatened end of its monopoly in 1813 would undo the achievements of men like Robert Clive and Lord Wellesley through whose "gigantic efforts" India had been secured for Britain in the first place.[38]

But it would be erroneous to see scientific activity under the company as solely motivated by, or unwaveringly supportive of, imperial ideology. Science also served economic needs. British interest in India was at heart commercial and one of the pragmatic functions of science was to further this material interest. In mercantilist thinking, India was to provide goods that Britain could not itself produce but which were valuable to its trade and manufacturing or which might otherwise remain the monopoly of its rivals. As the eminent naturalist and Company adviser Joseph Banks remarked of India in 1788: "A country like this, blessed with advantages of soil, climate and population so eminently above its mother country," seemed "by nature intended" to supply it with abundant raw materials. He added in a more political vein, that a colony yielding "that kind of tribute binds itself to the mother country by the strongest and most indissoluble of human ties, that of common interest and mutual advantage."[39] For Banks and many surgeon-naturalists in India, science was, to quote John Gascoigne, "above all applied science—a source of knowledge which could transform society for the benefit of humanity."[40] What Banks had in mind for India was the cultivation of such tropical commodities as cotton, coffee, sugar, cochineal, indigo, and vanilla.[41] Despite the dwindling value of the spice trade by the early nineteenth century, one of the ambitions of the Company and its naturalists was either to annex those localities where spices were currently produced, mostly under Dutch control (cinnamon in Ceylon, mace and nutmeg in the Indonesian islands), or alternatively to create botanic gardens and plantations in India to replace those sources of supply.[42] Another high priority, again reflecting military and naval needs, was timber—teak for shipbuilding and other Indian hardwoods, such as *sal* and *sissoo*, for the construction of army gun carriages.[43]

In this economic reconnaissance of India, botany—that is, applied or "economic" botany—was crucial. Surgeons sought to locate medicinal plants, whether for their greater efficacy, or to replace expensive imported drugs; they investigated timbers, dyes for staining cotton cloth, plant fibers suitable for making ropes and cordage, and food plants and commercial crops that might be introduced into India to reduce the threat of famine and to stimulate agricultural improvement. In part, India was thought of as a "tropical estate," meeting some of the same functions as the West Indies had hitherto for the British and French or the Indonesian islands for the Dutch. While drawing pragmatically upon "native" usages and sources of information, the quest for plant knowledge was, therefore, often impelled by needs and informed by comparisons external to India. As British power penetrated into the more northerly parts of the subcontinent, following the Anglo-Gurkha War and the annexation of the sub-Himalayan ranges, so did the possibility increase of developing more temperate crops like tea and cinchona.[44]

Two considerations were at work here. One was the capitalistic spirit that lay behind the colonial drive for improvement—to increase crops yields, promote better animal husbandry, expand the production of commercial commodities, drain marshes, and turn jungle "wastes" into lush pastures, well-managed plantations, and timber-rich forests. This was as much a moral as a commercial agenda, a position captured by the leading role that the Baptist missionary William Carey played in setting up the Agricultural and Horticultural Society of India in 1820.[45] But alongside this confident affirmation of colonialism's role as the agent of capitalist improvement (and so of a superior moral order and material civilization) ran a second strand—one of imperial anxiety. At a time of intense military conflict and global warfare, there was deep anxiety about possessing adequate supplies of timber for building warships and gun carriages—hence the need for creating plantations (as the French were said to have done) to ensure future timber supplies.[46] Anxiety pervaded other matters as well. There was the secrecy or belligerence of neighboring powers whose military strength, material resources, and technological skills might rival those of the Company or erode its supremacy.[47] The devastating impact of India's climate and diseases challenged European constitutions and so questioned the very ability of Britons to inhabit, let alone rule, the subcontinent or maintain a healthy military presence there.[48] In addition were the recurrent droughts and famines that from 1770 onward further threatened agriculture, revenues, and rural subsistence.[49] In response to these multiple and interlinked anxieties over India's environment, natural resources, and their human impact, men of science were expected to provide expert advice and muster scientific solutions.

Science before Humboldt

Given the diversity of scientific activity in the late eighteenth and early nineteenth centuries, it would be rash to propose a single way of characterizing it in its entirety. Science was not readily separated from the wider colonial milieu in which it was situated and, to a large degree, subsumed. But that is not to say that everything that happened in the name of science simply served the immediate needs of the Company and the nascent colonial state. Certainly, almost everyone who could be described as a scientist in this period was either, like Francis Buchanan or Colin Mackenzie, employed by the Company or, like the Baptist missionary and agricultural improver, William Carey, allied with it for many ideological and practical purposes. It has been part of the argument of this chapter that those who followed science made a significant contribution to the way in which the Company state functioned and presented a public rationale for its dominion over India. They may not have had much influence over the way in which campaigns were fought or the policies by which provinces were annexed (though when it came to forest resources in particular their views were not entirely without significance), but they helped provide a scientific legitimacy for British rule. They assisted in seeking the means to preserve the health of soldiers and civilians, to improve agriculture and harvest timber, and in surveying and mapping the Company's ever-growing territories.

But we can further evaluate the nature of this age of emerging scientific endeavor in the light of the kind of science that later emerged in association with Alexander von Humboldt—a figure well known in scientific circles in British India by the 1820s and 1830s, who came to serve as a touchstone for colonial scientific aspirations.[50] What, then, was India's pre- or un-Humboldtian science like?

Its diversity offers one clue. There was little attempt by Company naturalists and engineers to theorize and systematize at an elevated, abstract level. Much of the science of the period was practical, observational, and descriptive—concerned with surveying and measuring India, with treating sick soldiers and ailing Europeans, with establishing the properties and utility of particular timbers, drugs, and dyes. Even if there was time, there was rarely the disposition to be more philosophical about science. This can be illustrated anecdotally by referring to a "Statistical and Geological Memoir" published in the *Journal of the Asiatic Society* in 1835. Written by a doctor in the Bombay service, the author confessed to making his "hasty observations" in the course of a "rapid march" across central and southern India several years earlier. And yet, despite the haste and the lapse of time since making his observations, the physician believed that his personal observations might still be of value, observing:

> Connected as this portion of the globe now is with the political and commercial prosperity of Great Britain, the resources of the country, the mineral treasures of its rocks, the capabilities and productions of its soil, the conditions of the inhabitants, and their prospect and means

> of attaining a higher scale of civilization, deserve the attention of the legislator, merchant, philanthropist, and man of science.[51]

Despite its late date of publication, this article helps exemplify a continuing trend. Drawing on individual experiences, and in an age before many systematic scientific surveys had been carried out, European travelers offered to the service of science "an interesting accession of knowledge" about a country "never before explored."[52] Knowledge—not least geographical knowledge—was often seen as a "desirable object" in itself, without the need for further theorization and synthesis.[53] It could be said, though, even without a Humboldt, science in India had begun to change by the early 1830s. Discoveries were made, like those of the Tertiary era Siwalik fossils in northern India, or the delivery to London by Nathaniel Wallich of the accumulated "treasures" of Indian plant botany, that attracted recognition in scientific circles in London and beyond.[54] Partly inspired by Humboldtian biogeography, Company botanists like J. F. Royle began to think more systematically about the relationship between plants and their location and about the implications this might have for the introduction of exotic species.[55] Science in India was beginning to seem less insular and more connected by the 1830s than it had done 15 or 20 years earlier. The decades after 1820 saw, too, a greater distancing of Western science from Indian knowledge systems and the beginning of increasing specialization and differentiation between the leading fields of scientific enquiry, including botany, geology, meteorology, zoology, and ethnography, many of which, in time, acquired their own specialist, nonmilitary, personnel.

It could perhaps be argued, though, that one of the areas where company science did assume a more philosophical stance was in its relation to indigenous knowledge. Thus the work on botany by Sir William Jones, the founder of the Asiatic Society in Bengal and India's premier Orientalist until his death in 1794, undertook a critical engagement with Sanskrit texts and with the problems of plant taxonomy and nomenclature that they posed. Discussions of medicine and astronomy were often marked by a wonderment at the antiquity, complexity, and sophistication of Indian (specifically Hindu) systems of knowledge and a desire to connect this "lost" Eastern knowledge with biblical authority and with Europe's folk and classical traditions.[56] But little of this philosophical speculation and literary enquiry took the form of an engagement with the practicalities of science, and there was a feeling in India after Jones's death that the opportunity had been lost to cultivate a more exacting knowledge of natural history and the physical sciences.[57]

Conclusion

Some recent scholars have emphasized the "dialogic" nature of the relationship between indigenous and Western knowledge.[58] Others have seen science as part of a process of "circulation" between India and the West extending well into the nineteenth century.[59] But there are grounds for being skeptical. There was a European appropriation of what was considered to be of direct and practical use, but this was tempered (as in the work of Buchanan and Heyne) by repeated, often passionate, criticism of Indian backwardness, ignorance, and superstition. There was extensive use of Indians as subordinate agents: both Mackenzie and Buchanan were heavily reliant in their scientific surveys on "intelligent" Brahmin interpreters and informants.[60] Botanists like Wallich used Indians as plant collectors and botanical illustrators; for other investigators they served as a source of artisanal knowledge. But the British seldom regarded such Indians as their partners, as their equals in science. And, as we have seen, much of the science of the period presented at learned bodies like the Asiatic Society of Bengal or published in journals like *Asiatick Researches* was expressly intended for European consumption. In many cases, at least before the 1830s, Company science failed even to find an audience outside India. All this looks more like appropriation than exchange, a colonization of information rather than a genuine (and reciprocal) circulation of knowledge. In the end, science seemed intent on garnering tribute to the superior civilization of Britain, and of the West more generally, and in providing evidence, broadly supportive of Company rule and the colonial project, of Indian backwardness and barbarity.

On the other hand, with Humboldtian science in mind, we should recognize that the pursuit of science in India in the late eighteenth and early nineteenth centuries was not merely part of ongoing ideological warfare or mere amateurism and provincialism compared with what was happening in Europe. Growing beyond immediate political needs and military imperatives, careful attention was paid to observable detail, to quantification and measurement, and to the nature and utility of Indian tools and implements and Western scientific instruments.[61] There was the systematic examination of human bodies, alive and dead, and serious efforts to classify plants and extend knowledge of their characteristics and properties. Statistics became a central part of thinking about health and disease, about climate, demography, and agriculture. Statistics were critical in trying to put medicine and science on a "rational footing."[62] The outbreak of cholera in India in 1817, and the subsequent pandemic, gave fresh impetus to local enquiry and causal explanation.[63] But in many ways the science of the period in India appears in retrospect like a science in waiting—if not for Humboldt, then for a clearer sense of its own nature, purpose, and function. This only really emerged, scientifically speaking, under the Crown rule after 1858.

Notes

1. Richard B. Barnett, ed., *Rethinking Early Modern India* (New Delhi: Manohar, 2002); Ian J. Barrow and Douglas E. Haynes, "The Colonial Transition: South Asia, 1780–1840," *Modern Asian Studies* 38 (2004): 469–78; P. J. Marshall, *The Making and Unmaking of Empires: Britain, India, and America, c. 1750–1783* (Oxford: Oxford University Press, 2005). For an account that more closely aligns medicine and empire, see Pratik Chakrabarti, *Materials and Medicine: Trade, Conquest and Therapeutics in the Eighteenth Century* (Manchester: Manchester University Press, 2010).
2. See Bernard S. Cohn, *Colonialism and Its Forms of Knowledge: The British in India* (Princeton: Princeton University Press, 1996); cf. C. A. Bayly, *Empire and Information: Intelligence Gathering and Social Communication in India, 1780–1870* (Cambridge: Cambridge University Press, 1996); Kapil Raj, *Relocating Modern Science: Circulation and the Constitution of Scientific Knowledge in South Asia and Europe, 1650–1900* (Basingstoke: Palgrave Macmillan, 2007).
3. Marshall, *Making and Unmaking*, 129; *Minutes of Evidence taken before the Select Committee on the Affairs of the East India Company, V: Military* (Cmd 735), 1832, xix.
4. C. A. Bayly, *Imperial Meridian: The British Empire and the World, 1780–1830* (Harlow: Longman, 1989), 128; *Minutes of Evidence* 1832, vii, xv.
5. C. A. Bayly, *Indian Society and the Making of the British Empire* (Cambridge: Cambridge University Press, 1988), 87. On the Company's military regime, see, too, Douglas M. Peers, *Between Mars and Mammon: Colonial Armies and the Garrison State in Early Nineteenth-Century India* (London: I. B. Tauris, 1995).
6. Philip Lawson, *The East India Company: A History* (Harlow: Longman, 1993), 134–36.
7. Wellesley to Lord Harris, Governor of Madras, June 26, 1798; Wellesley, minute, August 12, 1798, in *The Despatches, Minutes, and Correspondence of the Marquess Wellesley during His Administration in India*, ed. Montgomery Martin (London: John Murray, 1836), 1: 62, 82, 205.

8. Wellesley to Court of Directors, August 3, 1799, in Martin, *Despatches*, 2: 85. For a contrasting view of Mysore, see Irfan Habib, ed., *Confronting Colonialism: Resistance and Modernisation under Haidar Ali and Tipu Sultan* (London: Anthem Press, 2002).
9. David Arnold, "Agriculture and 'Improvement' in Early Colonial India: A Pre-History of Development," *Journal of Agrarian Change* 5 (2005): 505–25. As Richard Drayton has pointed out, the doctrine of "improvement" formed part of the wider relationship between science and empire: Richard Drayton, *Nature's Government: Science, Imperial Britain and the "Improvement" of the World* (New Haven: Yale University Press, 2000). It is important to note, though, that the drive for improvement was generated and given institutional form by the British in India in response to local conditions and not simply as a consequence of imperial policies and scientific practices emanating from Britain.
10. See Wellesley's instructions to Buchanan, February 24, 1800, in Francis Buchanan, *A Journey from Madras through the Countries of Mysore, Canara, and Malabar* (London: T. Cadell & W. Davies, 1807), 1: vii–xiii; Marika Vicziany, "Imperialism, Botany and Statistics in Early Nineteenth-Century India: The Surveys of Francis Buchanan (1762–1829)," *Modern Asian Studies* 20 (1986): 625–60.
11. Cohn, *Colonialism*, 7–8.
12. [Colin Mackenzie], "Biographical Sketch of the Literary Career of the Late Colonel Colin Mackenzie, Surveyor-General of India," *Journal of the Royal Asiatic Society* 1 (1834): 336–38.
13. Montgomery Martin, ed., *The History, Antiquities, Topography, and Statistics of Eastern India*, (London: W. H. Allen, 1838), 1: vii–x. "A more comprehensive programme than this was probably never entrusted to a single officer in or out of India," according to D. Prain, "A Sketch of the Life of Francis Hamilton (once Buchanan)," *Annals of the Royal Botanic Garden, Calcutta* 10 (1905), xix.
14. For example, William Roxburgh to secretary, Public Department, February 1, 1809, Bengal Public Consultations, India Office Records, British Library, London [hereafter IOR]. On the "natural history of man," see Francis Hamilton, *An Account of the Kingdom of Nepal* (Edinburgh: Archibald Constable & Co, 1819), 60. One could date the explicit identification with "science" to the founding of the Calcutta journal *Gleanings in Science* in 1829, but references to the "arts and sciences" were common before that.
15. Buchanan's original journals provide examples of this subjectivity: see C. E. Oldham, ed., *Journal of Francis Buchanan Kept during the Survey of the District of Shahabad in 1812–1813*, (Patna: Superintendent of Government Printing, Bihar and Orissa, 1926).
16. "Notes with Respect to the Formation of a College at Fort William," July 10, 1800, and "Regulations for Fort William College," in Martin, *Despatches*, 2: 359.

17. Wellesley, minute, July 26, 1804, in Ray Desmond, *Wonders of Creation: Natural History Drawings in the British Library* (London: British Library, 1986), 122. On Wellesley's remarkable collection of natural history paintings and drawings from India and Southeast Asia, see Mildred Archer, *Natural History Drawings in the India Office Library* (London: Her Majesty's Stationery Office, 1962), 6.
18. D. G. Crawford, *A History of the Indian Medical Service, 1600–1913* (London: W. Thacker & Co., 1914), 2: 208, 214; Chakrabarti, *Materials and Medicine*, Chapter 3.
19. Jean Théodoridès, "Humboldt and England," *British Journal for the History of Science* 3 (1966): 43–44.
20. George Ballingall, *Practical Observations on Fever, Dysentery, and Liver Complaints as They Occur amongst the European Troops in India* (Edinburgh: David Brown & A. Constable, 1818). On Ballingall, see also Mark Harrison, *Medicine in an Age of Commerce and Empire: Britain and Its Tropical Colonies, 1660–1830* (Oxford: Oxford University Press, 2010), 189–91.
21. For example, John Fleming, "A Catalogue of Indian Medicinal Plants and Drugs, with Their Names in the Hindustani and Sanscrit Languages," *Asiatick Researches* 11 (1810): 153–96; Hamilton, *Account*, 85–100.
22. On Western dissections and the contrast with "Hindu medicine," see Ballingall, *Practical Observations*, 59–63; James Annesley, *Sketches of the Most Prevalent Diseases of India* (London: Thomas & George Underwood, 1825), 1: 45–46; Whitelaw Ainslie, *Materia Indica* (London: Longman, Rees, Orme, Brown & Green, 1826), 2: vii.
23. Patrick Russell, "Preface," William Roxburgh, *Plants of the Coast of Coromandel* (London: George Nicol, 1795), 1: i.
24. For example, Roxburgh, *Plants*, 1: 3–7, 10–11, 21–22, 57. For a more decorative view of botany, see N. Wallich, "Descriptions of Some Rare Indian Plants," *Asiatick Researches* 13 (1820): 369–415; David Arnold, *The Tropics and the Traveling Gaze: India, Landscape and Science, 1800–1856* (Seattle: University of Washington Press, 2005), Chapter 5.
25. The description was that of Sir James Mackintosh in Bombay: Prashant Kidambi, *The Making of an Indian Metropolis: Colonial Governance and Public Culture in Bombay, 1890–1920* (Aldershot: Ashgate, 2007), 159.
26. For the "natural sciences" articles published by the Asiatic Society, see *Centenary Review of the Asiatic Society of Bengal, from 1784 to 1883* (Calcutta: Thacker, Spink & Co., 1885), Part 3.
27. "Introduction," *Asiatick Researches* 18 (1833): i–ii.
28. Andrew Grout, "Geology and India, 1770–1851: A Study in the Methods and Motivations of a Colonial Science," (PhD diss., School of Oriental and African Studies, London, 1995), 89, 104.

29. See Ray Desmond, *The European Discovery of the Indian Flora* (Oxford: Oxford University Press, 1992), Chapters 6–9; J. F. Royle, "Account of the Honorable Company's Botanic Garden at Saharanpur," *Journal of the Asiatic Society of Bengal* 1 (1832): 41–58.
30. For example, Wallich's expedition to Nepal in 1820–1821 while superintendent of the botanic garden in Calcutta: see Board's Collections F/4/655: 18040 and F/4/712: 19459, IOR.
31. Wallich, "Descriptions"; I. H. Burkill, "William Jack's Letters to Nathaniel Wallich, 1819–1821," *Journal of the Straits Branch of the Royal Asiatic Society* 73 (1916), 147–268; Mackenzie, "General View of the Results of Investigations into Geography, History, Antiquities, and Literature, in the Island of Java," in [Mackenzie], "Biographical Sketch," 353–60.
32. On the military connection with science, see Douglas M. Peers, "Colonial Knowledge and the Military in India, 1780–1860," *Modern Asian Studies* 33 (2005): 157–80.
33. Kalipada Biswas, *The Original Correspondence of Sir Joseph Banks Relating to the Foundation of the Royal Botanic Garden, Calcutta* (Calcutta: Royal Asiatic Society of Bengal, 1950), 3–10.
34. Matthew H. Edney, 1997, *Mapping an Empire: The Geographical Construction of British India, 1765–1843* (Chicago: University of Chicago Press, 1997).
35. William Lambton, "An Account of a Method for Extending a Geographical Survey across the Peninsula of India," *Asiatick Researches* 7 (1808): 312.
36. William Lambton, "An Account of the Measurement of an Arc of the Meridian," *Asiatick Researches* 12 (1818): 1–2.
37. William Lambton, "Account of the Measurement of an Arc on the Meridian", *Asiatick Researches* 13 (1820): 7–8.
38. Benjamin Heyne, *Tracts, Historical and Statistical, on India* (London: Robert Baldwin, 1814), 437–38.
39. Quoted in Desmond, *European Discovery*, 202.
40. John Gascoigne, *Joseph Banks and the English Enlightenment: Useful Knowledge and Polite Culture* (Cambridge: Cambridge University Press, 1994), 4.
41. For the example of indigo cultivation, see Prakash Kumar, *Indigo Plantations and Science in Colonial India* (Cambridge: Cambridge University Press, 2012), Chapter 2.
42. Minute of Robert Kyd, June 1, 1786, in Biswas, *Original Correspondence*, 6–7; Heyne, *Tracts*, 365–66.
43. Michael Mann, "German Expertise in India? Early Forest Management on the Malabar Coast, 1792–1805," in *Explorations in the History of South Asia: Essays in Honour of Dietmar Rothermund*, ed. Georg Berkemer et al. (Delhi: Manohar, 2001), 9–26; David Arnold, "Plant Capitalism and Company Science: The Indian Career of Nathaniel Wallich," *Modern Asian Studies* 42 (2008): 913–14.

44. Royle, "Account," 42–44.
45. Arnold, *Tropics*, 106–7.
46. W. Carey, "Remarks on the State of Agriculture, in the District of Dinajpur," *Asiatick Researches* 10 (1812): 22–23.
47. For an earlier example of this with respect to cotton, see A. P. Hove, *Tours for Scientific and Economical Research made in Guzerat, Kattiawar, and the Conkuns, in 1787–88* (Bombay: Bombay Government Press, 1855).
48. On British climatic anxieties in India, see Philip D. Curtin, *Death by Migration: Europe's Encounter with the Tropical World in the Nineteenth Century* (Cambridge: Cambridge University Press, 1989), Chapter 1; Mark Harrison, *Climates and Constitutions: Health, Race, Environment and British Imperialism in India, 1600–1850* (New Delhi: Oxford University Press, 1999), Chapters 2–3.
49. For these factors in the establishment of the Calcutta botanic garden, see Biswas, *Original Correspondence*; Richard H. Grove, *Green Imperialism: Colonial Expansion, Tropical Island Edens and the Origins of Environmentalism, 1600–1860* (Cambridge: Cambridge University Press, 1995), 332–48.
50. On Humboldt, see J. A. Hodgson and J. D. Herbert, "An Account of Trigonometrical and Astronomical Observations for Determining the Heights and Position of the Principal Peaks of the Himalaya Mountains," *Asiatick Researches* 14 (1822): 211; James Annesley, *Researches into the Causes, Nature, and Treatment of the Most Prevalent Disease of India, and of Warm Climates Generally* (London: Longman, Rees, Orme, Brown & Green, 1828), 1: 77. On Humboldtian science, see Margarita Bowen, *Empiricism and Geographical Thought from Francis Bacon to Alexander von Humboldt* (Cambridge: Cambridge University Press, 1981); Michael Dettelbach, "Global Physics and Aesthetic Empire: Humboldt's Physical Portrait of the Tropics," in *Visions of Empire: Voyages, Botany, and Representations of Nature*, ed. David Philip Miller and Peter Hanns Reill (Cambridge: Cambridge University Press, 1996), 258–92.
51. James Bird, "A Statistical and Geological Memoir of the Country from Punah to Kittor," *Journal of the Asiatic Society of Bengal* 2 (1835): 65.
52. H. T. Colebrooke, "Introductory Note" to William Moorcroft, "A Journey to Lake Manasarovara in Undes, a Province of Little Tibet," *Asiatick Researches* 12 (1818), 380.
53. Hodgson and Herbert, "Account," 188.
54. Grout, "Geology," 83; Roger de Candolle and Alan Radcliffe-Smith, "Nathaniel Wallich and the Herbarium of the Honourable East India Company," *Botanical Journal of the Linnean Society* 83 (1981): 325–48.
55. Royle, "Account."
56. Thomas R. Trautmann, *Aryans and British India* (Berkeley: University of California Press, 1997), Chapter 2; Michael J. Franklin, *Orientalist*

Jones: Sir William Jones, Poet, Lawyer, and Linguist, 1746–1794 (Oxford: Oxford University Press, 2011).
57. "Analyses of Books," *Gleanings in Science* 1 (1829): 211.
58. Eugene F. Irschick, *Dialogue and History: Constructing South India, 1795–1895* (Berkeley: University of California Press, 1994); Bayly, *Empire and Information.*
59. Raj, *Relocating Modern Science.*
60. "Biographical Sketch," 334; Hamilton, *Account*, 1.
61. See the description of scientific instruments in Hodgson and Herbert, "An Account," 200–10. On "instruments of tillage," see Carey, "Remarks," 3–7; Heyne, *Tracts*, 74–75. For Buchanan's statistical approach, see Vicziany, "Imperialism."
62. Annesley, *Researches*, 2: 255.
63. James Jameson, *Report of the Epidemick Cholera Morbus, as It Visited the Territories Subject to the Presidency of Bengal in the Years 1817, 1818, and 1819* (Calcutta: Government Gazette Press, 1820).

Bibliography

Ainslie, Whitelaw. *Materia Indica.* 2 vols. London: Longman, Rees, Orme, Brown & Green, 1826.

"Analyses of Books." *Gleanings in Science* 1 (1829): 211–30.

Annesley, James. *Researches into the Causes, Nature, and Treatment of the Most Prevalent Disease of India, and of Warm Climates Generally.* 2 vols. London: Longman, Rees, Orme, Brown & Green, 1828.

Annesley, James. *Sketches of the Most Prevalent Diseases of India.* 2 vols. London: Thomas & George Underwood, 1825.

Archer, Mildred. *Natural History Drawings in the India Office Library.* London: Her Majesty's Stationery Office, 1962.

Arnold, David. "Agriculture and 'Improvement' in Early Colonial India: A Pre-History of Development." *Journal of Agrarian Change* 5 (2005): 505–25.

———. *The Tropics and the Traveling Gaze: India, Landscape and Science, 1800–1856.* Seattle: University of Washington Press, 2006.

———. "Plant Capitalism and Company Science: The Indian Career of Nathaniel Wallich." *Modern Asian Studies* 42 (2008): 899–928.

Ballingall, George. *Practical Observations on Fever, Dysentery, and Liver Complaints as They Occur amongst the European Troops in India.* Edinburgh: David Brown & A. Constable, 1818.

Barnett, Richard B., ed. *Rethinking Early Modern India.* New Delhi: Manohar, 2002.

Barrow, Ian J., and Douglas E. Haynes. "The Colonial Transition: South Asia, 1780–1840." *Modern Asian Studies* 38 (2004): 469–78.

Bayly, C. A. *Indian Society and the Making of the British Empire.* Cambridge: Cambridge University Press, 1988.

———. *Imperial Meridian: The British Empire and the World, 1780–1830.* Harlow: Longman, 1989.

Bayly, C. A. *Empire and Information: Intelligence Gathering and Social Communication in India, 1780–1870*. Cambridge: Cambridge University Press, 1996.

Bird, James. "A Statistical and Geological Memoir of the Country from Punah to Kittor." *Journal of the Asiatic Society of Bengal* 2 (1835): 65–80.

Biswas, Kalipada. *The Original Correspondence of Sir Joseph Banks Relating to the Foundation of the Royal Botanic Garden, Calcutta*. Calcutta: Royal Asiatic Society of Bengal, 1950.

Bowen, Margarita. *Empiricism and Geographical Thought from Francis Bacon to Alexander von Humboldt*. Cambridge: Cambridge University Press, 1981.

Buchanan, Francis. *A Journey from Madras through the Countries of Mysore, Canara, and Malabar*. 3 vols. London: T. Cadell & W. Davies, 1807.

Burkill, I. H. "William Jack's Letters to Nathaniel Wallich, 1819–1821." *Journal of the Straits Branch of the Royal Asiatic Society* 73 (1916): 147–268.

Carey, W. "Remarks on the State of Agriculture, in the District of Dinajpur." *Asiatick Researches* 10 (1812): 1–26.

Centenary Review of the Asiatic Society of Bengal, from 1784 to 1883. Calcutta: Thacker, Spink & Co., 1885.

Chakrabarti, Pratik. *Materials and Medicine: Trade, Conquest and Therapeutics in the Eighteenth Century*. Manchester: Manchester University Press, 2010.

Cohn, Bernard S. *Colonialism and Its Forms of Knowledge: The British in India*. Princeton: Princeton University Press, 1996.

Colebrooke, H. T. "Introductory Note" to William Moorcroft, "A Journey to Lake Manasarovara in Undes, a Province of Little Tibet." *Asiatick Researches* 12 (1818): 380.

Crawford, D. G. *A History of the Indian Medical Service, 1600–1913*. 2 vols. London: W. Thacker & Co., 1914.

Curtin, Philip D. *Death by Migration: Europe's Encounter with the Tropical World in the Nineteenth Century*. Cambridge: Cambridge University Press, 1989.

de Candolle, Roger, and Alan Radcliffe-Smith. "Nathaniel Wallich and the Herbarium of the Honourable East India Company." *Botanical Journal of the Linnean Society* 83 (1981): 325–48.

Desmond, Ray. *Wonders of Creation: Natural History Drawings in the British Library*. London: British Library, 1986.

———. *The European Discovery of the Indian Flora*. Oxford: Oxford University Press, 1992.

Dettelbach, Michael. "Global Physics and Aesthetic Empire: Humboldt's Physical Portrait of the Tropics." In *Visions of Empire: Voyages, Botany, and Representations of Nature*, edited by David Philip Miller and Peter Haans Reill, 258–92.Cambridge: Cambridge University Press, 1996:

Drayton, Richard. *Nature's Government: Science, Imperial Britain and the "Improvement" of the World*. New Haven: Yale University Press, 2000.

Edney, Matthew H. *Mapping an Empire: The Geographical Construction of British India, 1765–1843*. Chicago: University of Chicago Press, 1997.

Fleming, John. "A Catalogue of Indian Medicinal Plants and Drugs, with Their Names in the Hindustani and Sanscrit Languages." *Asiatick Researches* 11 (1810): 153–96.

Franklin, Michael J. *Orientalist Jones: Sir William Jones, Poet, Lawyer, and Linguist, 1746–1794.* Oxford: Oxford University Press, 2011.

Gascoigne, John. *Joseph Banks and the English Enlightenment: Useful Knowledge and Polite Culture.* Cambridge: Cambridge University Press, 1994.

Grout, Andrew. "Geology and India, 1770–1851: A Study in the Methods and Motivations of a Colonial Science." PhD diss., School of Oriental and African Studies, London, 1995.

Grove, Richard H. *Green Imperialism: Colonial Expansion, Tropical Island Edens and the Origins of Environmentalism, 1600–1860.* Cambridge: Cambridge University Press, 1995.

Habib, Irfan, ed. *Confronting Colonialism: Resistance and Modernisation under Haidar Ali and Tipu Sultan.* London: Anthem Press, 2002.

Hamilton, Francis. *An Account of the Kingdom of Nepal.* Edinburgh, Archibald Constable & Co., 1819.

Harrison, Mark. *Climates and Constitutions: Health, Race, Environment and British Imperialism in India, 1600–1850.* New Delhi: Oxford University Press, 1999.

———. *Medicine in an Age of Commerce and Empire: Britain and Its Tropical Colonies, 1660–1830.* Oxford: Oxford University Press, 2010.

Heyne, Benjamin. *Tracts, Historical and Statistical, on India.* London: Robert Baldwin, 1814.

Hodgson, J. A., and J. D. Herbert. "An Account of Trigonometrical and Astronomical Observations for Determining the Heights and Position of the Principal Peaks of the Himalaya Mountains." *Asiatick Researches* 14 (1822): 187–363.

Hove, A. P. *Tours for Scientific and Economical Research made in Guzerat, Kattiawar, and the Conkuns, in 1787–88.* Bombay: Bombay Government Press, 1855.

Irschick, Eugene F. *Dialogue and History: Constructing South India, 1795–1895.* Berkeley: University of California Press, 1994.

Jameson, James. *Report of the Epidemick Cholera Morbus, as It Visited the Territories Subject to the Presidency of Bengal in the Years 1817, 1818, and 1819.* Calcutta: Government Gazette Press, 1820.

Kidambi, Prashant. *The Making of an Indian Metropolis: Colonial Governance and Public Culture in Bombay, 1890–1920.* Aldershot: Ashgate, 2007.

Kumar, Prakash. *Indigo Plantations and Science in Colonial India.* Cambridge: Cambridge University Press, 2012.

Lambton, William. "An Account of a Method for Extending a Geographical Survey across the Peninsula of India." *Asiatick Researches* 7 (1808): 312–35.

Lambton, William. "An Account of the Measurement of an Arc of the Meridian." *Asiatick Researches* 12 (1818): 1–101.

Lambton, William. "Account of the Measurement of an Arc on the Meridian." *Asiatick Researches* 13 (1820): 1–127.

Lawson, Philip. *The East India Company: A History*. Harlow: Longman, 1993.

[Mackenzie, Colin]. "Biographical Sketch of the Literary Career of the Late Colonel Colin Mackenzie, Surveyor-General of India." *Journal of the Royal Asiatic Society* 1 (1834): 333–64.

Mann, Michael. "German Expertise in India? Early Forest Management on the Malabar Coast, 1792–1805." In *Explorations in the History of South Asia: Essays in Honour of Dietmar Rothermund*, edited by Georg Berkemer et al., 9–26. Delhi: Manohar, 2001.

Marshall, P. J. *The Making and Unmaking of Empires: Britain, India, and America, c. 1750–1783*. Oxford: Oxford University Press, 2005.

Martin, Montgomery, ed. *The Despatches, Minutes, and Correspondence of the Marquess Wellesley during his Administration in India*. 5 vols. London: John Murray, 1836.

———. *The History, Antiquities, Topography, and Statistics of Eastern India*. 3 vols. London: W. H. Allen, 1838.

Minutes of Evidence Taken before the Select Committee on the Affairs of the East India Company, V: Military (Cmd 735), 1832.

Oldham, C. E., ed. *Journal of Francis Buchanan Kept during the Survey of the District of Shahabad in 1812–1813*. Patna: Superintendent of Government Printing, Bihar and Orissa, 1926.

Peers, Douglas M. *Between Mars and Mammon: Colonial Armies and the Garrison State in Early Nineteenth-Century India*. London: I. B. Tauris, 1995.

———. "Colonial Knowledge and the Military in India, 1780–1860." *Modern Asian Studies* 33 (2005): 157–180.

Prain, D. "A Sketch of the Life of Francis Hamilton (once Buchanan)." *Annals of the Royal Botanic Garden, Calcutta* 10 (1905): i–lxxv.

Raj, Kapil. *Relocating Modern Science: Circulation and the Constitution of Scientific Knowledge in South Asia and Europe, 1650–1900*. Basingstoke: Palgrave Macmillan, 2007.

Roxburgh, William. *Plants of the Coast of Coromandel*. 3 vols. London: George Nicol, 1795–1819.

Royle, J. F. "Account of the Honorable Company's Botanic Garden at Saharanpur." *Journal of the Asiatic Society of Bengal* 1 (1832): 41–58.

Théodoridès, Jean. "Humboldt and England." *British Journal for the History of Science* 3 (1966): 39–55.

Trautmann, Thomas R. *Aryans and British India*. Berkeley: University of California Press, 1997.

Vicziany, Marika. "Imperialism, Botany and Statistics in Early Nineteenth-Century India: The Surveys of Francis Buchanan (1762–1829)." *Modern Asian Studies* 20 (1986): 625–60.

Wallich, N. "Descriptions of Some Rare Indian Plants." *Asiatick Researches* 13 (1820): 369–415.

Chapter 2

Collecting and the Pursuit of Scientific Accuracy: The Malaspina Expedition in the Philippines, 1792

Raquel A. G. Reyes

Introduction

This chapter focuses on botanical collecting, an important aspect of Spanish scientific voyaging in the Enlightenment and one of the most significant achievements of the Malaspina expedition. I will discuss the expedition's collecting in the Philippine Islands, Spain's Pacific possession. The information collected by the expedition, I will show, was distinguished by scientific accuracy, the pursuit of which profoundly influenced all the expedition's data-gathering tasks, from surveying and mapping to botanizing and drawing.

Toward the end of July 1789, Alejandro Malaspina, a brilliant 34-year-old Italian-born naval officer in the employ of the Spanish Crown, set sail from Cádiz in command of an ambitious expedition organized by the Spanish Crown and what would be one of the best examples of expeditionary science in the Spanish Enlightenment. The chief officer of the expedition was his friend and fellow officer José Bustamante y Guerra, a capable and courageous Spaniard, several years Malaspina's junior. Their ships were the *Atrevida* and the *Descubierta*, three-masted corvettes, each 120 feet long, each carrying 102 men, supplied with books, manuscripts, maps, and the latest scientific instruments; the accompanying personnel comprised naturalists, artists, draughtsmen, and a cartographer—men of exceptional learning, experience, and possessed of eclectic interests. The voyage, it was estimated, would take three-and-a-half years, and would attempt a comprehensive scientific exploration of Spanish territories in the Americas and the Pacific. Ultimately the journey took five years, following routes that explored the west coast of the Americas from Cape Horn to the Gulf of Alaska, several Pacific islands, the Philippines, New Zealand, Australia, and Atlantic Patagonia. Though celebrated on his return to Spain in 1794, Malaspina soon after became embroiled in political intrigues and he incurred the displeasure of Manuel de Godoy, a high-ranking minister at the court of Carlos VI. The entanglement cost him dearly. He was arrested for treason, imprisoned for six years (from 1796 to 1802) and, upon his release, was exiled for life in Italy, where he died in obscurity in 1810. As a consequence of Malaspina's fall from grace, the vast amount of valuable information collected by the expedition was impounded and suppressed, remaining unpublished until the late nineteenth century.[1]

The scale of this tragedy moved Alexander von Humboldt to remark that Malaspina was more famous for his misfortunes than for his discoveries.[2] The significant scholarly attention the expedition has lately received, however, has brought some vindication. The interests of Malaspina's select team of natural historians, artists, and hydrographers were so many and diverse that scholars have described the scope of the expedition as encyclopedic.[3] Dolores Higueras, curator of the data amassed by the expedition currently housed in the Museo Naval in Madrid, has shown that the expedition produced over 300 journals and logbooks; 450 albums of astronomical observations, 1500 hydrographic reports, 183 charts, 361 views of coastal elevations, and almost 1000 botanical and ethnographic drawings. The expedition made a contribution to science that, to paraphrase Felipe Fernandez-Armesto, "enhanced the world-picture."[4]

The expedition's sojourn in the Philippines in 1792 is an aspect of the voyage largely ignored by scholars.[5] The Islands were important to Spain at the time mainly because Manila functioned as an *entrepôt* for trade. From the city, the famed "Manila galleons" plied between Acapulco on the west coast of Mexico and Manila bringing shipments of silver bullion and minted coin to be exchanged for cargoes of Chinese goods, mainly silk.[6] Malaspina had been to the Philippines twice previously. In 1777, captaining the ship *Astrea,* he had sailed to and from the archipelago via the Cape of Good Hope, and in 1784, in the service of the Royal Philippine Company, he embarked on a circumnavigation of the globe traveling through ports on the east coast of South America, around Cape Horn before sailing north to Lima, and crossing the Pacific to the Philippines before returning to Spain via the Cape of Good Hope. It was these voyages that opened his eyes to the possibilities of new discoveries, and, inspired by the journeys of James Cook and Jean-François de La Pérouse, he put forward, in 1788, his ambitious "Plan of a Scientific and Political Voyage Around the World."[7]

Malaspina allotted a period of six months for the visit, or at least while favorable northeast winds blew, during which time he would collect hydrographic measurements and data on the depth of near-shore waters, survey and map coastlines, and collect botanical and animal specimens. The expedition stayed for nine months and for almost each day of those months, Malaspina kept a maritime diary describing in detail, and often in technical language, his activities—how tasks were undertaken, the logistical challenges faced, and the encounters with local peoples.

Science and the Collection of Plants and Animals

Since the early Renaissance, Europeans invested considerable time and energy in collecting objects from different cultures and distant places. Botanicals, things from nature (*naturalia*), curiosities, and rarities—the marvelous (*mirabilia*), those that originated from strange places (*exotica*), and human-made artifacts (*artificialia*) intellectually delighted, intrigued, and materially profited European audiences. As objects of fascination and appreciation, wondrous natural exotica and their collection and display in European museums and private curiosity cabinets have garnered the lion's share of scholarly attention.[8] To some extent, the imbalance has been redressed in recent years by an upsurge in works that have examined European collecting in India, the Atlantic world, colonial Latin America, and the Dutch East Indies.[9] These studies demonstrate the need to examine the connective tissues that linked intellectual, medical, and mercantile interests with imperialism, patriotism, and early globalization, factors which shaped thinking about nature and the so-called tropics. They richly describe the reliance of Europeans on local informants in the production and translation of scientific information, the strategies employed in the dissemination and circulation of knowledge, and the contributions to the Scientific Revolution made by exploring and collecting expeditions. Generally speaking, however, the early modern quest for, and collection of, medicinal plants and naturalia in what is today known as Southeast Asia, is less well studied. In the case of the Philippines under Spanish rule, the field has been almost entirely neglected.[10]

Spain's commercial rivalry with the British, French, and Dutch in the location, exploitation, and trade in valuable natural commodities such as pepper, spices, and tea is well known.[11] Historical research on Spanish scientific expeditions in the Spanish Empire has rightly emphasized Spain's competitive drive to collect botanicals for medicines and to profit from the global trade in spices and other luxuries. Like other European powers in the Enlightenment, Spain sought to explore the natural resources of far-flung regions for economic, political, and scientific gain. It was a utilitarian mind-set that was often allied to an equally fervent state intent on inventorizing resources and producing highly detailed information on Spain's possessions.[12] From out of 56 expeditions organized by the Spanish Crown from the mid-eighteenth century and prior to the nineteenth century, 5 were solely focused on plant collecting.[13] It is also likely that, as in the case of Malaspina's expedition, naturalist-botanists joined expeditions whose aims were not specifically botanical. From 1735 to 1805, 9 of those expeditions were specifically sent to the Philippines.[14] Naturalists, botanists, and artists, of diverse nationalities, joined these expeditions and utilized specific practices and technologies to collect and visualize the flora and fauna of Spanish possessions.[15]

The production of images based on careful observation and representation of people, plants, and animals encountered in new lands was a cornerstone of early modern European scientific practices in relation to natural history. However, in the Spanish Empire, visual representations of nature functioned on a host of other levels. Daniela Bleichmar has recently argued that a "way of knowing based on visuality" or what she has termed "visual epistemology" became an integral part of the Spanish imperial apparatus. Drawings of flora and fauna did not just serve natural history, they carried the imperial purpose "of making the New World knowable and governable [which] involved making it visible."[16] Paula De Vos has explored the extent to which the Spanish Crown also gave importance in the early eighteenth century to the collection of natural curiosities and rarities.[17] For instance, out of the 335 shipments of commercially useful plants that were sent to Madrid from all over the Empire in the latter half of the eighteenth century, 25 boxes contained curiosities. Those collected from the Philippines included live and preserved animals, shells, corals, and minerals, and an assortment of "monsters." These specimens, destined for the Bourbon royal cabinet and botanical garden, were important markers of power, status, and prestige.[18]

The recruitment of three naturalist-botanists for the Malaspina expedition reflected Spain's appetite for global botanizing. Malaspina's appointments are notable for their diverse nationalities and the wide-ranging accomplishments and expertise of those selected. Antonio

Pineda y Ramírez, appointed the expedition's chief of natural history, was a Spanish army officer born in Guatemala in 1753. He had served in the Royal Spanish Guards and had been assigned at the Royal Museum of Natural Science where he finished a work on the museum's collection of birds. Malaspina described him as intelligent, experienced in all branches of the natural sciences, and a man of "admirable energy and disposition."[19] The second naturalist of the expedition was a Frenchman, Luis Neé, who was born in Paris but had worked for most of his life in Spain. Neé had been employed for a number of years by the Royal Botanical Garden in Madrid. He had collected plants in the Basque region sending over 1,200 species from Navarre to Madrid in one year and had completed a study on the vegetation of southern Spain. As Malaspina noted, Neé was "well-versed in theoretical and practical botany, was indefatigable in his investigations."[20] The third recruit was the 28-year-old brilliant, resourceful Bohemian adventurer Tadeo Haenke. A polyglot, Haenke was fluent in Spanish, French, Latin, Italian, and German; he held a doctorate from the University of Prague, and had visited Syria, Corinth, and Tyrol. Arriving on board the expedition in spectacular fashion, Haenke had narrowly missed the expedition's departure from Cadiz and attempted to catch up in Buenos Aires. There, the boat he was on was shipwrecked and he managed to swim ashore bringing with him only a copy of Linnaeus. Missing the expedition again, he made his way to Santiago de Chile, the expedition's next stop, walking overland and collecting 2,500 plants along the way. He caught up with the expedition in 1790.[21] Finally, the expedition's physician and surgeon aboard the *Atrevida,* Pedro María González, was also a naturalist and contributed by describing a quantity of zoological specimens collected in Acapulco.[22]

Accompanying the work of the collectors were several artists who joined the expedition at separate phases of the journey. José Guió, a young artist and taxidermist, stayed with the expedition until Acapulco where they arrived in 1791; the highly educated and talented artist of perspective serving the *Descubierta,* José del Pozo was replaced by Tomás de Suría on account of the former's indolence; initially joining as a servant, the Spanish draughtsman José Cardero was recognized as an official artist of the expedition by the time they arrived in Acapulco, industriously working throughout the time in Alaska drawing the Tlingit before departing to join another expedition. The artists responsible for drawings undertaken in the Philippines were two Italians, the 28-year-old Milanese Fernando Brambila, and Juan Ravenet, the latter being Malaspina's family acquaintance.[23]

Political Contexts, National and Personal Motivations

The Malaspina expedition was assigned three tasks: (1) to map coastlines and trace navigation routes; (2) to inventory, classify, and evaluate the natural resources of Spanish territories; and (3) to undertake an investigation of the Empire's border zones. Broadly, the Spanish Bourbon monarchy attached the utmost importance to maintaining maritime control in the Pacific and reinforcing internal government within its viceroyalties. The expedition to Peru in 1735 by La Condamine, accompanied by Jorge Juan and Antonio de Ulloa, marks the first of 63 scientific expeditions that were organized by the Spanish monarchy throughout the eighteenth century until the early nineteenth century.[24] During this period, Spain allocated a bigger budget for scientific expeditionary purposes than any other European state. From 1765 onward, Charles III, renowned in his interest in and support for the development of natural sciences, sponsored a number of important expeditions whose range of objectives embraced global, hydrographic, botanical, cartographic, geostrategical, and astronomical aims. Between 1775 and 1800, nine expeditions set out for America with the particular purpose of studying its flora, fauna, and natural resources.[25] Of the expeditions sent out specifically to the Philippines, almost all were planned as hydrographic missions indicative of the crown's interest in defending shipping routes and the safety of navigation.[26] Spain sought, firstly, to ensure the defense and security of trade and navigation routes; secondly, to defend national honor; thirdly, to pursue scientific interests; and fourthly, political gain.[27] Similarly, the motives of French and English exploration of the Pacific in the 1760s and 1770s have often been identified as prosaic—motives of pride, rivalry, and national defense. The English desired to maintain their naval preeminence, a position which they were justly proud, and both countries were prepared to invest heavily in commercial prospects, expansion and the development of new arenas for trade.[28] In the age of Enlightenment, exploratory voyages made manifest the elevation of science, deeper philosophical interests in human societies and their development, and the concept that scientific achievement was closely linked to human progress. That said, Spain was, to be sure, also stimulated into action by the scientific activities of the English, French, and Russians in the Pacific. Spain's alarm at their activities stemmed from her belief that the area was exclusively reserved for her own trade and navigation. Spain's indignation clearly surfaced when British whalers were sighted off South American coasts and British fur traders at Nootka Sound.[29] Certainly, the Spanish did not attribute solely scientific aims to British exploration. Captain Cook's voyage in 1788, for instance, was perceived in terms of not only enriching scientific knowledge, but also strategically acquiring land. Speaking about the British settlement in New South Wales, one Spaniard noted: "The endeavours of the energetic Cook... have placed his Nation in a position to compensate itself for the loss of North America by the acquisition of a country almost as vast."[30]

But the threat to trade routes was not the only factor that angered Spain. Spanish expeditionary activity was also set within a political context that saw the country defending her beleaguered national honor. In examining eighteenth-century Spanish political economy, Jorge Canizares-Esguerra has recently shown that a characteristic of the Spanish Enlightenment was the tension between the patriotic repudiation of foreign criticism and confronting the reality of decline.[31] It was claimed that Spain had turned inward, inclined to keep itself, and others, from knowing more about the Pacific. During the first two decades of the eighteenth century, France and England had been busy exploring the west coast of South America for exploitable wealth. Mocking Spain for her alleged ignorance and backwardness, French and British intellectuals of the European Enlightenment would talk of Spain's "decline," pointing to a number of purported faults—from a mismanaged economy, indulgent and emasculated courtiers, to an idle population. Spain was censured for failing to accept new sciences and technologies produced by the Scientific Revolution, for her lack of knowledge about the New World, of the region's plants and natural history, and the general absence of scientific interest. To counter these accusations, Spain heavily funded and launched scientific expeditions and established botanical gardens.

Malaspina's project was commissioned by a Spanish monarchy partly in response to the threat made on its security and trade in the Pacific, and partly because Spain was defending its reputation against allegations of national decline. The expedition's purpose was to make a systematic investigation of Spain's possessions and survey imperial strengths and weaknesses. Its commitment to science was always linked to politics, commerce, and national honor.[32] In keeping with Enlightenment ideals, motives that aimed to advance sea navigation, geography, and the progress of humanity ostensibly did much to replace old sixteenth-century imperialist attitudes of conquest, religious conversion, and economic exploitation with which the Indies were typically regarded.[33] There was the expectation that the expedition would make: "new discoveries, careful cartographic surveys, important geodesic experiments in gravity and magnetism, botanical collections, and descriptions of each region's geography, mineral resources, commercial possibilities, political status, native peoples and customs."[34] To ensure the efficient organization and successful implementation of scientific tasks, the expedition followed a core set of procedures, or principles. Scientists were carefully chosen; learned institutions and scientists were consulted; and accumulated materials were to be systematically sent back to Spain. Moreover, following established practice, information would be obtained through carefully drafted questionnaires that would be completed with the cooperation of local scientists and institutions.[35]

But political gain remained an important subtext, and Malaspina's mission brought together scientific interests with the business of reporting on the political and economic state of the Spanish Empire, or, as it was spelled out: "to explore, examine, and knit together Madrid's far flung empire, report on problems and possible reforms, and counter the efforts of rivals to obtain colonial possessions at Spain's expense."[36] The aim was to obtain reliable data in order to gain a clear idea of Spain's overseas dominions, and the resources that could be harnessed for securing and maintaining Spanish hegemony in the Pacific.[37]

Did Malaspina go beyond these directives? Disdain for the old motives of imperial acquisitiveness, commerce, and power was not uncommon, and European intellectuals were wont to show both imperial and scientific high-mindedness.[38] Malaspina, however, appears to have emphasized the importance of science and politics in equal measure. Juan Pimentel has recently portrayed the Tuscan-born navigator and explorer as a "social Newtonian," whose moral standpoint, global vision, and Enlightenment ideals imprinted a universal character upon the expedition. Malaspina believed in a fundamental set of truths or what he called ten "axioms" that he deemed necessary for Spain's regeneration. Elaborated in *Axiomas políticos sobre la América* (1789), a work written before the departure of the expedition, they included a call for the liberalization of trade and local political institutions, the reform of Hispanic bureaucratic systems, and opposition to conquest and commercial monopolies.[39]

Malaspina in the Philippines

By the time the expedition arrived in the Philippines, the voyage was in its third year and had journeyed through Montevideo; the Rio de la Plata, the Falklands, and Tierra del Fuego; crossed South America and the Andes on foot; visited Lima, Panama, Prince William Sound off the Gulf of Alaska; Nootka Sound in present-day British Columbia; Acapulco; and Guam. "The nearer we came to the Philippines," wrote Malaspina on 22 February 1792, "the greater was my preoccupation regarding good planning to ensure that the expedition produced the best results without risking failure by being excessively adventurous.". Written en route to Port Palapag in the Philippines from Guam, the diary entry offers a glimpse of Malaspina's anxious mood. He was hoping to make charts of the country's principal islands. While in the Philippines, the corvettes would be refitted and the crew, exhausted from illness, suffering from fevers and "very severe attacks of bilious colic" and fearful of pirate attacks, would rest. The corvettes themselves were infested with cockroaches, so severe that cabins had to be dismantled to completely rid them of insects.[40]

Malaspina noted his first impressions of the country: "The scenes that meet the eye are so many and varied," he wrote, and, scanning the landscape of Sorsogon, observed, "a uniformly luxuriant vegetation, the terrain... gently rounded or steeply rising to volcanoes and other higher mountains, the watchtowers of each of the towns." Throughout the visit, the expedition was welcomed by Spanish officials who were attentive, provided them with food, horses, and armed escorts.[41] Upon landing, Malaspina sought the aid of missionaries while the crew collected seashells and bought provisions from Filipinos, making a point to buy cockerels which were used both as food and as amusement in cockfights. Indeed, Malaspina was particularly grateful that cockfights were frequent throughout the country because they served to distract his crew away from sexual temptations or "worse pastimes."[42] It was noticed that in some of the crew of both corvettes, the "dreadful symptoms of venereal disease" had appeared and had to be treated with mercury.[43] To prevent further infections, Malaspina imposed several disciplinary measures and enforced routines to keep behavior in check: "The marines were ordered to perform firearms drill twice a week and to have a clothing inspection once a week. Any seamen who missed roll call without leave for three days was to be considered a deserter... Anyone who transgressed these very moderate regulations... would be condemned to the Cavite galley."[44] What proved to be more effective however, were other distractions especially cockfights:

> A sailor's cravings are inexhaustible... there was an abundance of domestic fowls, and the cockerels, which are the object of the natives' constant attention... It had to be considered an advantage, therefore, that the useful diversion of the cock-fights continued throughout our

> stay in this port, distracting the men from other vices and providing a quantity of healthy food for the cooking pot.[45]

The plan was for the corvettes to separate, the *Atrevida* to sail onto Macau, while the *Descubierta* would remain. Malaspina was to send parties by land and sea to undertake soundings, chart harbors and survey coastlines covering eastern Luzon, the passage around the northeast extremity of the island of Samar and the various ports scattered from east to west at the southern end of Luzon. In Manila, Malaspina sought out local officials to gain information on the country's political conditions and natural history; artists sketched and painted ethnological portraits, vistas, and buildings. Antonio Pineda, Luis Neé, and Tadeo Haenke set out on inland excursions around the country to make observations and collect specimens.[46]

Output from the visit to the Philippines was substantial. Malaspina produced the first accurate charts of Luzon, Mindanao, Mindoro, and Negros; Felipe Bauza, the expedition's cartographer, drew coastline views and various maps of ports; documents in Filipino archives were copied and, in the field of botany, Neé discovered a variety of unknown plants and wrote a comparative study of plants found in Luzon and Mindanao.[47] Pineda's writings included his meteorological observations, research on volcanoes, minerals, commerce, and agriculture, and his observations of the Philippine peoples.[48]

In Pursuit of Scientific Accuracy

The corvettes were supplied with a library of predominantly British and French works on navigation, astronomical observations, and hydrographic charts ordered from London and Paris. It was equipped with the latest surveying and navigational instruments recommended by Sir Joseph Banks and purchased from the most important instrument-makers in London, with the help of Alexander Dalrymple.[49] Sixty-eight pieces of equipment were issued, of which sixty-one had been purchased from London.[50] Included in Malaspina's cache of precision instruments were a quantity of pocket chronometers (because they tended to break); the Ramsden astronomical quadrant, so famed because it was made by one of the most skilled instrument-makers in eighteenth-century England; a variety of telescopes and compasses; a eudiometer for measuring air quality on board the ship; at least two longitude chronometers; and, interestingly, a microscope that could have been the first microscope to arrive in the Philippines. This may well have been a solar microscope, invented in the 1750s, a useful apparatus for naturalists, which operated in a dark room much like a magic lantern, producing highly magnified images of very small objects with the aid of the sun's rays falling on a plane mirror outside the room, and reflecting toward a condensing lens.[51]

Malaspina's emphasis on accuracy was revealed by his habit of methodically comparing his data, and checking differences in mathematical calculations, with those obtained by previous expeditions. He regularly referred to Cook's writings, copies of which he kept on board the *Descubierta,* and Dalrymple's engraved charts, which he depended upon during storms and high winds.[52] In addition, he regularly compared his astronomical and hydrographic Philippine data with calculations taken by the Göttingen astronomer Johann Tobias Mayer (1723–62) (Mayer's star catalogue), and findings from at least two other expeditions. In the Philippines he consulted Le Gentil's longitudinal measurements obtained from the unfortunate Frenchman's hapless voyages to record the 1761 and 1768 transits of Venus, and astronomical data gathered by Jean-François de La Pérouse on his expedition to the Pacific from 1785 to 1788.[53]

Pineda's list of books carried by the expedition show a reliance on British and French secular accounts of Pacific voyages, and included the most authoritative works of the period on zoology, mineralogy, astronomy, and botany—from Linnaeus and Buffon to Plumier. He sought out titles from Paris and London, and works that detailed the latest methods in calculating distances and charting coastlines.[54] Works by local clerics, on the other hand, were conspicuous by their absence within the expedition's library.[55] At the time of Malaspina's arrival, much of what was already known about the Philippines was through works by the religious orders. Priests were well placed within a community, resided in the country for lengthy periods, and were at the vanguard of botanical and natural history writing in the Philippines. From the sixteenth century to the end of the eighteenth century, *Historias* and *Relaciónes,* written by friar and Jesuit scholars, offered the most detailed sources of information on the archipelago. Works such as Pedro Murillo Velarde's *Historia de la provincial de Philipinas de la Compania de Jesus* (1749), and the multivolume *Historia General de Philipinas* by the Augustinian Juan de la Concepción published in 1788, one year before Malaspina's departure, described local customs, flora, and fauna, and the geography and climate of the archipelago, with the aim of bringing science into the service of God.[56] The botanical drawings, detailed annotations, and collections of plant and animal specimens by the Czech Jesuit lay brother Georg Josef Camel, for example, were the first serious inquiry into the botanic and zoologic life found in the Philippine archipelago; and the two-volume *Historia* of the Visayan peoples, written in 1668 by the Jesuit Ignacio Alcina, was a richly detailed ethnography, the result of a 42-year residency in the Philippines and a strong grasp of local languages.[57]

Although Malaspina may not have consulted these clerical writings, he certainly did seek out Spanish missionary friars for information and practical support in his travels around the archipelago. Frequently, they were his first port of call, as when he immediately went to see the resident Franciscan friar when he arrived in Port Palapag.[58] While surveying the coasts of Pangasinan, Francisco Viana, an ensign on the voyage, drew on the hospitality of Dominican and Augustinian friars. Writing flatteringly of the religious orders, whom he says showed "the extent of their influence over the natives, undoubtedly won by gentleness, love and kindness," Viana compared the work of Spanish missionaries to Filipino secular priests. The former had brought civilization and order, as shown in clean and orderly towns, well-dressed and hardworking natives; the latter, he perceived, were just the same as the "mass of people" because they shared the customs of their fellow Filipinos; he charged them with idleness, bearing little influence over people, and administered towns where idleness, neglect, and laxity were widespread.[59]

The Art of Empirical Research

Mary Louise Pratt has argued that the exploration and documentation of continental interiors, from about 1750 to 1800, presented a new orientation in travel writing and a "major object of expansionist energies and imaginings."[60] Expedition writings relating to the interior exploration of the Philippines reflect an eclecticism of interests and give a forceful impression of industrious Europeans moving about the country in all directions, searching, observing, and collecting. Tadeo Haenke botanized (*herborizando*) his way across dense forests and sandy plains[61]; Luis Neé wrote with enthusiasm and curiosity toward all that he encountered—whether the subject was the plants he collected (in Acapulco, during one month alone, Neé had collected and described 30 new specimens of plants), or ethnography, "Indian" societies he came across, particularly the Negrito peoples with whom he stayed simply out of curiosity, or the physical terrain of the places he visited. Neé was also interested in zoology and conchology; he scaled the Albay volcano, noting the surrounding landscape of gypsum rocks, fissures, and gullies; he looked into the production of commercial crops and surveyed the cultivation and plantations of cocoa, mulberries, indigo, pepper, sugar, and cotton.[62] Antonio Pineda's account, written by Neé from notes and journals saved after his death, appears as a narrative of physical suffering, dogged perseverance, and industriousness under intolerable conditions. Juan Ravenet sketched Negrito peoples. Witnessing a demonstration of their dances and exchanging adornments, Ravenet saw them as innocents, showing him no distrust or ill will[63]. Commenting on the work of the naturalists in Sorsogon, Malaspina noted the return of Pineda and Haenke who had been away for several weeks collecting, bringing back " a vast number of acquisitions and much useful information." Impressed by their diligence under the difficult circumstances, he wrote:

> These industrious naturalists had, in the short space of eight days travelled through a considerable part of the province...made a close study of the volcano...and the molten rock...After this, botany and zoology had been the principal object of their observations, and although the rain, the rough tracks and the short time available were obstacles to the success of their excursion, their survey could nonetheless be considered complete, particularly as Don Luis Neé had explored the area surrounding the harbour with his usual zeal.[64]

Neé was collecting in the Bicol region, in Sorsogon, the southernmost province of southern Luzon, through the provinces of Albay and Camarines before returning to Manila and the environs. He was careful to take note of useful plants such as *abaca,* fibers obtained from a species of banana that could be used for rope and cordage; forests "full of excellent timber for construction," planting of *bonga* palms, indigo, and cocoa; and indigenous uses of plants especially in relation to medicines and textiles.[65]

Meanwhile, Haenke made his way from Manila to Ilocos Norte, the northernmost region of Luzon. En route he was aided by many priests, one of whom, Father Fray Manuel del Barrio, a poet, Haenke praised for his "vast erudition and love of science."[66] Like his colleagues, Haenke kept an eye out for the profitable, taking note of gold mines worked by the *igorots*, a term given by the Spaniard to the indigenous peoples of the Cordillera region, and their cultivation of various cottons. He remarked on the fertility of the area, relentlessly collecting all plants that he considered important, new, and precious. Over a period of two weeks he had amassed a "large collection of well-dried plants of about 2000 species" taking care that the collections were well protected from damage and humidity. On his return journey to Manila, he collected the flowers of plants that had come into season and, despite a brief bout of illness, was able to write his observations "on the little known fruition of [the] palms."[67]

Antonio Pineda set out on 11th April, a few days ahead of Haenke. He tarried awhile in the rural environs of Manila, examining first the cinnamon plantations and the thermal waters of Laguna, the latter accomplished by following the methods of the renowned Swedish scientist Torbern Olof Bergmann (1735–83).[68] He observed how the springs of Laguna were used to irrigate over 5,000 cinnamon trees and 500,000 trees of various species, and tested the waters for their purity.[69] Accompanied on the early leg of his journey by Juan de Cuellár, a naturalist who had been commissioned to collect plants by the Real Compañia,[70] Pineda intended, eventually, to travel north, by way of Pampanga, in central Luzon, and follow the entire length of the Cagayan river that flowed through northeastern Luzon, passing through settlements established by the Dominicans. He carried with him a barometer, a hydrometer, several other boxes, a few provisions, and a single blanket. The hostilities of the *igorots* toward Spaniards had compelled him to travel with 20 Filipinos armed with spears and arrows; he, his European servant, and the painter, possibly Ravenet, carried muskets. In northern Luzon, the party's progress was slow due to the difficult terrain—steep, slippery mountain slopes, tortuous tracks, and thickly forested areas impassable by horseback.[71] Throughout the journey, Pineda made the acquaintance of priests who sheltered and shared with him their knowledge. In Cagayan, for instance, a learned Dominican friar Fray Antonio Lobato shared his observations on the Kalingas, the rice farmers of the Cordilleras; in Badoc he was tended to by Augustinians.[72] Pineda himself was attentive: he noted the fine quality of the cocoa that grew on the rocky soils of mountain ridges; he observed how the Spanish tobacco monopoly had impoverished villages and diminished original trades such as the collection of wax; and he praised the industriousness of women weavers.[73]

Neé's retelling of Pineda's journey was inclined to emphasize the arduousness and peril Pineda encountered, thus deterministically presenting the journey, and Pineda's death, as most importantly distinguished by heroism. Pineda, Neé recounted, suffered innumerable deprivations and hardships that took a toll on his health. Already beleaguered by fevers, most likely malaria, his condition took a turn for the worse after drinking from the waters of the Cagayan river, though he knew the water to be impure. He died on 23rd June at the age of 38.

Conclusion

The Malaspina expedition was driven by a host of purposes that were to directly benefit the Spanish Crown—from the making of hydrographic charts to assist in mercantile navigation within even the remotest regions of Spain's Empire, to the execution of a systematic and comprehensive survey of American politics, commerce, natural resources, manufacturing, and defense, in relation to Spain. But Malaspina also infused the undertaking with an expansive enlightened thinking that sought to gather information and collect all manner of exotica, flora and fauna, for the Royal Museum and Botanical Garden, where everything could be studied and exhibited for the benefit of science and humanity. The careful and internationally diverse choice of officers on the Malaspina expedition, men chosen for their scientific breadth, depth, eclecticism, and experience, appears to have been a case of obtaining the best men to carry out the expedition's special scientific aims. State support for Malaspina's international choices seems to indicate a distinct lack of fear of espionage, a confidence though difficult to explain was rewarded by Malaspina and others. The Czech naturalist Tadeo Haenke, for instance, had been appointed on the recommendation of the Austrian ambassador to Spain, and continued to serve the Spanish Crown as a naturalist until his death in Alto Perú, what is today Bolivia, on November 4, 1816.[74]

"Science," as Jorge Cañizares-Esguerra has written and other scholars have also observed, "was the handmaiden of the Iberian empires."[75] The eighteenth century was a favorable moment for Spanish science. The period saw the establishment of a number of important scientific institutions in Madrid (the Royal College of Medicine, the Astronomical Observatory, for instance); and Spanish scientific exploration excelled most notably in the field of botany. Naturalist-botanists on board expeditions sponsored by the Crown sent innumerable floral and faunal specimens to Spain; completed thousands upon thousands of folio botanical illustrations; and wrote detailed descriptions of new specimens, valuable woods, precious resins and gums, medicinals, curiosities, and fish and other marine life. Yet, conspiracies at court and imprudence were able to consign the considerable achievements of one of the most extensive and far-ranging expeditions launched by Spain to obscurity until the late twentieth century. Two volumes of hydrographic and astronomical information were published in 1809, but this was only a small fraction of the expedition's work.[76] Why had the expedition's findings been permitted to languish for so long? Scholars have pointed to the enduring "bureaucratic culture of secrecy," *arcane imperii,* long upheld by the Spanish monarchy.[77] Knowledge, it was thought, could only be circulated in manuscript form among the trusted, a strategy that would effectively obstruct the ambitions of imperial rivals, or so it was reasoned. Perhaps combined with this culture of secrecy, there might also have been a general distrust of print and print culture, though knowledge was allowed to circulate in the form of images—masses and masses of drawings that might have served as mnemonic aids so that plants and knowledge of distant places were experienced visually.[78] What is certain is that the failure to publish information from the Malaspina expedition greatly deprived Spain of national prestige. In the Philippine case, the information gathered by the botanists Pineda, Haenke, and Neé would have substantially added to the knowledge already gathered by priests. Had their findings been published, the authority of clerical writings would surely have been challenged long before the late nineteenth century.

Notes

1. The Malaspina expedition and the life and political thought of Alejandro Malaspina are well documented. A few important works are Maria Dolores Higueras Rodriguez, *Catálogo Critic de los Documentos de la Expedición Malaspina en el Museo Naval,* 3 vols. (Madrid: Museo Naval, 1989–1994); Andrew David, Felipe Fernandez-Armesto, Carlos Novi, and Glyndwr Williams, with an introduction by Donald C. Cutter. *The Malaspina Expedition 1789–1794: Journal of the Voyage by Alejandro Malaspina,* trans. Sylvia Jamieson, 3 vols. (London and Madrid: The Hakluyt Society and the Museo Naval, Madrid, 2004); Iris H. W. Engstrand, *Spanish Scientists in the New World: The Eighteenth Century Expeditions* (Seattle: University of Washington Press, 1981); Juan Pimentel, *La Física de la Monarquía: Ciencia y Política en el Pensamiento Colonial de Alejandro Malaspina (1754–1810)*, Colección de Historia Natural *Theatrum Naturae* (Aranjuez: Doce Calles, 1998); Juan Pimentel, *Testigos del Mundo: Ciencia, Literatura y Viajes en la Ilustración* (Madrid: Marcial Pons Historia, 2003).
2. Alexander von Humboldt, *Political Essay on the Kingdom of New Spain*, vol.II, trans. John Black (London: Longman, Hurst, Rees, Orme and Brown, 1811), 376.
3. Juan Pimentel, *Viajeros Cientificos: Tres Grandes Expediciones al Nuevo Mundo; Jorge Juan, José Celestino Mutis y Alejandro Malaspina* (Madrid: Nivola, 2001); Virginia González Claverán, *La Expedición Científica de Malaspina en Nueva España, 1789–1794* (Mexico: El Colegio de México, 1988).
4. Felipe Fernandez-Armesto, *Pathfinders: A Global History of Exploration* (Oxford: Oxford University Press, 2006), 306.

5. Exceptionally, for a chronological survey of the expedition including the Philippines, see Emilio Soler Pascual, *La Aventura de Malaspina: la Gran Expedición Científica del Siglo XVIII por las Costas de América, las Filipinas y las Islas del Pacífico* (Barcelona: Ediciones B, 1999); José María A. Cariño and Sonia Pinto Ner, *Álbum: Islas Filipinas 1663–1888* (Hong Kong; Ars Mundi Philippinae, 2004), esp. 42–72.
6. William Lytle Schurz, *The Manila Galleon: The History of the Spanish Galleons Trading between Acapulco and Manila* (New York: Dutton, 1959).
7. Engstrand, *Spanish Scientists in the New World*, 45.
8. Work on European collecting and collecting practices is vast. Some prominent examples are: Thomas DaCosta Kaufmann, *The Mastery of Nature: Aspects of Art, Science, and Humanism in the Renaissance* (Princeton, NJ: Princeton University Press, 1993); John Elsner and Roger Cardinal, eds., *The Cultures of Collecting* (Cambridge, MA: Harvard University Press, 1994); Lorraine Daston and Katherine Park, *Wonders and the Order of Nature 1150–1750* (New York: Zone Books, 1998); Paula Findlen, *Possessing Nature: Museums, Collecting, and Scientific Culture in Early Modern Italy* (Berkeley: University of California Press, 1994); Nicholas Jardine, Emma Spary, and J. A. Secord, eds., *Cultures of Natural History* (Cambridge: Cambridge University Press, 1996); Anthony Grafton and Nancy Siraisi, *Natural Particulars: Nature and the Disciplines in Europe* (Cambridge, MA: MIT Press, 1999); Pamela H. Smith and Paula Findlen, eds., *Merchants and Marvels: Commerce, Science and Art in Early Modern Europe* (London: Routledge, 2002); Eddy Stols, Werner Thomas, and Johan Verberckmoes, eds., *Naturalia, Mirabilia e Monstrosa en los Imperios Ibéricos* (Leuven: Leuven University Press, 2007); also Part 3, chapters 9, 10, and 11 in Claire Farago, ed., *Reframing the Renaissance: Visual Culture in Europe and Latin America 1450–1650* (New Haven and London: Yale University Press, 1995). I thank Thomas DaCosta Kaufmann for bringing this book to my attention.
9. David Arnold, *Science, Technology and Medicine in Colonial India* (Cambridge: Cambridge University Press, 2000); several art historical studies on the Mughal empire, for instance, Som Prakash Verma, *Flora and Fauna in Mughal Art* (Bombay: Marg Publications, 1999); Harold J. Cook, *Matters of Exchange: Commerce, Medicine, and Science in the Dutch Golden Age* (New Haven and London: Yale University Press, 2007); Londa Schiebinger, *Plants and Empire: Colonial Bioprospecting in the Atlantic World* (Cambridge, MA: Harvard University Press, 2004); Londa Schiebinger and Claudia Swan, eds., *Colonial Botany: Science, Commerce, and Politics in the Early Modern World* (Philadelphia: University of Pennsylvania Press, 2007); Londa Schiebinger, "Forum Introduction: The European Colonial Science Complex," *Isis* [Special Issue] 96 (2005): 52–55; Jorge Cañizares-Esguerra, *Nature, Empire and Nation: Explorations of the History*

of Science in the Iberian World (Stanford: Stanford University Press, 2006); Roy MacLeod, ed., "Nature and Empire: Science and the Colonial Enterprise," *Osiris* [Special Issue] (2000). See also selected essays in the Variorum volumes: Ursula Lamb, ed., *The Globe Encircled and the World Revealed*, vol. 3 (Ashgate: Variorum, 1995); Felipe Fernández-Armesto, ed., *The Global Opportunity*, vol. 1 (Ashgate: Variorum, 1995); and Sanjay Subrahmanyam, ed., *Merchant Networks in the Early Modern World, 1450–1800*, vol. 8 (Ashgate: Variorum, 1996).

10. Scholarly work is scattered and fragmentary. For recent treatments on Jesuit botanizing with mention of the Philippines, see Sabine Anagnostou, "Jesuits in Spanish America and Their Contribution to the Exploration of the American Materia Medica," *Pharmacy in History* 47 (2005): 3–17; Sabine Anagnostu, "International Transfer of Medicinal Drugs by the Society of Jesus (Sixteenth to Eighteenth Centuries) and Connections with the Work of Carolus Clusius," in *Carolus Clusius in a New Context: Cultural Histories of Renaissance Natural Science*, ed. Florike Egmond, Paul Hoftijzer, and Rob Visser (Amsterdam: Edita Pubs. Royal Dutch Academy, 2006), 293–312; and Raquel A. G. Reyes, "Botany and Zoology in the Late Seventeenth-Century Philippines: The Work of Georg Josef Camel, SJ (1661–1706)," *Archives of Natural History* 36 (2009): 262–76. On the early-seventeenth-century Franciscan botanist Father Blas de la Madre de Dios, see Francisco Guerra and María Carmen Sánchez Tellez, *El Libro de Medicinas Caseras de FR. Blas de la Madre de Dios: Manila, 1611* (Madrid: Editorial Cultura Hispánica, Fac de Medicina Alcalá de Henares, 1986); Evelyn Ansay-Miranda, "Positive Contributions of Spanish Religious Scientists in Philippine Society," in *Imperios y Naciones en el Pacífico*, vol. 1, ed. Maria Dolores Elizalde, Josep M. Fradera, and Luis Alonso (Madrid: Biblioteca de Historia, CSIC, 2001), 585–95; Mercedes Planta, "Traditional Medicine and Pharmacopoeia in the Philippines, 16th and 17th Centuries," *Historical Perspectives on East Asian Science, Technology and Medicine* (Singapore: Singapore University Press, 1999). For a brief mention of the role of Juan de Cuéllar and the cultivation of pepper, coffee, cacao, cinnamon, indigo, and mulberry trees for silkworms in the Philippines, see Daniela Bleichmar, *Visible Empire: Botanical Expeditions and Visual Culture in the Hispanic Enlightenment* (Chicago and London: University of Chicago Press, 2012), 135–37.

11. For a recent survey of early modern European commercial and imperial rivalry, see Ronald Findlay and Kevin H. O'Rourke, *Power and Plenty: Trade, War, and the World Economy in the Second Millennium* (Princeton, NJ: Princeton University Press, 2007).

12. Antonio Barrera-Osorio, *Experiencing Nature: The Spanish American Empire and the Early Scientific Revolution* (Austin: University of Texas Press, 2006); Daniela Bleichmar et al. eds., *Science in the Spanish and*

Portuguese Empires 1500–1800 (Stanford: Stanford University Press, 2009).

13. Chiyo Ishikawa ed., *Spain in the Age of Exploration 1492–1819* (Seattle: University of Nebraska Press, 2004), 171–72.
14. Ibid.
15. Daniela Bleichmar, "A Visible and Useful Empire: Visual Culture and Colonial Natural History in the Eighteenth-Century Spanish World," *Science in the Spanish and Portuguese Empires,* 290–309.
16. Bleichmar, *Visible Empire*, 10. On the application of "visuality" to the study of cultural interaction in the early modern period see Claire Farago, "Understanding Visuality," in *Seeing Across Cultures in the Early Modern World,* ed. Dana Leibsohn and Jeanette Favrot Peterson (Farnham: Ashgate, 2012), 239–57.
17. Paula De Vos, "The Rare, the Singular, and the Extraordinary: Natural History and the Collection of Curiosities in the Spanish Empire," *Science in the Spanish and Portuguese Empires,* 271–90.
18. Ibid. On botanical collecting and the culture of gift exchange and patronage during the Hispanic Enlightenment see Daniela Bleichmar, "Visible Empire: Scientific Expeditions and Visual Culture in the Hispanic Enlightenment," *Journal of Postcolonial Studies,* 12 (2009): 441–66. On the origins of the Royal Natural History Cabinet, see María Angeles Calatayud, *Pedro Franco Dávila y el Real Gabinete de H.a Natural* (Madrid: CSIC, 1988).
19. Engstrand, *Spanish Scientists in the New World,* 46.
20. Ibid., 47; John Kendrick, *Alejandro Malaspina: Portrait of a Visionary* (McGill: Queen's University Press, 1999), 41.
21. Engstrand, *Spanish Scientists in the New World,* 48.
22. Ibid., 52; Iris H. Wilson Engstrand, " Of Fish and Men: Spanish Marine Science during the Late Eighteenth Century," *Pacific Historical Review* 69 (2000): 3–30.
23. For a comprehensive catalogue of the expedition's artists, see Carmen Sotos Serrano, *Los Pintores de la Expedicion de Alejandro Malaspina* (Madrid: Real Academia de la Historia, 1982). Also, José Torre Revello, *Los Artistas Pintores de la Expedición Malaspina*, vol. 2 (Buenos Aires: Estudios y Documentos para la Historia del Arte Colonial Universidad de Buenos Aires, 1944).
24. Scholarship on early modern Spanish colonial science in the Philippines is extremely limited. While tentative research on science in nineteenth-century Philippines has begun, literature is slight on earlier periods and is frequently found in the form of translations, annotations, and reprints of manuscripts. Exceptionally, some work has been done in the fields of cartography, navigation, and astronomy in the mid to late sixteenth century, most notably on the charts, maps, and sketches of Southeast Asia made by the Portuguese cartographer Francisco Rodrigues in 1515 and the Augustinian friar Andrés de Urdaneta, well known for discovering a fast route to Acapulco from

Manila. See Felipe-Fernandez Armesto, *Pathfinders*; the Filipino historian Horacio de la Costa closely examined Jesuit activities in the Philippines, including some of their scientific contributions: Horacio de la Costa, *The Jesuits in the Philippines, 1581–1768* (Cambridge, MA; Harvard University Press, 1961); in the history of botany, see Maria Belen Bañas Llanos, *Una Historia Natural de Filipinas: Juan de Cuéllar* (Madrid: Ediciones del Serbal, S.A., 2000).

25. Maria Higueras, "The Sources; The Malaspina and Bustamante Expedition: A Spanish State enterprise," appendix 6 to *The Malaspina Expedition 1789–1794: Manila to Cadiz*, vol. 3 (London: Hakluyt Society, 2004), 372.
26. Ishikawa, ed., *Spain in the Age of Exploration, 1492–1819*, 185–86.
27. J. H. Elliot, *Spain, Europe and the Wider World 1500–1800* (New Haven and London: Yale University Press, 2009).
28. Daniel Baugh, "Seapower and Science: The Motives for Pacific Exploration," in *Scientific Aspects of European Expansion*, ed. William K. Storey (London: Ashgate Variorum, 1996), 85–139.
29. See Alan Frost, *The Global Reach of Empire: Britain's Maritime Expansion in the Indian and Pacific Oceans 1764–1815* (Carlton, VIC: Miegunyah Press, 2003); Alan Frost and Jane Samson, eds., *Pacific Empires: Essays in Honour of Glyndwr Williams* (Vancouver: University of British Columbia Press, 1999).
30. Alan Frost, "Science for Political Purposes: European Explorations of the Pacific Ocean, 1764–1806," *Scientific Aspects of European Expansion*, 37.
31. Jorge Canizares-Esguerra, *Nature, Empire, and Nation: Explorations of the History of Science in the Iberian World* (California:Stanford University Press, 2006), 96.
32. Scholars have portrayed the Malaspina expedition as particularly representative of a "new phase" in Spain's relationship with the New World. Richard L. Kagan and Benjamin Schmidt have emphasized the exploratory drive, and the scientific objectives underpinning the expedition.
33. Ishikawa, ed., *Spain in the Age of Exploration, 1492–1819.*
34. Frost, "Science for Political Purposes," 38.
35. The questionnaires that circulated in the 1570s and 1580s came to be known as the *Relacíoneses de Indias.* Antonio Barrera-Osorio has carefully elucidated Spain's information-gathering strategies in relation to the Spanish administration and domination of the Americas, and the management of natural resources: Barrera-Osorio, *Experiencing Nature,* 82–83.
36. Frost, "Science for Political Purposes," 39.
37. Higueras, "The Sources," 371.
38. Baugh, "Sea Power and Science," 40.
39. Juan Pimentel, "Alessandro Malaspina (1754–1810): In Search of the Nature of Empire" (lecture, Malaspina University College).

40. *The Malaspina Expedition 1789–1794 Panama to the Philippines*, vol. 2, 319.
41. See for instance the treatment the crew received in Zamboanga, described in *The Malaspina Expedition 1789–1794, Manila to Cadiz*, vol. 3, 16.
42. Ibid., 276.
43. *The Malaspina Expedition 1789–1794, Panama to the Philippines*, vol. 2, 324.
44. Ibid., 322.
45. Ibid., 289.
46. Ibid., 273–74.
47. Luis Neé, "Observaciones Botanicas Islas Philipinas 1792," unpublished manuscript, Archivo de Real Jardín Botanico, Madrid.
48. Cariño and Ner, *Album*, 47.
49. Andrew David, "Surveying and Navigational Instruments and Related Books Supplied to the *Descubierta* and *Atrevida*," appendix 5 to *The Malaspina Expedition 1789–1794, Panama to the Philippines*, vol. 3, 361.
50. Ibid.
51. Gerard L'Estrange Turner, *Scientific Instruments, 1500–1900: An Introduction* (Berkeley: University of California Press, 1998), 96; Robert S. Whipple, "Scientific Instruments in the Eighteenth Century," *The Philosophical Magazine* (1948), 113–21.
52. *The Malaspina Expedition 1789–1794, Panama to the Philippines*, vol. 3, 4.
53. Ibid., 302–3.
54. Engstrand, *Spanish Scientists in the New World*, 48.
55. David, "Surveying and Navigational Instruments," 365–70.
56. Others include Juan de Plasencia, *Relacíon de las Costumbres los Tagalos* (1589); Francisco Colin, *Labor Evangelíca* (1604), Pedro Chirino, *Relacíon de las Islas Filipinas* (1663); and Francisco Combes *Historia de las Islas de Mindanao, Jolo y sus Adjacentes* (1667).
57. Ignacio Francisco Alcina, *History of the Bisayan People in the Philippine Islands (1668)*, ed. Cantius J. Kobak and Lucio Gutierrez (Manila: University of Santo Tomas Publishing House, 2005), 19.
58. *The Malaspina Expedition 1789–1794, Panama to the Philippines* vol. 2, 275.
59. Ibid., 361.
60. Mary Louise Pratt, *Imperial Eyes,* 23.
61. *The Malaspina Expedition 1789–1794, Panama to the Philippines* vol. 2, ff.7, 399. Notes by Tadeo Haenke on Philippine plants are in Div. VI Serie, Tadeo Haenke, Leg.2–3, Archivo de Real Jardin Botanico, Madrid.
62. *The Malaspina Expedition 1789–1794, Panama to the Philippines,* vol. 2, 389.

63. Ibid., 324.
64. Ibid., 290.
65. Ibid., 390.
66. Ibid., 400.
67. Ibid., 402.
68. Ibid., 404.
69. Ibid. The cinnamon from the Philippines was not the true cinnamon, *Cinnamomum Zeylanicum*, which was then found only in Sri Lanka.
70. Lourdes Díaz-Trechuelo, Susana Pinar, Ma. Belén Bañas Llanos, Juana Molia Nortes, and Domingo A. Madulid eds., *La Expedición de Juan de Cuéllar a Filipinas* (Madrid: Caja Madrid Real Jardín Botánico, 1997); María Belén Bañas Llanos, *Una Historia Natural de Filipinas: Juan de Cuéllar, 1739?–1801* (Barcelona: Ediciones del Serbal, 2000); María Pilar de San Pío Aladrén, ed., *La Expedición de Juan de Cuéllar a Filipinas* (Madrid: Lunwerg, 1997).
71. *The Malaspina Expedition 1789–1794, Panama to the Philippines*, vol. 2, 404.
72. Ibid., 409–11. For a detailed analysis of the expedition's ethnographic interests see David J. Weber, *Bárbaros: Spaniards and Their Savages in the Age of Enlightenment* (New Haven and London: Yale University Press, 2005).
73. *The Malaspina Expedition 1789–1794, Panama to the Philippines*, vol. 2, 408–10.
74. José Ignacio González-Aller Hierro, "Officers and Supernumeraries on the Malaspina Expedition," appendix 3 to *The Malaspina Expedition 1789–1794, Panama to the Philippines*, vol. 2, 345.
75. Jorge Cañizares-Esguerra, "Introduction," *Science in the Spanish and Portuguese Empire*, 1.
76. Carlos Novi, "The Road to San Antón: Malaspina and Godoy," appendix 2 to *The Malaspina Expedition 1789–1794, Panama to the Philippines*, vol. 3, 330–31.
77. Jorge Cañizares-Esguerra, "Introduction," 3.
78. Ibid., 4–5.

Bibliography

Alcina, Ignacio Francisco. *History of the Bisayan People in the Philippine Islands (1668)*, edited by Cantius J. Kobak and Lucio Gutierrez. Manila: University of Santo Tomas Publishing House, 2005.

Anagnostou, Sabine. "Jesuits in Spanish America and Their Contribution to the Exploration of the American Materia Medica." *Pharmacy in History* 47 (2005): 3–17.

Aladren, María Pilar de San Pío, ed. *La expedición de Juan de Cuéllar a Filipinas*. Madrid: Lunwerg, 1997.

Ansay-Miranda, Evelyn. "Positive Contributions of Spanish Religious Scientists in Philippine Society." In *Imperios y Naciones en el Pacífico*, edited by Maria Dolores Elizalde, Josep M. Fradera, and Luis Alonso, vol. 1, 585–95. Madrid: Biblioteca de Historia, CSIC, 2001.

Arnold, David. *Science, Technology and Medicine in Colonial India.* Cambridge: Cambridge University Press, 2000.

Barrera-Osorio, Antonio. *Experiencing Nature: The Spanish American Empire and the Early Scientific Revolution.* Austin: University of Texas Press, 2006.

Baugh, Daniel. "Seapower and Science: The Motives for Pacific exploration." In *Scientific Aspects of European Expansion,* edited by William K. Storey, 85–139. London: Ashgate Variorum, 1996.

Bleichmar, Daniela. "Visible Empire: Scientific Expeditions and Visual Culture in the Hispanic Enlightenment." *Journal of Postcolonial Studies* 12 (2009): 441–66.

Bleichmar, Daniela, *Visible Empire: Botanical Expeditions and Visual Culture in the Hispanic Enlightenment.* Chicago and London: University of Chicago Press, 2012.

Bleichmar, Daniela, et al., eds. *Science in the Spanish and Portuguese Empires 1500–1800.* Stanford: Stanford University Press, 2009.

Calatuyod, María Angeles. Pedro Franco Dávila y el Real Gabinete de H.a Natural. Madrid: CSIC, 1988.

Canizares-Esguerra, Jorge. *Nature, Empire and Nation: Explorations of the History of Science in the Iberian World.* Stanford: Stanford University Press, 2006.

Carino, José María A. and Sonia Pinto Ner. *Álbum: Islas Filipinas 1663–1888.* Hong Kong: Ars Mundi Philippinae, 2004.

Claveran, Virginia González. *La Expedición Científica de Malaspina en Nueva España, 1789–1794.* Mexico: El Colegio de México, 1988.

Cook, Harold J. *Matters of Exchange: Commerce, Medicine, and Science in the Dutch Golden Age.* New Haven and London: Yale University Press, 2007.

Daston, Lorraine, and Katherine Park. *Wonders and the Order of Nature 1150–1750.* New York: Zone Books, 1998.

David, Andrew, Felipe Fernandez-Armesto, Carlos Novi, and Glyndwr Williams, with an introduction by Donald C. Cutter. *The Malaspina Expedition 1789–1794: Journal of the Voyage by Alejandro Malaspina.* Translated by Sylvia Jamieson. 3 vols. London and Madrid: The Hakluyt Society and the Museo Naval, Madrid, 2004.

De la Costa, Horacio. *The Jesuits in the Philippines, 1581–1768.* Cambridge, MA: Harvard University Press, 1961.

De Vos, Paula. "The Rare, the Singular, and the Extraordinary: Natural History and the Collection of Curiosities in the Spanish Empire." In *Science in the Spanish and Portuguese Empires,* edited by D. Bleichmar et al., 271–90. California: Stanford University Press, 2009.

Diaz-Trechuelo, Lourdes, Susana Pinar, Ma. Belén Bañas Llanos, Juana Molia Nortes, and Domingo A. Madulid, eds. *La Expedición de Juan de Cuéllar a Filipinas.* Madrid: Caja Madrid Real Jardín Botánico, 1997.

Egmond, Florike, Paul Hoftijzer, and Rob Visser, eds. *Carolus Clusius in a New Context: Cultural Histories of Renaissance Natural Science*, 293–312. Amsterdam: Edita Pubs. Royal Dutch Academy, 2006.
Elliot, J. H. *Spain, Europe and the Wider World 1500–1800*. New Haven and London: Yale University Press, 2009.
Elsmer, John, and Roger Cardinal, eds. *The Cultures of Collecting*. Cambridge, MA: Harvard University Press, 1994.
Engstrand, Iris H. W. *Spanish Scientists in the New World: The Eighteenth Century Expeditions*. Seattle: University of Washington Press, 1981.
Engstrand, Iris H. Wilson. "Of Fish and Men: Spanish Marine Science during the Late Eighteenth Century." *Pacific Historical Review* 69 (2000): 3–30.
Farago, Claire, ed. *Reframing the Renaissance: Visual Culture in Europe and Latin America 1450–1650*. New Haven and London: Yale University Press, 1995.
Fernandez-Armesto, Felipe. *Pathfinders: A Global History of Exploration*. Oxford: Oxford University Press, 2006.
———, ed. *The Global Opportunity*. Vol. 1. Ashgate: Variorum, 1995.
Findlay, Ronald, and Kevin H. O'Rourke. *Power and Plenty: Trade, War, and the World Economy in the Second Millennium*. Princeton, NJ: Princeton University Press, 2007.
Findlen, Paula. *Possessing Nature: Museums, Collecting, and Scientific Culture in Early Modern Italy*. Berkeley: University of California Press, 1994.
Frost, Alan. *The Global Reach of Empire: Britain's Maritime Expansion in the Indian and Pacific Oceans 1764–1815*. Carlton, VIC: Miegunyah Press, 2003.
Frost, Alan, and Jane Samson, eds. *Pacific Empires: Essays in Honour of Glyndwr Williams*. Vancouver: University of British Columbia Press, 1999.
Guerra, Francisco and María Carmen Sánchez Tellez. *El Libro de Medicinas Caseras de FR. Blas de la Madre de Dios: Manila, 1611*. Madrid: Editorial Cultura Hispánica, Fac de Medicina Alcalá de Henares, 1986.
Ishikawa, Chiyo, ed. *Spain in the Age of Exploration 1492–1819*. Seattle: University of Nebraska Press, 2004.
Jardine, Nicholas, Emma Spary, and J. A. Secord, eds. *Cultures of Natural History*. Cambridge: Cambridge University Press, 1996.
Kaufmann, Thomas DaCosta. *The Mastery of Nature: Aspects of Art, Science, and Humanism in the Renaissance*. Princeton, NJ: Princeton University Press, 1993.
Kendrick, John. *Alejandro Malaspina: Portrait of a Visionary*. McGill: Queen's University Press, 1999.
Lamb, Ursula, ed. *The Globe Encircled and the World Revealed*. Vol. 3. Ashgate: Variorum, 1995.
Llanos, María Belén Bañas. *Una Historia Natural de Filipinas: Juan de Cuéllar*. Barcelona: Ediciones del Serbal, 2000.
MacLeod, Roy, ed. "Nature and Empire: Science and the Colonial Enterprise." *Osiris* [Special edition] 15 (2000).

Neé, Luis. "Observaciones Botanicas Islas Philipinas 1792." Unpublished manuscript. Archivo de Real Jardín Botanico, Madrid, 1792.
Pascual, Emilio Soler. *La Aventura de Malaspina: la Gran Expedición Científica del Siglo XVIII por las Costas de América, las Filipinas y las Islas del Pacífico.* Barcelona: Ediciones B, 1999.
Pimentel, Juan. *La Física de la Monarquía: Ciencia y Política en el Pensamiento Colonial de Alejandro Malaspina (1754–1810).* Aranjuez: Doce Calles, 1998.
———. *Viajeros Cientificos: Tres Grandes Expediciones al Nuevo Mundo; Jorge Juan, José Celestino Mutis y Alejandro Malaspina.* Madrid: Nivola, 2001.
———. *Testigos del Mundo: Ciencia, Literatura y Viajes en la Ilustración.* Madrid: Marcial Pons Historia, 2003.
Planta, Mercedes. "Traditional Medicine and Pharmacopoeia in the Philippines, 16th and 17th Centuries." In *Historical Perspectives on East Asian Science, Technology and Medicine,* edited by Alan K. L. Chan et al., 157–71. Singapore: Singapore University Press, 1999.
Revello, José Torre. *Los Artistas Pintores de la Expedición Malaspina.* Buenos Aires: Estudios y Documentos para la Historia del Arte Colonial Universidad de Buenos Aires, 1944.
Reyes, Raquel A. G. "Botany and Zoology in the Late Seventeenth-Century Philippines: The Work of Georg Josef Camel, SJ (1661–1706)." *Archives of Natural History* 36 (2009): 262–76.
Rodriguez, Maria Dolores Higueras. *Catálogo Critic de los Documentos de la Expedición Malaspina en el Museo Naval.* 3 vols. Madrid: Museo Naval, 1989–94.
Schiebinger, Londa. *Plants and Empire: Colonial Bioprospecting in the Atlantic World.* Cambridge, MA: Harvard University Press, 2004.
———. "Forum Introduction: The European Colonial Science Complex." *Isis* [Special Issue] 96 (2005): 52–55.
Schiebinger, Londa, and Claudia Swan, eds. *Colonial Botany: Science, Commerce, and Politics in the Early Modern World.* Philadelphia: University of Pennsylvania Press, 2007.
Schurz, William Lytle. *The Manila Galleon: The History of the Spanish Galleons Trading between Acapulco and Manila.* New York: Dutton, 1959.
Serrano, Carmen Sotos. *Los Pintores de la Expedición de Alejandro Malaspina.* Madrid: Real Academia de la Historia, 1982.
Siraisi, Nancy. *Natural Particulars: Nature and the Disciplines in Europe.* Cambridge, MA: MIT Press, 1999.
Smith, Pamela H., and Paula Findlen, eds. *Merchants and Marvels: Commerce, Science and Art in Early Modern Europe.* London: Routledge, 2002.
Stols, Eddy, Werner Thomas, and Johan Verberckmoes, eds. *Naturalia, Mirabilia e Monstrosa en los Imperios Ibéricos.* Leuven: Leuven University Press, 2007.
Subrahmanyam, Sanjay, ed. *Merchant Networks in the Early Modern World, 1450–1800.* Vol. 8. Ashgate: Variorum, 1996.

Turner, Gerard L'Estrange. *Scientific Instruments, 1500–1900: An Introduction*. Berkeley: University of California Press, 1998.

Verma, Som Prakash. *Flora and Fauna in Mughal Art*. Bombay: Marg Publications, 1999.

Von Humboldt, Alexander. *Political Essay on the Kingdom of New Spain*. Translated by John Black. London: Longman, Hurst, Rees, Orme and Brown, 1811.

Weber, David J. *Bárbaros: Spaniards and Their Savages in the Age of Enlightenment*. New Haven and London: Yale University Press, 2005.

Whipple, Robert S. "Scientific Instruments in the Eighteenth Century." *The Philosophical Magazine: Natural Philosophy through the Eighteenth Century and Allied Topics* (Commemoration Number to Mark the 150th Anniversary of the Foundation of the Magazine), 113–21. London: Taylor and Francis, 1948.

Chapter 3

Empire without Science? The Dutch Scholarly World and Colonial Science around 1800

Klaas van Berkel

Introduction: The Absentee Dutch Scientists

The year 1761 promised to be a very rewarding year for the scientific community in Europe. On June 6, a conjunction would occur between the Sun and the planet Venus. On that day, anyone with a good telescope might see a tiny dark spot crossing the disc of the Sun, like a fly walking across a lamp; at least as long as the weather was nice, since a few clouds could easily spoil everything. Yet, observation of the so-called transit of Venus was important for the scientific community. The exact measurement of the time the planet needed for this transit would enable astronomers to calculate the distance between the Sun and the Earth—a fundamental astronomical standard.[1]

All over the world, astronomers were preparing for this major scientific event. Despite the Seven Years War, astronomers from almost all European countries traveled to distant places on earth to observe the transit of Venus. A few years earlier, the French astronomer Joseph-Nicolas Delisle had drawn a so-called *mappemonde*, a map of the earth indicating where one might be able to see the transit of Venus. This map proved to be very helpful, and astronomers raised their telescopes in places as far away as Irkoetsk in Siberia, the island of St. Helens in the southern Atlantic, and the trading post of Pondicherry in India, as well as in Paris, London, and St. Petersburg. Never before had such a complicated and well-coordinated scientific expedition taken place.

Due to bad weather conditions, the results were meager, to say the least. No results of any significance were published. Yet all the costly expeditions were not without some kind of reward. The fact that astronomers from so many countries had been able to set up an expedition of this magnitude was perhaps much more important than the actual results. And besides, in 1769, astronomers had a second chance to observe the transit of Venus. These transits do not occur very often, but when they do, they do so in couples with only a few years in between. As soon as the 1761 expeditions were over, astronomers began planning for the next transit.

While in 1761 the French had taken the lead, now English astronomers were leading the way. No less than four English expeditions were sent out to places such as the North Cape, nearby Hammerfest, the Hudson Bay in northern America, and the southern Pacific (this was the famous expedition led by Captain James Cook). But the French were not far behind; after consultation with the British, they sent astronomers to the Philippines, southern California, and the Caribbean island of Santo Domingo. There were, of course, other expeditions by the Swedes, the Russians, and even the British colonists in northern America. In total, some 151 observers, in at least 77 places across the world, were preparing for the event—significantly more than in 1761. The amount of money spent on these expeditions far exceeded the amount of money spent on the 1761 expeditions. Once again the astronomical results were somewhat disappointing, but the increase of knowledge in botany and geography was a reward in itself. During his trip to the southern Pacific, Cook for instance finally disproved the existence of the mythical Southern Continent or Terra Australis. Furthermore, on account of the preparations for this expedition, the ties between the state and the scientific community were strengthened, to the benefit of both parties.

Among the many astronomers who observed the transits of Venus in 1761 and 1769, the Dutch were surprisingly absent. Not completely so, since a few scattered observations took place, but compared to what other colonial powers had done, the contribution of the Dutch was negligible. In 1761 and 1769, the best observations the Dutch could boast of were actually made by Johann Moritz Mohr, a German minister in Batavia, the capital of the Dutch colonial empire in Asia. In 1761 Mohr had observed the transit of Venus on his estate outside Batavia, and in 1769 he was able to do his observations from a recently constructed astronomical observatory on the roof of his new house in Batavia. Not only was the building itself amazing, it was stuffed with the most modern equipment one could think of. Mohr had married a wealthy widow and therefore money was not a major problem for him. Still, his observations and calculations were unimpressive, and for all his money his contribution to the advancement of astronomy in the eighteenth century was practically zero.[2]

The absence of substantial Dutch participation in the observation of the transits of Venus is remarkable, since in the preceding century the Dutch Republic had been one of the leading nations as far as astronomy was concerned. In the beginning of the seventeenth century, the telescope was more or less invented in the Netherlands by the Middelburg spectacle maker Hans Lipperhey (although, of course, Galileo had been the first to use the telescope as an astronomical instrument). During the second half of the century, Christiaan Huygens was, without a doubt, one of the best astronomers in Europe, embedding some sort of astronomical tradition in Holland. Furthermore, the Dutch had a huge colonial empire, stretching from Surinam in the West Indies and the Cape of Good Hope on the southern tip of Africa all the way to southern India, Sri Lanka, Indonesia, and finally, the tiny island of Deshima in Japan. So the question remains, why were there no Dutch expeditions to Indonesia or southern Africa to observe the transits of Venus? Why did it have to be Mohr—a lonely scientist, a minister of German descent—who saved the honor of the Dutch in 1761 and 1767?

Remarkable as the absence of a serious contribution of the Dutch to eighteenth-century astronomy may be, it was by no means exceptional. In disciplines like botany, medicine, and linguistics, the Dutch lagged behind as well. It is not at all difficult to find major Dutch contributions to colonial science in the seventeenth century. Jacob Bontius improved our botanical knowledge of the Indies, Hendrik van Reede tot Drakestein did the same for the Malabar Coast in India, and George Rumphius meticulously documented the natural world of Ambon and other islands in the eastern part of the Indonesian archipelago. Willem Piso and Georg Markgraf published on the natural history of Dutch Brazil, and Maria Sibylla Merian carefully studied and painted birds and insects in Surinam.[3] These are just a few of the medical doctors, colonial officials, and amateur scientists whose names still figure in modern histories of seventeenth-century colonial science. In contrast, it would be very difficult, not to say impossible, to find an equally impressive string of names from the eighteenth century, even though the Dutch still controlled a large colonial empire. Again and again one wonders why the Dutch did not exploit their colonies scientifically in the same way the British and the French did.

Questions like these have been asked before. In the historical literature, explanations of the so-called decline of Dutch culture, including science, abound. I will not deal with this larger issue; instead I will concentrate on explaining the near invisibility of the Dutch in colonial science in the eighteenth century. I will argue that there were several structural deficiencies in the Dutch scientific system that made it impossible for them to participate in colonial science in the manner in which France, England, or even Russia did. In the preceding century, the way in which science in the Netherlands was organized fitted reasonably well with the way in which science was practiced. In the late seventeenth and early eighteenth centuries, however, science changed, but the system, at least in the Netherlands, did not change accordingly. In the final paragraph I will also show that as soon as the system did change, during and after the Napoleonic period, the interest in colonial science in Holland manifested itself again.

Changing Patterns in Eighteenth-Century Science

A major change in the nature of science in the eighteenth century was the introduction of the scientific expedition. In the years around 1800, expeditions like the ones to observe the transits of Venus in such faraway places as Pondicherry and the Hudson Bay became an integral part of science. In the eighteenth century, however, this way of practicing science was relatively new. In 1735, the first major international scientific expedition was launched, in a joint effort to determine once and for all the exact shape of the earth. The French had taken the initiative and wanted to determine whether the earth was a sphere, as French, meaning Cartesian, geography would have it, or a spheroid that was flattened at the poles, like the Newtonians in Britain claimed. One team of scientists and geographers, led by the French mathematician Mauptertuis, went north to Lapland, to measure a longitudinal degree at the Meridian. Another group of scientists headed to South America to take the same measurement at the equator near Quito. Officially, this expedition was led by the mathematician Louis Godin, but it has gone down in history under the name of one of its few survivors, the geographer Charles de la Condamine, simply because it was La Condamine who reported to the Academy of Sciences.

Scientific expeditions are expensive of course and because of their complicated organization they can only be launched by major scientific institutions. In the case of the 1735 expedition, the French Academy of Sciences set up the entire enterprise. Likewise, the 1761 and 1768 expeditions to observe the transits of Venus were fitted out by the larger scientific academies in Europe, for instance, the Académie des Sciences in Paris, the Royal Society in London, and the Imperial Academy in St. Petersburg. The results of these expeditions, meager though they were, were also published in the transactions and memoirs of these academies. In the eighteenth century, when science began to require substantial organizational and financial backing, only scientific academies were able to provide these. In disciplines like mathematics or linguistics, the solitary university professor or the wealthy amateur was still able to contribute substantially, but in colonial science the demands of logistics and diplomatic support made the intervention of an academy that was founded or chartered by the king or the government indispensable.

Keeping this in mind, it becomes clear why the Dutch lagged far behind in the domain of colonial science in the eighteenth century. The Dutch Republic could boast several good universities, but scientific academies were nonexistent until the second half of the eighteenth century, and those that were founded after 1750 were of a purely private and in fact, though not in theory, local or at the most regional character. The Dutch had every reason to be proud of their universities.[4] Each of the seven provinces that together constituted the Dutch Republic had the authority to found a university, and most of them had done so in the course of the late sixteenth and early seventeenth centuries. Leiden was founded in 1575 as the university for the provinces of Holland and Zeeland. The province of Friesland followed in 1585 with the University of Franeker, while the province of Groningen instituted its university in 1614. Later on, other provinces founded universities in Utrecht, Harderwijk, and Nijmegen (the last one for a short time only). There also existed the so-called illustrious schools—institutions that provided academic instruction without having the authority to issue degrees. Amsterdam, for instance, had no university, but it did have an illustrious school that, during at least part of its history, functioned like a small university. Taken together, these universities and illustrious schools constituted an impressive system of higher education in the Dutch Republic. In the seventeenth century, the universities—particularly Leiden—attracted a large number of foreign students. In the eighteenth century, however, when most of the foreigners went to other places or found opportunities for higher education nearer home, the Dutch universities continued to provide adequate education for the inhabitants of their own provinces, as they were meant to from the very start. For aspiring young scholars and scientists, a job as a university professor was a reasonable career perspective. It paid well, one had a lot of time for research or other occupations, and in polite society the professor was a well-respected man. Of course for someone like Christiaan Huygens, who came from an aristocratic background, a university career was unthinkable, but in general the university provided a hospitable environment for scientists and scholars. It is therefore no coincidence that in the seventeenth century, Dutch university professors, even though they did not have the obligation to do research, participated more in the enterprise of science than in surrounding countries. Even in the early eighteenth century, Leiden professors, like the physician Herman Boerhaave and the physicist Willem Jacob 's Gravesande, were scientists with an international reputation, not unlike the anatomist Petrus Camper, who in the second half of the eighteenth century was a professor of medicine in Franeker, Amsterdam, and Groningen.

The existence of an excellent network of universities made the foundation of an academy of science less urgent, at least in the eyes of the scholars. The local authorities, meaning the city governments and the provincial estates, did not feel the need to found a scientific academy either. As the governing body of a city or a province they only felt responsible for education, not for science, and since they already spent a lot of money on the provincial universities, the idea of founding a scientific academy did not even cross their minds. And why should it, when all the major scientific academies were the creation not of local regents or wealthy individuals, but of ruling monarchs? In the case of academies in Paris, St. Petersburg, Leipzig, and later in Brussels and several German principalities, it was the prince who took the initiative to create state-financed academies of science. Although the situation in London was more complicated, the Royal Society being a private affair with private funding and only a formal charter given by the king, the mere fact that it was a *Royal* Society made a big difference. In the Dutch Republic there was no monarch. The princes of the House of Orange and Nassau were only stadholders and therefore, in theory, the servants of the provincial estates. Some of the provinces did not even have a stadholder for considerable stretches of time. Besides, the stadholders did not have the financial resources to incorporate an academy of science into their tiny courts in The Hague and Leeuwarden. Thus it is not surprising that scientific academies did not exist in the Dutch Republic. The federal structure of the country, the relatively flourishing state of the universities, and the absence of a prince with the wealth and the power to create a scientific academy can easily explain why the example of the Royal Society in London and the Academy of Sciences in Paris was not followed.

Now what about the Dutch East India Company? In the seventeenth century, the East India Company had been, at times at least, very helpful in furthering the scientific and scholarly exploration of the Indies.[5] For those interested in the natural history of the tropics, a job with the East India Company was the only way to get there and to find time for research. A surprisingly large number of employees of the Company did indeed spent some time doing botanical or medical research in the Indies. We also know the names of several leading members of the board of the East India Company—men like Nicolaes Witsen and Johan Huydecooper van Maarsseveen—who used the facilities of the Company to further their own interests in natural history or the history of the people living overseas.[6] Even later in the eighteenth century the Company was still profitable enough to continue supporting scientific research in the colonies. An example is provided by the governor of Ceylon, Joan Gideon Loten (1710–1789), who contributed to the knowledge of nature by initiating a collection of detailed drawings of East Indian animals and plants, and communicating his findings to the naturalists in the Dutch Republic and England (Loten was not a naturalist himself, but rather a connoisseur and a dilettante).[7] In a sense, therefore, the East India Company functioned as a scientific institution in itself. Furthermore, besides stuffing the cabinets of the wealthy with exotic objects and providing naturalists and other scholars with opportunities to do research in the tropics, the Company also influenced science by favoring a new way of assessing and evaluating knowledge in general. The rise of this new science was in some way related to the new appraisal of the methods through which the merchants, who ruled companies like the East India Company, gathered their information about spices, trade routes, and potential competitors. Commercial interests definitively stimulated accurate description, objective information, and the recognition of the value of accumulated knowledge—whether in medicine, geography, or botany. Although it is an overstatement to say that "the intellectual activities we call science emerged from the ways of knowing valued most highly by the merchant-rulers of urban Europe," the connection is unmistakable.[8]

Yet there is reason to be careful. The Dutch East India Company was not founded to stimulate science; it just proved to be the only vehicle for those who wanted to investigate nature in the colonies. The Company provided some facilities, but hardly ever actively encouraged scientific research. On the contrary, there are several instances in which the board of the Company or its local officials actively discouraged or even tried to forbid scientific inquiry in the districts under their control, mostly because too much time was lost in doing research, or because it was feared that the Company's commercial and political rivals might take advantage of the outcomes. For this reason, much of Rumphius's pathbreaking studies of the fish, plants, and shells of

Ambon and the Maluku Islands was kept under lock and key by the governing board of the East India Company for many years. Certainly in the eighteenth century, the Company could in no way make up for the absence of modern scientific institutions in the Dutch Republic.

The Learned Societies

But then, in 1752, an informal group of amateur scientists in the city of Haarlem, not far from Amsterdam, was transformed into the official Hollandsche Maatschappij der Wetenschappen—the Holland Society of Sciences ("Holland" referring to the province of Holland, not to the Dutch Republic as a whole).[9] The society was soon given a charter by the estates of the province of Holland, thereby acquiring a semiofficial status. The founding members or directors (most of them wealthy regents with no particular scientific background) attracted new scientific members and appointed paid officials. They then proceeded to collect scientific objects and made plans to publish scientific treatises and issue prize competitions—just like the larger scientific societies in surrounding countries. Soon the example of Haarlem was followed in other cities and provinces within the Dutch Republic. Restricting ourselves to societies at least partially devoted to natural science, those worthy of at least a mention are: the Zeeland Society of Sciences at Vlissingen, founded in 1767, the Batavian Society for Experimental Philosophy in Rotterdam, also founded in 1767, the Provincial Utrecht Society for Arts and Sciences, founded in 1778, and finally, in the very same year, the Batavia Society of Arts and Sciences, founded in the city of Batavia in the Dutch East Indies (present-day Jakarta). Around the same time, Haarlem saw the creation of a second scientific society within its walls when Teylers Society was established after the death and on the request of a wealthy Mennonite merchant, Pieter Teyler van der Hulst.

Could these societies, or at least some of them, provide the institutional backing that modern science in the eighteenth century, and certainly colonial science, needed? Before answering this question, let us first look at what these societies actually did to promote colonial science. The Batavia Society of Arts and Sciences was explicitly created to further research in the Indonesian archipelago. But what about the others; what about the most prestigious of all Dutch societies, the Holland Society of Sciences in Haarlem? What did the society do with its money; was there a genuine interest in colonial science? Did the topics proposed for the prize essays that were issued each year touch on colonial science and did the so-called *Verhandelingen* [*Memoirs*] of the society contain letters, essays, and treatises that deal with the natural history of the colonies?

Let us first take a look at the prize competitions that were on the society's program from the mid-eighteenth century until the beginning of the twentieth century.[10] During roughly the first half-century of the existence of the society, almost one hundred prize competitions were advertised, of which only seven could in any way be labeled as "colonial." It started in 1758 with the question of what was known of the diseases of the crews on board the ships that went to the West Indies, and what could be done to remedy those diseases. At first no answer was received, but after repeating the prize competition in 1759, two entries were received: one came from Edinburgh and the other was written by the Rotterdam doctor Salomon de Monchy. Dr. de Monchy was awarded a gold medal and the essay was published in the sixth volume of the *Memoirs* of the society.[11] The second prize contest dates from a few years later. In 1774, the society asked for essays discussing the possibilities of introducing new crops in the West Indies, besides the already existing cultures of coffee, sugar, cacao, and cotton. Preferably, the essay was to be based on real experiments, executed in the Dutch West Indies or in other colonies with a comparable climate. Although the prize competition was repeated in 1776, no one took the trouble to answer the call.[12] Of course, this will not have encouraged the directors and scientific members of the society to devise other prize contests specifically devoted to the colonies.

A few years later, however, the society received an offer from the newly founded Batavia Society in the East Indies.[13] The young colonial society had transferred 315 guilders to the Haarlem Society to enable it to issue prize contests "for the benefit of our country or its colonies." A committee was created to formulate a number of prize competitions, and from 1779 onward a colonial prize contest was advertised almost every year. The first competition dealt with an urgent problem encountered by the inhabitants of Batavia, to wit, the stagnant water in its canals, which was believed to be a major cause of the declining health conditions in the city. The contest was repeated in 1785 and 1790 and indeed three entries were received, but none of them was judged to be good enough for a gold medal.[14] The results of the other colonial prize contests were not encouraging either. The topics were attractive enough: the improvement of secondary education in Batavia, the introduction of the Dutch language in the colonies in the East so as to strengthen the ties between the Dutch and the indigenous peoples, the acclimatization of Europeans in the tropical climate, and finally Indonesian medicine, including medicine as practiced by the Chinese. Yet apparently no Dutch scholar was able or interested in colonial topics and so none of these last four prize contests led to prize essays, let alone to an award for one of them. After 1785 the colonial prize contest was quietly dropped from the program.

When we look at the contents of the *Memoirs* of the Haarlem Society, the situation is hardly better. In the first volume, published in 1754, only one report is indirectly related to colonial science. It is a list of meteorological observations made by the crew of a ship that had arrived in Batavia in October 1752. Most of these observations deal with temperature and were executed with a newly invented mercury thermometer. Nothing, however, is done with these observations. They had simply been recorded and sent to Haarlem, hoping that perhaps someone might later find them useful.[15] Meteorological observations are also the subject of some reports in the second volume of the *Memoirs*, published in 1755. The Leiden professor of philosophy and natural philosophy Jean Nicolas Sébastien Allamand (1713–1789) contributed a translation of meteorological observations made at the Cape of Good Hope by the Frenchman Abbé de la Caille. Somewhat later, Dr. M. Seymens, an alderman from the small shipping port of Enkhuizen, reported on the winds that were blowing in and around the city of Batavia and about a curious meteorological event that had happened way back in 1730—presumably around the time Seymens had himself been in Batavia as a young merchant.[16] One does not have to be an expert in the history of meteorology to see that these scattered observations have in no way furthered the science of meteorology. In the fourth volume of the *Memoirs*, the same magistrate from Enkhuizen published a treatise on the natural causes of the trade winds and monsoon rains in the Indian Ocean and the seas between Japan, China, and Indonesia. He even inserted a map of this region in which he indicated the normal directions of the winds at several places. This map is more interesting than the long lists of meteorological data that were printed in almost every volume of the *Memoirs* of the Haarlem Society of Sciences.

Also interesting is a report, in the second volume of the *Memoirs*, about an electrical fish in South America.[17] Allamand, the aforementioned Leiden professor, relates how two years earlier he had received the remains of an electric eel, sent to him by a gentlemen called 's Gravesande. Allamand, who is sometimes credited with being the coinventor of the Leiden jar alongside Petrus van Musschenbroek, and who was interested in electrical phenomena—more specifically in the presumed electrical nature of neurological phenomena—immediately became interested and asked for more details, which he got in a second letter. On the basis of this letter, he drew up a report that was published in the *Memoirs*. Allamand's correspondent was Laurens Storm van 's Gravesande (1704–1775), who for many years was the director of several Dutch plantations on the banks of the Rio Issiquibo, now known as Essequebo in Guyana, some 400 kilometers to the west of Paramaribo in Surinam and at that time still a Dutch colony. Allamand introduces 's Gravesande as a trustworthy observer. He tells us that his correspondent is the director of a large plantation, and that he spends a lot of time, at least when he is not running the business of the plantation, thoroughly getting to know everything that is worthy of study in the places that are under his command. We soon understand why Allamand trusts this colonial official so much. 's Gravesande describes the sensation of touching this remarkable eel. It is the sensation of electricity; the same sensation that he felt, so he says, "when I was with you and held in my hand a bottle that was connected, by way of an iron wire, to an electrical machine." Here he refers to visits to Holland in 1750 and 1751, during which he also went to Leiden, where his uncle, Willem Jacob's Gravesande, was a professor of natural philosophy. The professor must have introduced his nephew to his colleague Allamand, who taught natural history, but was also an expert in modern physics and apparently had an electrical machine at his disposal. In his report, Storm van 's Gravesande admits that others have written about this electrical fish before. Allamand confirms this, but he nevertheless calls upon his readers to provide him with better specimens of this fish, preferably alive, so that he can dissect it and find out what causes these electrical shocks. There is, according to Allamand, a specimen in the Leiden Cabinet of Curiosities (of which he was the curator), but since this object was so highly valued—it seems to have been one of the showpieces of the cabinet—the board of the university would not allow him to dissect it. He also writes that he hopes for fresh specimens of this creature and even details how to keep the animal alive during the journey from the West Indies back to Holland.

Five years later a second letter concerning the electric eel was published in the *Memoirs* of the Holland Society.[18] In this letter, which was communicated to the society by an unnamed member, a Dutch surgeon called Frans van der Lott, who like Storm van 's Gravesande was located in the Dutch colony of Essequibo, gives a detailed description of the effects that an electric eel could have on people and how the shock one gets on touching the eel might also cure various forms of neurological diseases. Van der Lott, for instance, tells his correspondent that when five persons were put on in row, hand in hand, with the one at the far left touching the eel, the person at the far right would also feel the shock. Several slaves on the plantation who had fallen ill had been cured by the electric shock drawn from the eel. Some of these experiments were executed in the presence of Storm van 's Gravesande, some even at his home. One is inclined to think that 's Gravesande, who had no medical background, had stimulated Van der Lott to further investigate the enigmatic animal and that the correspondent who communicated the letter to the society once again was Allamand.

These letters form an interesting example of the exchange between metropolitan and colonial science. Allamand was not interested in the electric eel as such. The taxonomy of the different eels and fish was relatively unimportant to him. What mattered to him was the possibility to prove, first, that what happened in nerves was an electrical and not a mechanical phenomenon and, second, that it was possible to cure specific diseases by administering an electrical shock. The paralyzing effects of the South American eels as described by Storm van 's Gravesande and Van der Lott had not been unknown before these letters were written. In 1714, in a report to the Academy of Sciences in Paris, René-Antoine Ferchault de Réaumur had interpreted these effects (produced not by the eel, but by the animal that is now known as the torpedo or electric ray) as being mechanical by nature, that is, as being an effect of the fast movement of the muscles of the animal. Yet the more people became acquainted with electrical phenomena, the more it was doubted whether this interpretation was correct and whether electricity might have something to do with it. However, even La Condamine, who reported on the effects of the torpedo in his 1745 memoir of his journey through South America, did not express his views on this issue. For Allamand, the news from Essequibo therefore offered an opportunity to finally prove that neurological effects were indeed electrical by nature. Both the memoir by Allamand and the report by Van der Lott were written in Dutch, however, which partially explains why these interesting communications had so little effect on the development of eighteenth-century neurology.

Interesting though these communications are, they are the exception proving the rule that in the *Memoirs* of the Holland Society, colonial science does not figure prominently, to put it mildly. Only occasionally there is a report that is of some interest. For instance, in 1773 Johan Moritz Mohr reported from Batavia on his observations of the transit of Venus in 1761, earning himself membership of the

Haarlem Society. Yet Mohr's much more interesting report on the 1769 transit of Venus was published in the *Philosophical Transactions* of the Royal Society and not in the *Memoirs* of the Holland Society (this was because Cook, on visiting Batavia in 1770, had practically stolen the report from the ailing minister and sent it to London). The overwhelming impression one gets from reading the volumes of the *Memoirs* is that colonial science hardly existed. In the seventeenth century, Dutch scholars and amateurs had been highly interested in the natural history of the colonies in the East and the West. The cabinets of curiosities were stuffed with objects from all the kingdoms of nature and huge prizes were paid for rare or particularly beautiful specimens of birds, shells, or crocodiles. Yet in the eighteenth century, at least in the second half of the century, this interest in colonial objects seems to have faded and in the volumes of the *Memoirs* of the Holland Society the colonial element is negligible.

The situation in Haarlem was in no way exceptional. The *Memoirs* of other learned societies contain even fewer articles and reports concerning colonial affairs. The Zeeland Society did indeed pay some attention to the colonies, but for the Utrecht Society, the Rotterdam Society, and the Teylers Foundation, the colonies to the east or the West Indies apparently did not exist. Scholars, directors, and readers in the Dutch Republic appear to have been interested almost exclusively in topics closer to home, like the dikes along the rivers, the canals, the dunes, the rains, and storms in Holland. They were very much preoccupied with finding ways of improving both commerce and industry, and even in this context hardly ever thought of the colonies. The late eighteenth century was a time during which the social and political elite seemed almost obsessed by the idea of the decline of the Dutch Republic, searching for ways to redress the situation, restore the Republic to its former glory, and make it powerful and prosperous again. In the closing decades of the eighteenth century, many of the learned societies in the Netherlands were established with the explicit purpose of countering the putative decline of the Republic. The colonies seemed not to have played a large role in these plans. This was the main reason why a country with such a vast colonial empire had so little interest in colonial matters. Despite its enormous colonial possessions, Holland had become an inward-looking country. This, combined with the fact that colonial science in the age of the first scientific expeditions required so much more coordination and money than in the seventeenth century, explains why the Dutch were practically absent in this field in the eighteenth century.

State-Sponsored Science since 1800

The situation changed for the better only in the first two decades of the nineteenth century. Much had happened in between. The French had occupied—or, as some said, "liberated"—the country in 1795 and in the next year the tottering Dutch East Asia Company was dissolved. After the disastrous war against England in the early 1780s, the once profitable Company had run into serious problems and finally in 1796 its debts and possessions were taken over by the newly created Dutch revolutionary government. Over the next 15 years, however, all of the colonies of the former East India Company fell into the hands of the British. In 1796 they already occupied the Dutch colony at the Cape of Good Hope and finally took over Java in 1811. In the meantime, the former Republic of the Seven United Provinces had become a unitary state, first as the Batavian Republic and after 1806, as the Kingdom of Holland under King Louis Napoleon, the emperor's brother. Louis Napoleon's reign was—at least in the eyes of his little brother—a disaster, and in 1810 the kingdom was finally annexed to the French Empire. Not for long though, for in November 1813 the French withdrew from Holland and a new sovereign was installed: the son of the last stadholder, the Prince of Orange. William I soon became king, not only of the old Republic, but also of the former Austrian Netherlands and the bishopric of Liège, comprising present-day Belgium. Part of the colonial empire of the East India Company was given back to the Dutch state, but southern Africa, Ceylon (Sri Lanka), and trading posts in India were permanently lost to the British. Even more than before the Napoleonic Wars, Java became the focal point of the Dutch colonial empire.

It soon became evident that the upheavals during the French period had not been unfavorable for the Dutch scientific system. To begin with, Louis Napoleon could not imagine a modern state not having a national academy of sciences and when he discovered that the existing learned societies, and especially the Haarlem Society of Sciences, were not prepared to give up their local autonomy, he created, from scratch, a new academy of arts and sciences in Amsterdam: the so-called Royal Institute of Sciences, Letters, and Fine Arts. He also started to reform the universities and even though he did not have the time to finish it, during the period of French annexation the process continued and within two years after the beginning of the reign of William I, the king had fundamentally transformed the system. All universities were freed from their provincial restraints and reconstituted as part of a national system of higher education. Important impediments to a flourishing colonial science had now been removed and, as it turned out, within a few years scientific research of the Indies would become an integral part of the Dutch scholarly world.

The new interest in the Indies did not restrict itself to the investigation of the natural world. The languages spoken by the indigenous people in the Indies also began to attract the attention of scholars in the Netherlands. As soon as the Royal Institute of Sciences, Letters, and Fine Arts, established by Louis Napoleon and re-instituted by William I, resumed its meetings after the liberation from French occupation, its members urged the king to promote the study of Indonesian languages. On March 30, 1814, the king visited a combined meeting of all the divisions of the institute and on that occasion—the first time a ruling sovereign visited the institute—presentations were made by members of all four divisions, including the so-called Third Class of ancient and foreign languages and history (the languages of the Far East included). The well-known Orientalist Joannes Willmet spoke about the way in which commerce and state government in the Indies could benefit from the study of Oriental languages. Every time a representative of the Dutch government negotiated with indigenous princes, so Willmet said, it would be a huge advantage not to have to rely on interpreters and other intermediaries, because as everyone knew these people were not to be trusted. Some of them are even worse than the Jews, Willmet confessed. What was needed was to train Dutch colonial officials in the languages of the districts they were supervising, and in order to do so the king should take the initiative to found an Oriental Institute in Batavia, linked to the Batavia Society. Willmet said he had already discussed this plan with the former king Louis Napoleon, but nothing had come of it then. Now he thought it was the right time to relaunch his plan and so he prompted the king to do what the French, the English, the Prussians, the Russians, and even the Austrians had done before. Willmet explicitly referred to the French Oriental Institute in Cairo, more particularly to the Society for Eastern Literature in Calcutta, established in 1783, and to the Oriental Society in London, established in 1800. This argument for a practically oriented study of the Indonesian languages did not fall on deaf ears, and it was the beginning of a strong Dutch tradition in the linguistic studies of Indonesia.[19]

Still, the king put his money mostly in researching the natural history of the Indonesian archipelago. In 1815, he dispatched a committee to the East to take over the colonial possessions from the British. The advisor to this committee was the former professor of botany at Harderwijk, the German-born Caspar Georg Carl Reinwardt.[20] He had been a favorite of Louis Napoleon and was a prominent member, and temporarily even chairperson, of the reestablished Royal Institute. Reinwardt was ordered to go to the East Indies and to collect as much information about the natural world as he could. During the short period in which the British had occupied Dutch possessions in the East (1811–1816), Lieutenant-General Thomas Stamford Raffles had started to investigate these dominions, and his example may have inspired the king to send Reinwardt to the Indies.[21] From the instructions Reinwardt received, it is clear however that Dutch officials were very much aware of the rudimentary nature of their knowledge of the Dutch colonies. "It is time that we compensate for the harm we have inflicted on ourselves and the learned world," the king's secretary Anton R. Falck wrote to Reinwardt in December 1814. "We should no longer miss the merits of knowing our colonies as thoroughly as our neighbors do."[22] Also Reinwardt was sent to the East with a very detailed list of questions, the elementary nature of which demonstrates the lack of real knowledge of the Indies. For instance: "Are there already factories and manufacturers, and if that is the case, which kind of factories can be found?" or "Can diamonds and other precious stones be found there?"[23] Whatever the case, Reinwardt's journey to the Indies was the first scientific expedition organized by the Dutch. Reinwardt stayed in Indonesia until 1822 and, among other things, founded a botanical garden in Buitenzorg (Bogor) near Batavia, soon to become one of the major centers of scientific research in the Dutch East Indies. Reinwardt returned to the Netherlands with a wealth of material—so much that he never managed to write the final report about his expedition to the Indies. Yet others followed in his footsteps, and soon Java and the other islands of the Indonesian archipelago became a place of very special interest for the Dutch and other scientists.

Conclusion

The call on the new king to establish a school of Indonesian languages and Reinwardt's expedition to Java are just two examples of a new interest in the study of the colonial world in the beginning of the nineteenth century, but they both show that something important had changed around 1800. In both cases, science and scholarship were now marshalled in the service of the Dutch state, which from the start took its colonial mission in the East seriously. Although the East India Company had an official charter, the Company had only been a private enterprise and was as such not really interested in the scientific exploration of the colonies; the Company showed itself even rather suspicious when individual scholars and scientists showed interest in the natural history of their possessions. In the Dutch Republic, because of the fragmented, mostly provincial nature of the learned world and the lack of a central coordinating institute, scholars and scientists only now and then showed an interest in colonial affairs. In the eighteenth century, the preoccupation with the theme of national decline made things even worse (from the perspective of colonial science of course). In the early nineteenth century, there finally emerged a unitary state headed by a king with unprecedented powers, a national system of higher education, and a central institution for the promotion of science and scholarship, the Royal Institute. There is no denying that William I spoiled much of what he could have achieved by letting the confrontation with the Belgians in 1830 slip out of hand and of course the Royal Institute had yet to become the nodal point of science and scholarship in the Netherlands. Still the basic outlines of a new scientific system and a new arrangement between state and science were there. Colonial science would profit from these new arrangements enormously.

Notes

1. Harry Woolf, *The Transits of Venus: A Study of Eighteenth-Century Science* (Princeton: Princeton University Press, 1959).
2. Mohr is not mentioned by Woolf in his study of the transits of Venus (see previous note). For a detailed description and analysis of his observations and those of other Dutch observers: Huibert Jan Zuidervaart, *Van "konstgenoten" en hemelse fenomenen. Nederlandse sterrenkunde in de achttiende eeuw* (Rotterdam: Erasmus Publishing, 1999), 279–333; in this volume, see also chapter 5, Boomgaard on Indonesia.
3. There are several surveys of Dutch colonial science in Dutch, beginning with M. J. Sirks, *Indisch natuuronderzoek: Een beknopte geschiedenis van de beoefening der natuurwetenschappen in de Nederlandsche koloniën* (Amsterdam: Koloniaal Instituut, 1915). For a survey in English, see Klaas van Berkel, "The Natural Sciences in the Colonies," in *A History of Science in The Netherlands: Survey, Themes and Reference*, ed. Klaas van Berkel, Albert van Helden and Lodewijk C. Palm (Leiden: Brill, 1999), 210–28.
4. W. Th. M. Frijhoff, *La société néerlandaise et ses gradués, 1575–1814* (Amsterdam: APA/Holland University Press, 1981); Van Berkel, Van Helden, Palm, *A History of Science in The Netherlands*, 29–36, 43–52, 69–76.
5. Klaas van Berkel, "Een onwillige mecenas? De VOC en het Indische natuuronderzoek," in *Citaten uit het boek der natuur: Opstellen*

over Nederlandse wetenschapsgeschiedenis (Amsterdam: Bert Bakker, 1998), 131–46; Leonard Blussé and Ilonka Ooms, eds., *Kennis en Compagnie: De Verenigde Oost-Indische Compagnie en de moderne Wetenschap* (Amsterdam: Balans, 2002).

6. On Witsen: Marion Peters, *De wijze koopman: Het wereldwijde onderzoek van Nicolaes Witsen (1641–1717), burgemeester en VOC-bewindhebber van Amsterdam* (Amsterdam: Bert Bakker, 2010).
7. A. J. P. Raat, *The Life of Governor Joan Gideon Loten (1710–1789): A Personal History of a Dutch Virtuoso* (Hilversum: Verloren 2010).
8. Harold J. Cook, *Matters of Exchange: Commerce, Medicine, and Science in the Dutch Golden Age* (New Haven: Yale University Press 2007), 40.
9. On the Holland Society and other learned societies in the Dutch Republic: W. W. Mijnhardt, *Tot Heil van 't Menschdom: Culturele genootschappen in Nederland, 1750–1815* (Amsterdam: Rodopi, 1987). For a more up-to-date survey: Joost Kloek and W. W. Mijnhardt, *1800: Blueprints for a National Community* (Assen: Van Gorcum/ Basingstoke: Palgrave Macmillan, 2004).
10. Jan Gerrit de Bruijn, *Inventaris van de prijsvragen uitgeschreven door de Hollandsche Maatschappij der Wetenschappen 1753–1917* (Haarlem: H. D. Tjeenk Willink en Zoon, 1977).
11. De Bruijn, *Inventaris*, 34. The Scottish entry was left out of consideration because of the language in which it was written.
12. Ibid., 47–48.
13. On the prize competitions initiated by the Batavia Society see: J. A. Bierens de Haan, *De Hollandsche Maatschappij der Wetenschappen, 1752–1952* (Haarlem: H. D.Tjeenk Willink & Zoon, 1952), 206–7; on the Batavia Society: J. P. M. Groot, *Van de Grote Rivier naar het Koningsplein: Het Bataviaasch Genootschap van Kunsten en Wetenschappen 1778–1867* (Leiden: Published by the author, 2006).
14. De Bruijn, *Inventaris*, 54–55.
15. "Waarnemingen, gedaan op 't Oostindiesch Comp. schip de Hoop, met een origineelen Quik-thermometer van Prins, geplaatst in de open lugt, in de schaduwe van de Zon 's middags ten 12 uuren," *Verhandelingen uitgegeeven door de Hollandse Maatschappy der Weetenschappen te Haarlem* 1 (1754), 775–78.
16. *Verhandelingen* 2 (1755), 413–18.
17. J. N. S. Allamand, "Kort verhaal van de uitwerkzelen welke een Amerikaanse Vis veroorzaakt op de geenen die hem aanraaken," *Verhandelingen* 2 (1755), 372–79. For a more extensive analysis of this particular memoir (and the related one by Van der Lott, see the next note): P. J. Koehler, S. Finger, "De Zuid-Amerikaanse aal: Twee vroege brieven uit de Nederlandse koloniën over dierlijke electriciteit," *Studium* 3 (2008), 185–94.
18. F. van der Lott, "Kort bericht van den conger-aal, ofte drilvisch," *Verhandelingen* 6 (1762), 87–94.

19. Klaas van Berkel, *De stem van de wetenschap: Geschiedenis van de Koninklijke Nederlandse Akademie van Wetenschappen, deel I, 1808–1914* (Amsterdam: Bert Bakker 2008), 96.
20. Andreas Weber, "Encountering the Netherlands Indies: Caspar G. C. Reinwardt's Field Trip to the East (1815–1822)," *Itinerario* 33 (2009), 45–60; Andreas Weber, *Hybrid Ambitions: Science, Governance, and Empire in the Career of Caspar G. C. Reinwardt (1773–1854)* (Leiden: Leiden University Press, 2012). Weber excellently shows how Reinwardt refused to act as a malleable tool of metropolitan science in Leiden and Amsterdam, and instead pictured himself as a heroic figure who had encountered the wild and fertile nature of the Indies all by himself, thereby proving that natural history can only be created in the field and not in the secluded rooms of the metropolitan museums.
21. Raffles published his *History of Java* in two volumes in 1817.
22. As quoted in Weber, "Encountering the Netherlands Indies," 49.
23. Ibid.

Bibliography

Allamand, J. N. S. "Kort verhaal van de uitwerkzelen welke een Amerikaanse Vis veroorzaakt op de geenen die hem aanraaken." *Verhandelingen uitgegeeven door de Hollandse Maatschappy der Weetenschappen te Haarlem* 2 (1755): 372–79.

Berkel, Klaas van. "Een onwillige mecenas? De VOC en het Indisch natuuronderzoek." In *Citaten uit het boek der natuur: Opstellen over Nederlandse wetenschapsgeschiedenis*, 131–46. Amsterdam: Bert Bakker, 1998.

———. "The Natural Sciences in the Colonies." In *A History of Science in The Netherlands: Survey, Themes and Reference*, edited by Klaas van Berkel, Albert van Helden, and Lodewijk C. Palm, 210–28. Leiden: Brill, 1999.

———. *De stem van de wetenschap: Geschiedenis van de Koninklijke Nederlandse Akademie van Wetenschappen, deel I, 1808–1914.* Amsterdam: Bert Bakker, 2008.

Bierens de Haan, J. A. *De Hollandsche Maatschappij der Wetenschappen, 1752–1952.* Haarlem: H. D. Tjeenk Willink & Zoon, 1952.

Blussé, Leonard and Ilonka Ooms, eds. *Kennis en Compagnie: De Verenigde Oost-Indische Compagnie en de moderne Wetenschap.* Amsterdam: Balans, 2002.

Bruijn, Jan Gerrit de. *Inventaris van de prijsvragen uitgeschreven door de Hollandsche Maatschappij der Wetenschappen 1753–1917.* Haarlem: H. D. Tjeenk Willink & Zoon, 1977.

Cook, Harold J. *Matters of Exchange: Commerce, Medicine, and Science in the Dutch Golden Age.* New Haven: Yale University Press, 2007.

Frijhoff, Willem Th. M. *La société néerlandaise et ses gradués, 1575–1814.* Amsterdam: APA/Holland University Press, 1981.

Groot, J. P. M. *Van de Grote Rivier naar het Koningsplein: Het Bataviaasch Genootschap van Kunsten en Wetenschappen 1778–1867*. Leiden: Published by the author, 2006.

Kloek, Joost J. and W. W. Mijnhardt. *1800: Blueprints for a National Community*. Assen: Van Gorcum/Basingstoke: Palgrave Macmillan, 2004.

Koehler, P. J. and S. Finger. "De Zuid-Amerikaanse aal: Twee vroege brieven uit de Nederlandse koloniën over dierlijke elektriciteit." *Studium* 3 (2008): 185–94.

Lott, F. van der. "Kort bericht over den conger-aal, ofte drilvisch." *Verhandelingen uitgegeeven door de Hollandse Maatschappy der Weetenschappen te Haarlem* 6 (1762): 87–94.

Mijnhardt, W. W. *Tot Heil van 't Menschdom: Culturele genootschappen in Nederland, 1750–1815*. Amsterdam: Rodopi, 1987.

Peters, Marion. *De wijze koopman: Het wereldwijde onderzoek van Nicolaes Witsen (1641–1717), burgemeester en VOC-bewindhebber van Amsterdam*. Amsterdam: Bert Bakker, 2010.

Raat, A. J. P. *The Life of Governor Joan Gideon Loten (1710–1789): A Personal History of a Dutch Virtuoso*. Hilversum: Verloren, 2010.

Semeyns, M. "Bericht wegens de eigenschappen der Land- en Zee-winden, waaijende langs de kusten van 't eiland Java, en wel inzonderheid te Batavia volgens eene veel-jaarige Bevindinge." *Verhandelingen uitgegeeven door de Hollandse Maatschappy der Wetenschappen te Haarlem* 2 (1755): 413–18.

Sirks, M. J. *Indisch natuuronderzoek: Een beknopte geschiedenis van de beoefening der natuurwetenschappen in de Nederlandsche koloniën*. Amsterdam: Koloniaal Instituut, 1915.

"Waarnemingen, gedaan op 't Oostindiesch Comp. schip de Hoop, met een orgineelen Quik-thermometer van Prins, geplaatst in de open lugt, in de schaduwe van de Zon 's middags ten 12 uuren." *Verhandelingen uitgegeeven door de Hollandse Maatschappy der Weetenschappen te Haarlem* 1 (1754): 775–78.

Weber, Andreas. "Encountering the Netherlands Indies: Caspar G. C. Reinwardt's Field Trip to the East (1815–1822)." *Itinerario* 33 (2009): 45–60.

———. *Hybrid Ambitions: Science, Governance, and Empire in the Career of Caspar C. C. Reinwardt (1773–1854)*. Leiden: Leiden University Press, 2012.

Woolf, Harry. *The Transits of Venus: A Study of Eighteenth-Century Science*. Princeton: Princeton University Press, 1959.

Zuidervaart, Huibert Jan. *Van "konstgenoten" en hemelse fenomenen: Nederlandse sterrenkunde in de achttiende eeuw*. Rotterdam: Erasmus Publishing, 1999.

Chapter 4

Why Was There No Javanese Galileo?

*Gerry van Klinken**

The Dutch being scientific *and* colonial, this chapter poses two questions about the scientific interests of their colonial subjects in Java. First, did the Javanese have an evidence-based science of the natural world in 1808? Second, if not—and it seems they did not—then why not?

The nominal year at the focus of our enquiry is 1808, birth year of the Netherlands Academy of Sciences. How much did the Javanese know about the natural world at this time? European observers were just beginning to take an interest in the contours of Javanese knowledge at the beginning of the nineteenth century. The pioneering Batavia Society of Arts and Sciences (Bataviaasch Genootschap van Kunsten en Wetenschappen) published its first proceedings in 1779. Among the papers on practical colonial science were some on Javanese ethnography and history.[1] Later, the economic malaise around the turn of the century paralyzed scientific work. Whatever scholarship was produced in Batavia in the early years of the nineteenth century amid the vicissitudes of the Revolution and the British Interregnum seems to have been taken to London by the energetic departing lieutenant-governor Thomas Stamford Raffles in 1815. He incorporated it into his magisterial *History of Java*.[2] The little that was known appeared to indicate that the Javanese knew no systematic, evidence-based science. The inimitable John Crawfurd, the Scottish surgeon whom Raffles appointed resident of Yogyakarta in 1811, put his opinion on the matter about as crudely as anyone has ever done before or since. They were "wholly ignorant of arithmetic as a science, and indeed know nothing of the common rules of calculation," he wrote. Imprecision in weights and measures was characteristic of their "poverty and barbarism." About their navigational skills, he warned the reader: "It will be their ignorance rather than their knowledge that will constitute the principal matter of discussion."[3] But other European scholars developed a warm appreciation for the refined spirit that the courtly Javanese manuscripts breathed. It seemed to glow with the same romanticism that permeated their own century. Johannes Hageman even included the natural sciences when he wrote effusively that "Java possesses, in her sciences and her literature, all those characteristics that earned the homelands of the sciences, Egypt, India, Greece, and Rome, the highest praise."[4] A quote from Schiller's poem *Die Götter Griechenlands* headed up his lyrical chapter on the written knowledge of the Javanese. A lithographic image of a Javanese sultan he promoted to the Dutch public illustrates this romantic admiration.

The sense of wonder has fortunately prevailed among students of the literature of Java. They have been mainly interested in belles lettres, but they certainly kept an open eye for the vision on nature displayed there. P. J. Zoetmulder wrote that the Old Javanese poet was not only a "priest of literary magic," whose task it was to enhance the power and glory of the ruler, but also an aesthete dedicated to a love of the beauty in nature and in women, and who adopted a rich aesthetic language of rapture called *lengleng*. The typical introduction to a poem (*manggala*) would acknowledge that he worshipped gods whose domain was that of beauty, found through meditation and tantric yoga. "We seem to find in all this," Zoetmulder concluded, "the poetic expression of a basic element of the Old Javanese way of thinking: the unity of the cosmos and the interrelatedness of everything in it.... Thus all beauty is essentially one."[5] Theodore Pigeaud noted similarly that modern Javanese texts consistently revealed "the central concept of Javanese civilization: social, cosmic and religious Order."[6]

There still seems little reason to doubt the prevailing understanding that Java never adopted an observation-based science of the natural world. The Dutch historian of science Floris Cohen gave only two examples of civilizations completely without science, namely "Rome and the archipelago of the East Indies," adding: "If there was no science to begin with, we would gain little by asking why no Scientific Revolution occurred."[7] The historian of Java, Ann Kumar, confirmed: "Of theoretical science there was no sign."[8]

However, perhaps surprisingly, no one has yet made a systematic study of the evidence to conclude that the Javanese knew no science. This evidence can mainly be found scattered in the Javanese texts now kept in archives. Sources—finding, reading, dating, and localizing them—are a problem for scholars of Javanese history. Javanese manuscripts in public collections number about 19,000, with perhaps as many again in private ones. But the vast majority remains poorly described, and only a handful of scholars are working on them today. Defeating "our resounding ignorance of the greater part of Javanese literature as it developed over the better part of the past five centuries" will be the work of generations, wrote Tim Behrend.[9] Making any generalizations about their contents is therefore a hazardous undertaking, particularly when it concerns a topic apparently as marginal to the corpus as the natural world. Historians of natural science in the Arab, Indian, and Chinese worlds make the same complaint.

By far, the greater part of the extant Javanese manuscripts date from the nineteenth century. Village texts are rough, poorly preserved, heterogeneous, inexpert, and inadequately dated and localized. Court documents, the bulk of those we have, "originated under systems of patronage wherein the labor of a scribe was commissioned and directed by another individual."[10] The court manuscripts were produced in a royal scriptorium located within the palace, known in Java as the *kraton*. Professional scribes in the courts of Surakarta, Yogyakarta, and, to a lesser extent, in those of Cirebon and Madura mainly copied older texts, sometimes setting them to new meters. Production levels were low. The Kasultanan scriptorium in Yogyakarta, together with Surakarta's Kasunan, the biggest in Java, produced only three manuscripts a year in the period 1836–55, except for a big outpouring in 1846–47.[11] The works the Javanese scribes copied mostly had older origins. Many were products of the wave of literary (re) production that had swept the two main Javanese courts, the Kasunan in Surakarta and the Kasultanan in Yogyakarta, at the end of the eighteenth and the beginning of the nineteenth centuries. Most of these were in turn copies of older works, in the conservative tradition of the kraton's scriptorium.

Being unable to read the Javanese manuscripts myself, I relied largely on Pigeaud's magisterial survey of the collection held at Leiden University.[12] Although Pigeaud's interests were as literary as those of his predecessors, his vision on the written corpus of Javanese knowledge was broad and generous. He also took care to note references to the natural world. Let us start with a concrete subject, the stars. In all civilizations, the original interest in the stars is to determine the time—not chronological time but "kind of time," that is, whether a particular moment is good or bad for royal ceremonies, war, planting crops, sexual intercourse, and so on. Calendars classify kinds of time in a repetitive frame. The main object is thus divination. The Indian Vedas had already fixed a way of determining the calendar by about 1300 BC. Both the Chinese and the Indians divided the sky into 27 or 28 "lunar mansions," the star formations where the moon "rests" each night of the month. They are located near the plane that contains the earth's equator when extended into space. The Chinese first wrote about the lunar mansions (*xiu*) in the fifth century BC, and by the third century BC they were measuring the position of the sun, other stars, the moving planets, as well as astrologically significant transient events such as comets and supernovae, with respect to the central star in each *xiu* in degrees of right ascension. They had detailed star maps by 1000 AD. The Indians called their lunar mansions *naksatras* and the first written records of them measuring the position of the stars and planets date from the fifth century AD.[13] The Indian astronomer Aryabhata, born in 476 AD, took the massive stride forward of developing mathematical formulae to predict the orbits of the planets. He had made his own the theory of Ptolemy the Greek, and had improved its accuracy remarkably.

Javanese scholars, by contrast, appear not to have observed the stars. As Muslims, they did observe the moon to fix the lunar Islamic calendar, and particularly the fasting month Ramadan within it. Its strictly lunar character cut Quranic time off from the astrological powers and agricultural regularities tied to observation of the zodiac.[14] Pre-Islamic calendars were solar but based on Indic tables, not direct observation of the stars. Surviving Old Javanese time divination documents, known as *wariga*, deal little with the stars, and when they do, the stars are "literary" ones.

In 1925, Pigeaud described an unusual document, found among a pile of booty taken to Batavia by Dutch troops who occupied the Klungkung kingdom in Bali in 1908.[15] While in itself not so old, it may have had an ancient heritage as the last in a long line of copies. This manuscript was simply a replica (with some interesting modifications) of the *naksatras* of the oldest Indian calendric science, the Vedas, predating the work of Aryabhata. Names of constellations and planets were primarily those of divinities that rule the days of the calendar. This was literary astronomy, not indigenous observation of the stars. By the time the Vedic calendrical system reached Java, it was already millennia old and had lost all contact with actual observation of the stars. Where the needs of divination had elsewhere turned observing eyes to the night sky, in Java it had become a literary tradition that valued antiquity rather than conformity with real stars. The much more accurate system of Aryabhata, more than 12 centuries old by 1808, was apparently unknown or of no interest. Raffles noted in the early nineteenth century that the Javanese had "no pretensions to astronomy as a science."[16] John Crawfurd did learn that some Balinese scholars in the 1820s were using Indian tables to predict the occurrence of lunar eclipses. None of these tables are known to have survived today, but Gomperts believes they may have been derived from those made in the 1020s AD by the Ptolemaic Indian astronomer Al-Biruni on the basis of observation.[17]

The same loss of contact with observation occurred in the measurement of time. A basic Indian brahmin unit of time is the "stroke" or *muhurta*, equivalent to about 48 minutes, and this also occurs in Old Javanese manuscripts. Indian astronomic texts state that during the summer solstice the day is 18 muhurta long and the night 12. This is true at Indian latitudes but not in Java, which is closer to the equator and therefore has less seasonal variation in its daylight hours. Yet an Old Javanese text known as *Brahmandapurana* simply reproduces the brahminic 18:12 division without comment, clearly being more interested in the divinatory value of these times than in their duration.[18] Similarly, A. D. Cornets de Groot observed in the early nineteenth century that Javanese scholars calculated the Indo-Javanese calendar on the basis of Indian right ascensional parameters, apparently without realizing that these do not apply in Java.[19]

One of the oldest Javanese divinatory calendars is still determined entirely on rational principles, lacking any link to natural cycles. Known as *wuku,* it goes back at least to the tenth century AD.[20] It is not influenced by Hindu or Islamic knowledge. It has 30 seven-day weeks and is unrelated to the lunar or solar cycles. The *wuku* calendar is thus purely symbolic and rational. Cornets de Groot stated in his study of Gresik from 1822–23 that Java's founding myths are rooted in this calendar. The *wuku* calendar remains in use in mystical Javanese (*kejawen*) circles today.

Another source of science is an interest in numbers. Like the observation of stars, the contemplation of the properties of numbers always begins as a highly anthropocentric activity, shrouded in mystery. Pythagoras in the sixth century BC made discoveries about triangles in the secrecy of his brotherhood, and they revealed to him the mystery of the universe. By the tenth century AD, the Chinese possessed a dozen books devoted to topics like the determination of the value of *pi* and the solution of quadratic equations. They employed professional mathematical clerks—technicians, not theoreticians—who wrote their mathematics in prose or verse, as in European medieval mathematics. Aryabhata had systematized algebra, called *kuttaka*, to solve equations already in the fifth century AD. Neither the Chinese nor the Indians saw nature as essentially mathematical—Galileo first took that step. But they did regard algebra and geometry as important and useful intellectual endeavors.

When we turn to the Javanese literature, we do not see an interest in the inherent properties of numbers (nor of abstract space for that matter). Did Chinese, Indian Hindu, or Islamic texts on numbers not pass through some of the lively harbor towns on Java's north coast? If they did, they evidently left little impression. One rare exception is a sentence in the Old Javanese text *Brahmandapurana* reproducing, without comment, an Indian text crudely stating the value of *pi* as three.[21] As with a possibly lost knowledge of astronomy, some previous interest in mathematics may have been lost. The Borobudur funerary monument near Yogyakarta is built to within 1½ degrees of true geographic north. (The direction to the North Pole is indicated by the shadow of a vertical pole at mid-day; it deviates slightly from magnetic north.) It is difficult to imagine how this vast ninth-century AD structure could have been built without some kind of drawings using basic geometrical techniques. Sadly, almost nothing is known about the builders of the Borobudur.

Java's interest in numbers is not mathematical but totemistic. Chronograms (*candra sangkala*) are an esoteric form of numerology that links numbers to words.[22] Words are combined to form dates, which, like the calendar, contain predictions about the nature of certain times. The practice goes back to Old Javanese times. The number one goes with (among others) the earth, two with lips (and other things of which there are two), four with water, five with wind, nine with doorways of various kinds, and so on.[23] The historiography of Java's dynastic wars is based on a purely rational hundred-year cycle written down in *sangkala* lists.[24] The beloved late-nineteenth-century *Joyo Boyo* prophecy, which according to popular belief correctly predicted the Japanese Occupation, was based on *candra sangkala*. But later generations often failed to understand them.

The ancient Vedic Hindu concept of *rita*, or ritual order known only to the Brahmanic priests, provided the unity that bound together the universe. Where Chinese classifications mainly relied on the duality Yin-Yang, in Java, as in India, the number five was dominant. Five is always four plus one—the four cardinal points plus one point directly overhead. There are five aspects of Shiva, five Indian ascetics, five planets, five senses, five professions, five days of the week, and five elements (sky, heat, wind, water, and earth).[25] The correspondence between the microcosm within and the macrocosm without was also central. Islamic mystics drew diagrams known as *daerah* consisting of two concentric circles to represent the two worlds. At the center of both was *Dhat*, the divinity. Many of the correspondences between the outer and the inner were based simply on the sound of their names. When pairs of colors (red vs. white) were linked to pairs of directions (north vs. south), and to psychic characters (phlegmatic vs. sanguine), the European reader is likely reminded of heraldry. Pigeaud aptly called this "pre-logical" associative thinking, "a feeling of cosmic solidarity."[26] The deductive style of thinking, which assumes with Plato that the universe and the structure of the human mind correspond, is an important scientific impulse.[27] Java's feeling of cosmic solidarity was Platonic. But only in Greece did it lead to rigorous mathematical proof.

The disconnect between astronomical observation and the calculations underlying the *wuku* calendar exemplifies the conviction—a kind of Platonic rationalism—that only the inner world was real. Pigeaud wrote that the Javanese texts on science—mainly magic, medicine, and divination—were hardly separable from the main body of Javanese literature on religion and ethics. They were important as illustrations of the penetration of "the central concept of Javanese civilization: social, cosmic and religious Order" into all spheres of life; "their connection with religious ideas, ceremonies and ritual is in many cases very close."[28] The contrast is sharp with what Alfred North Whitehead once called the "through and through anti-intellectualist movement" that was the revolt of Galileo and Francis Bacon.[29] We read no skepticism in the Javanese manuscripts.

If deductive thought dominated the privileged world of the court scholar, the artisan obviously lived in a much more empirical one. The farmer, the healer, the gong-maker, and the horse trader depended for their livelihood on their ability to tame the natural forces. Here, not philosophy but magic provided the mental matrix for understanding the properties of the human body, of animals, or of metals. J. G. Frazer in *The Golden Bough* was the first to recognize the close connection between technology and magic.[30] Both are individualistic, specialized, concrete, practical, and everyday. Magic regards knowledge as power and attempts to systematize it. But the magician "does" very little, as images replace reality. Mauss wrote that magic is "the prefiguration of techniques . . . the most childish of skills."[31]

The Balinese trance-dances tourists love to photograph originated as a form of exorcism connected with illness.[32] A great number of Javanese-Balinese books of notes (*primbon*) contain unsystematic compilations of medical lore, with the names of medicines sometimes deliberately obfuscated to preserve their secrecy. Their social milieu is not courtly. They combine the fivefold divisions and the well-known micro- and macrocosm links, with long discussions on demons, on the mystical meaning of the afterbirth, and on the calendar ("kinds of time" again). Weck (1937), a medical officer who studied hundreds of Balinese manuscripts in the 1930s, observed that, mixed in with this strong interpretative framework was much careful observation on the symptoms of leprosy, cholera, and smallpox, and on pregnancy, the development of the fetus, and childbirth. Besides treatment using tantric formulae and psychic surgery, long sections described the preparation of healing herbs. Some of these herbal remedies derived from China, but most were considered indigenous to Indonesia.[33] In the nineteenth and twentieth centuries, this herbal knowledge began to shed its mystical associations and appear in printed form. One European wrote in the 1840s that Javanese medicines for gastric complaints were superior to European ones.[34]

Metallurgy too was suffused with magic. The manufacture of gongs and of daggers (*kris*) has always been surrounded by a mystical aura. Almost all Javanese men wore a kris up until World War II. Java exported them to China as long ago as the early tenth century ("short swords with hilts of rhinoceros-horn or gold").[35] The Javanese were casting their own cannon when foreigners first observed them in the sixteenth century, although gunpowder was imported from present-day Malaysia, as were muskets. Europeans marveled at the speed with which Javanese copied European gun designs. Today's Indonesian word for rifle, *senapan*, derives from a type of musket called the *snaphaen* that was briefly popular in the seventeenth century.[36] When homegrown expertise was insufficient, they took on foreign consultants. During the wars that the Islamic harbor town of Demak waged against the emergent kingdom of Mataram in Central Java during the early seventeenth century, they used Dutch prisoners of war to operate the guns.[37] When in 1629 Sultan Agung from Central Java attacked Jakarta, he shocked the Dutch by almost succeeding. He had transported his cannon right across Java on bullock carts. A road transport network, with permanent bridges passable to wheeled traffic in the dry season, has stretched across the island since the eight century. Yet in nonmilitary areas Java was technologically conservative. The bellows in use today in the average rural foundry are the same as those depicted on a relief of the fourteenth-century candi in Sukuh. Manuscripts continued to be scratched onto lontar-palm leaves until Islamic times despite the availability of Chinese paper.

Most dependent on the observation of nature were peasants, who needed to plant their crops *before* the rains started. The calendar they used to regulate the agricultural cycle, known as *mangsa*, was entirely different from the ritual calendar determined at the court (although it was eventually regulated by Sultan Paku Buwono VII with Dutch assistance in 1855). Several Western scholars have been fascinated by this artisanal knowledge of the stars in Javanese and other archipelagic cultures.[38] The *mangsa* calendar was calibrated by observing two key constellations at either sunup or sundown, namely the Pleiades and the belt of Orion. Completely unknown to the scholars at court, and thus never mentioned in their manuscripts, was the gnomonic method of determining the length of each of the 12 months of the year. This was based on direct observation. Muslim Chinese may have introduced it to the harbor towns along Java's north coast. Java is located about seven degrees south of the equator. At midday on any one day of the year the gnomon, a vertical pole known as a *bencet* in Javanese, casts a shadow that will lie either to its north or to its south, depending on the time of year. On the midwinter solstice, the midday shadow reaches its southern extremity, and on the midsummer solstice its northern extremity. Dividing the range between those two extremities into six divisions of equal length yields the boundary points for twelve months. Because of the geometry of the situation, the number of days in each month will not be equal. Indeed, the length of the traditional Javanese month varies between 23 and 43 days.

Animals were just as important to understand. Physiognomy is the art of judging character from physical features. The physiognomy of horses (*katuranggan*), of buffalo, cats, chickens, and birds was an important skill. The purpose was to discover whether an animal would bring good or ill luck to its owner. A horse with a yellow coat, if it is male, of moderate height, and has a cowlick patch of hair on his left cheek is auspicious. Its owner will surely grow rich, and if he rides it into battle will be victorious. On the other hand, a strand of hair from the horse's tail getting into the anus after defecating is a bad sign. The owner will have frequent fights with his wife.[39] Similar observations were made on the sounds of birds, on the color of soil, and of course on the bodies of men and women, where it often had an erotic loading. So for instance, a muscular twitch (*kedutan*) of the right eyebrow was a sign that lots of money was on the way, while a twitch in the right buttock indicated a relative on the way. A steel pan going "pop" when left in the sun was good or bad luck depending on the time of day.[40] The village aficionado knew (and knows) literally hundreds of these things.

Earthy as they are, however, these observations are far from a theoretical, evidence-based science of nature. At the risk of being proven wrong by the discovery of hitherto unknown scientific texts among those tens of thousands not yet examined, I am fairly confident in my suspicion that such texts hardly existed in Java in 1808.

Is it premature now to go on to the "why" question? Would this not simply confirm a prejudice rather than throw light on a fact? Perhaps. So let what follows be pleasant speculation, offered in the hope it may stimulate research. The question of Java's knowledge of the natural world was first put to me two decades ago by the Australian sinologist Stephen Fitzgerald, that country's first ambassador to communist China. I was at the time teaching physics to student physics teachers at a university in Central Java. We talked about Joseph Needham's vast and at that time still growing compendium on the science and civilization of China. When Needham began looking for evidence of Chinese scientific interests in the manuscripts, most experts expected he would find none. A summary of his early work was part of a course I was teaching on the history of physics. Needham's life has recently been made accessible by Simon Winchester's lively biography.[41] If the scientific revolution had occurred in China during the Sung Dynasty in the tenth century AD, as Needham had claimed, then, Fitzgerald asked me, what about Java? After all, Java had had centuries of trading and migratory contacts with China, India, and the Muslim Middle East, all areas where discoveries in the natural sciences had been made. It seemed worthwhile to ask the question at least. A certain naivety lurked behind the assumption that scientific knowledge would simply circulate through trading ports—the hydraulic metaphor—but I was myself engaged in "knowledge transfer" and was interested in its possible antecedents. Not long after that conversation, I dropped the subject and turned to studies of twentieth-century Indonesian political history, only to return to it afresh in 2008.

Part of my naivety lay in having an only hazy awareness of the fierce debates about the nature of scientific knowledge. Is there an "intrinsic" reality out there, waiting to be discovered, as I, like most practicing scientists, tended to believe? This was, after all, the knowledge I was "transferring" to my Javanese students in the 1980s. Or is what we know of the natural world as much a "social construction" as are the social sciences, as sociologists of knowledge believe who see no fundamental difference between the various branches of human knowledge? In that case there is little point comparing Galileo's Renaissance Italy with 1808 Java, for each culture creates its own incomparable knowledge. I now believe, with Ian Hacking, that both sides of this debate have much reason on their side.[42] On the one hand, the "local knowledge" approach to the history of science in non-Western societies correctly highlights the local contingency of what becomes known, and the politically negotiated process by which knowledge claims are adjudicated. On the other hand, the universality of the claims made by the natural sciences is unlike that in any other branch of human knowledge, and it is borne out every day in new technological applications. It still seems to me possible, indeed desirable, for the historian of non-Western science to make use of insights from both sides of this long debate without trying to resolve the debate itself.

The biochemist Joseph Needham spent a lifetime combing through thousands of Chinese manuscripts. He single-handedly established the field of the history of Chinese science. Only much later did I learn that mainstream historians of science such as Floris Cohen considered him a "maverick."[43] Needham's "Grand Question" was why the Scientific Revolution had occurred in Western Europe and not in China. The question is fascinating and valid. Needham answered that it very nearly *had* occurred in China, but that certain social formations had stood in the way. However, Cohen faulted both these answers. No one doubts that there were scholarly Chinese with a scientific (evidence-based, explanation-oriented) interest in the natural world. Indeed, along with the Indians, the Arabs, and the Europeans, the Chinese interest was far in advance of the more preliminary science practiced in premodern Japan, Korea, pre-Columbian America, and Byzantium. The science of China reached the equivalent of Leonardo da Vinci, and well before da Vinci. But it never approached that of Galileo, who took the first step beyond what Alexandre Koyre once called "the world of the more or less," the world of everyday weights and measures.[44] By undertaking exact measurement and rigorous proof, Galileo began the mathematization of nature that turned out to be so astonishingly fruitful. That, and not Leonardo da Vinci's inquisitive spirit, was the birth of the Scientific Revolution. For all Needham's praise of China's "organic" understanding of nature, Cohen insisted that this never offered the possibility of developing into a science of nature with real explanatory power.[45]

More importantly for our present purpose, Needham's critics also faulted him when he made claims about the social processes that gave rise, or failed to give rise, to China's scientific breakthroughs. They thought his respect for China, in itself admirable in an age of European self-centeredness, had tragically caused him to lose perspective. "Just about everything Needham has ever written to answer his Grand Question," wrote Nakayama and Sivin, "should be regarded as one giant projection of Joseph Needham's collected preconceptions upon China's society and world of thought."[46] The key ideas in Needham's explanation—that Taoists were science-minded (as opposed to the rationalist, antinaturalist Confucianists), that China's merchant class was too weak, that its feudal bureaucracy inhibited scientific work—were simply negative mirror images taken from Europe. They were not based on a good history of either Europe or China.[47] Concluding his demolition of Needham's lifework, which it had clearly pained him to write, Cohen suggested that Needham would have done better if he had adopted Weber's framework of comparison with respect to the birth of capitalism. "Elective affinities" between distinct social layers and their respective religious inclinations would have helped him move beyond his perennial merchants versus bureaucrats, he thought.[48]

As it happens, Goethe, at the very moment in 1808 that forms the focus of our present study, was in Weimar putting the finishing touches to a novella that was to become part of the answer. *Elective Affinities* (*Die wahlverwandtschaften*) places four people in an isolated summer villa, a husband and wife, an army captain, and a beautiful young woman, and watches their relationships change. Of course, the husband and the young woman fall for each other, as do the captain and the wife. Their after-dinner conversations during this social experiment make it clear that not whim drove these irrepressible affections but something akin to the "elective affinities" between chemical elements. Goethe was also (like Needham!) a chemist. The concept had recently been sharply formulated by the Swedish chemist Torborn Bergman. The novel's attack on free will made it hugely controversial throughout the nineteenth century. Max Weber read *Elective Affinities* as a schoolboy, under the desk during other lessons. Later he also read the chemistry that had provided the underlying notion, and of course he read the philosophy in which "affinity" had been a minor theme since the work of Kant. Although never defining it exactly, Weber turned the phrase "elective affinity" into one of his signature concepts, indeed the one notion that made all of social science possible.[49] If human action had been entirely contingent, then it would be impossible to observe any patterns in it, and there would be no sociology. At the same time, Weber could not accept Karl Marx's view that all values and their actions are determined by the material interests arising from the class structure. "Elective affinity" was the nondeterministic mutual interaction between two different areas of human action, the world of ideas and that of interests. He tried to make this plausible in *The Protestant Ethic and the Spirit of Capitalism*.[50] Calvinism with its strict ethic of hard work and saving was "adequate" to the emerging capitalism of Reformation Europe. Thus, ideas and interests interpenetrate. Ideas do not "express" interests, as Marx would have it. Certain ideas that have an elective affinity with material interests tend to flourish, while those that do not eventually wither away. Gerth and Mills put it this way in their introduction to a collection of Weber's essays:

> There is no pre-established correspondence between the content of an idea and the interests of those who follow from the first hour. But, in time, ideas are discredited in the face of history unless they point in the direction of conduct that various interests promote. Ideas, selected and reinterpreted from the original doctrine, do gain an affinity with the interests of certain members of special strata; if they do not gain such an affinity, they are abandoned...Both the ideas and their publics are seen as independent; by a selective process elements in both find their affinities.[51]

How does this help us solve the problem of the absence of a Javanese Galileo? It provides a firmer basis for studying the process of intellectual innovation than the alternatives in many histories of science. One of these inadequate alternatives is the idea that knowledge "circulates," like currents in an ocean. For Weber, knowledge is not a thing in itself, but develops out of human relationships. New ideas can have a powerful appeal in their own right, but the interests of those people who guard the gates of authentic knowledge tend to determine whether they spread.

Let us now return to Java and see whether Weber's insight throws light on the question of why the Javanese elite nurtured some ideas while resisting others. The first observation is that the year 1808 was not representative of Java's previous centuries. It was a bad moment to look for innovative thinking among scholarly Javanese. The fourteenth century might have been better. Despite the growing admiration for the fine arts of Java developing among a small group of nineteenth-century philologists, the reality was that Java in 1808 had, as Kumar put it with only slight exaggeration, "declined from a populous and glorious past to a fairly wretched, less populous and prosperous, present, subject also to the unchecked depredations of pirates and 'rabble.'"[52] A pristine precolonial Java lay almost 200 years in the past, with the defeat of Sultan Agung at the gates of Batavia. From that moment on, Java had increasingly felt the superior power of European capital, science, and technology. It is true that Dutch colonialism in the full sense did not begin until the Java War of 1825–30. Indeed the slow collapse of the VOC in the last half of the eighteenth century had given Yogyakartans the hope they would once more be able to breathe freely. On the other hand, in a way neither the Dutch nor the Javanese explicitly formulated, the history of the Dutch military intervention beginning in 1677 had left deep marks. The treaties of

1755 and 1757 by which the Javanese kingdom was divided made the threat of Dutch military action—despite waning Dutch prowess in subsequent years—so inescapable that it was no longer possible for the hitherto martial Javanese kings to resolve their differences by force of arms.[53] The freedom to make war that had defined their sovereignty was gone. A Dutch diplomatic and (largely ceremonial) military presence at both the courts had by 1808 been more or less permanent for nearly 130 years.[54] The Javanese kings had lost control over the trade on the north coast since they sold the rights to it to the Dutch in 1746. Moreover, in that very year 1808 politics took a turn for the worse in the court in Yogyakarta. The aggressive new governor-general in Batavia, Herman Willem Daendels, gave notice that the period of European weakness was over by asserting a diplomatic etiquette calculated to insult the Javanese royals. Four years later, another governor, Lieutenant-Governor Thomas Stamford Raffles, accompanied his troops into Yogyakarta to depose its sultan and loot its treasures and archives.[55]

Ricklefs thought it was this political impotence that turned the Javanese courts into "rather quaint, somewhat desiccated museum pieces."[56] Far from flourishing in a kind of "renaissance," as earlier scholars had maintained, he likened the literature of this period to Rabelais's "thick Gothic night."[57] It had lost touch with the delicate poetry and intricate chronograms of Old Javanese. Its concerns were mainly mystical.

Outside the core territories around the court, the Javanese kingdoms were "utterly ramshackle administrative entities."[58] They were hampered by poor communications, difficult mountain terrain, and scattered pockets of population that moreover held on to traditions of local autonomy. Even the bureaucracy, for example, to administer the irrigation network, was too loose to incubate the practical scholarship that can also be innovative. The scriptorium within the Javanese kraton was a far more confining institutional space than the public libraries and astronomical observatories that Islamic courts in the Middle East established to enhance their prestige. Moreover, the Javanese kingdoms had in past centuries been "predominantly military organizations."[59] Constant fighting (which always involved burning) and moving of the court had caused enormous losses to manuscript collections, particularly in the coastal towns.

The knowledge (re)produced at the courts was not intended to circulate widely. Manuscripts were copied out in tiny numbers onto expensive materials, with the result that hardly anyone read them—unlike in the Arab world, where the copyists deliberately employed cheap paper. The printing press was not used to make scholarship widely known. Indeed, the Dutch colonial government would have been quick to censor it if it was. A colonial government education estimate in Gresik in 1822 suggested that only 1 percent of the population had any degree of literacy in Javanese. This included native government officials. Elite Javanese acquired their knowledge from elders, friends, or relatives, rather than from schools.[60] Arabic-language religious schools for the nonelite, meanwhile, invested no time in a scientific and empirical education, unlike Europe in Galileo's time.[61]

The kraton's income was derived from taxation on rice cultivation in the royal domains and the appanage lands beyond their borders. The agricultural surplus was too small to encourage gratuitous consumption. The great bulk of it went on the soldiery, which in the Mangkunegaran in Surakarta in the late eighteenth century numbered around a thousand. The rest went on the courtly servants (*abdi*): "kris-makers, goldsmiths, grooms, riding-masters, *payung* [umbrella] bearers, and masters of traditional theatre."[62] Scholarship was a priority only as far as it served the urgent political needs of the court. The overwhelmingly "Platonic" nature of Javanese scholarship should therefore not be seen only as a sign of cultural decay. It also in some way reflected the interests of the elite over a longer duration. Institutionally, we must imagine a courtly writing establishment of limited extent and equally limited mandate, namely to ensure the survival of the court. In order to confront the constant likelihood of violent intrigue, courtly texts on history and religious ethics attempted to project an aura of divine inevitability, a theater of power (as Geertz and Anderson both wrote). The elective affinity between the Javanese courts, ruling precariously over a large agricultural population, and the Platonic convictions of "social, cosmic and religious Order" suffusing all their literature, lay in this survival imperative. It left no room for skepticism, intellectual rebellion, or uncertain experimentation with the physical world.

If this Weberian speculation linking intellectual inquiry to elite political interests has any value, it might also explain the change that began to occur after our nominal year of 1808. By this time, the court no longer faced a military imperative from rival kings, but it did confront the much more intellectual imperative of responding to the imperial Dutch. A Javanese intellectual elite began to emerge, whose world was not confined to the scriptorium. As the power of the courts began to wane, so too did elite Javanese reluctance to deal with the profane world. *Serat Centhini* is a rambling tale of several traveling students, edited in 1814 by the Surakarta court scholars Yasadipura II and Rangga Sutrasna, based on earlier accounts. A growing awareness of Java beyond the kraton expressed itself in an encyclopedic richness. Kumar saw in this manuscript "a shift in the intellectual tradition of Java from a largely a-temporal, a-specific, universalistic viewpoint to a more empirical and particular one."[63] Until *Serat Centhini,* Javanese geography was almost as literary and intellectual as Javanese astronomy. References to mountains and rivers in Javanese manuscripts had more often been to Indian, and thus metaphorical, mountains and rivers.[64] The author or authors of the story of *Ki Cabolang*, a subplot within the main Centhini cycle, were interested in the natural landscape as they had seen it with their own eyes.[65] There are woods, springs, caves, mountains, ravines, and strange phenomena such as a smoking pond, fire in an irrigated rice field, and big bones supposedly of a slain demon. We also read about real Javanese cultural monuments such as particular antiquities, graves, mosque-schools, and hermitages.

In 1819, the same Yasadipura II completed a moral-philosophical text entitled *Sasana Sunu*, to instruct the young aristocrat *(priyayi)* in such things as how to choose good friends and avoid bad ones, and how to discipline oneself through little eating, drinking, and sleeping and much meditating. Amid these moral concerns, the text conveys a new empirical spirit. The young priyayi assigned to collect taxes in a lowly village is told to learn all the agricultural tools—harrow, plow, sickle, crowbar, axes and hoes, and adzes and choppers—and to know about livestock.[66] It had been Galileo's intimate conversations with lowly gunners that led him to the solution of the trajectory of a projectile. A similar contact between the learned aristocrat and the peasant was about to occur in Java. Kumar saw here a refashioning of the intellectual world "from a Platonic to a Hegelian format."[67] This was a new intellectual force, arising within Java itself.

When the natural science teacher J. A. C. Oudemans in the second half of the nineteenth century wrote a set of school textbooks introducing indigenous students to the basics of astronomy such as the spherical earth, he did not refer to anything the Javanese may have already known.[68] He wrote it as a dialogue between a kindly Dutch gentleman-teacher and a polite, naïve young Javanese aristocrat called Abdullah. By then, upwardly mobile Javanese intellectuals were subscribing to their own Surakarta-published newspaper called *Bramartani,* which regularly lampooned Javanese "superstition" and praised "America, that land whose inventions of all kinds are ever greater and more amazing."[69]

By 1900, the first Javanese to follow a higher education in the natural sciences, medicine to be exact, had enrolled in a Dutch university. He was Raden Mas Ario Koesoema Joedha, son of Sultan Pakoe Alam V of Yogyakarta. His father was a Freemason, and had sent his son to the Netherlands in 1890 at the tender age of eight to have a complete Western education.[70] Abdul Rivai, from West Sumatra, became the first Indonesian to gain a PhD, also in medicine, in 1908. He was a *dokter djawa*, a graduate of the colonial medical education system that had been improving steadily since 1875.[71] He combined a brilliant career in medicine, specializing among others in ophthalmology, with a whirlwind of journalistic and literary activity.[72] He pleaded for more books "that push the indigenes to think and to observe."[73] The scientific urge was for him emancipatory, and anticolonial.

The same inspiration drove generations of Indonesian students to follow his example. My invitation to work at that Central Java university in the 1980s came from Liek Wilardjo, who completed his PhD in theoretical physics at the University of Michigan in 1970. Science has throughout his long life been an ethical calling. His prolific media columns are peppered with quotes from philosophers of science from Aristotle to Wittgenstein, and he famously resisted the introduction of nuclear energy in Indonesia (*nota bene* based on his own discipline!) as a "Faustian bargain."[74] The Indonesian astronomer Bambang Hidayat is also a historian of science. He writes about indigenous astronomical knowledge in the archipelago with the detachment that comes from living in a new era.[75] M. Sahari Besari, a vigorous emeritus professor of mechanical engineering at the Bandung Institute of Technology, in 2008 wrote a book with the somewhat masochistic title *Technology in Nusantara: 40 Centuries of Resistance to Innovation*. While the ancient Mesopotamians and Greeks were determining the value of *pi* to ever-increasing accuracy, he wrote, there was no record that the people of the archipelago contributed anything to the development of mathematics and science. He attributed this to the region's static culture. He even faulted those nineteenth-century Javanese aristocrats who had acquired Western values as "hypocrites," because they tried to combine feudal with progressive value systems. Indonesia's hope for modernity lay with a new elite born from the lower orders of society and then properly educated.[76]

The answers to the questions posed at the beginning of this chapter, therefore, appear to be as follows. The Javanese probably knew next to no evidence-based natural science around 1808. This was the conclusion of the first Dutch and British scholars who attempted to understand Javanese knowledge around this time, and the far more expert students who followed them have not seriously challenged it. However, we still cannot be sure. Many existing manuscripts remain unread. It is moreover not unlikely that the Javanese had known more about the natural world in previous, now scantily documented, centuries. We can only speculate about the reasons for the meager state of knowledge in this particular area of human endeavor in the early nineteenth century, which appears so contrary to the abundant opportunities for "knowledge transfer" at the time. Perhaps they had to do with the circumscribed function of the royal scriptorium—which was to build cosmic justifications for the political regime it served rather than to encourage skeptical inquiry. This guess finds some confirmation in the eagerness with which Javanese scholars embraced the natural sciences once the old regime had passed away.

Notes

*Senior researcher, Royal Netherlands Institute of Southeast Asian and Caribbean Studies (KITLV), www.kitlv.nl; email klinken@ kitlv.nl. I am grateful for critical comments from Amrit Gomperts, Ward Keeler, and Merle Ricklefs. Caveats are even more applicable here than is usual, firstly because their scholarship exceeds my own to an embarrassing degree, and secondly because I am afraid I have persisted in opinions that each of them, in their own way, perhaps finds inadvisable.

1. Hans Groot, *Van Batavia Naar Weltevreden: Het Bataviaasch Genootschap Van Kunsten En Wetenschappen, 1778–1867* (Leiden KITLV Uitgeverij, 2009), 115–23.
2. Ibid., 176.
3. John Crawfurd, *History of the Indian Archipelago: Containing an Account of the Manners, Arts, Languages, Religions, Institutions, and Commerce of Its Inhabitants*, 3 vols. (Edinburgh: Constable, 1820), I, 253, 81, 307.
4. Johannes Hageman Jcz, *Handleiding Tot De Kennis Der Geschiedenis, Aardrijkskunde, Fabelleer En Tijdrekenkunde Van Java*, 2 vols. (Batavia: Lange & Co, 1852), II, 396.
5. P. J. Zoetmulder, *Kalangwan: A Survey of Old Javanese Literature*, Translation Series, vol. 16 (The Hague: Martinus Nijhoff, 1974), 171–72, 214.
6. Theodore G. Th. Pigeaud, *Literature of Java: Catalogue Raisonné of Javanese Manuscripts in the Library of the University of Leiden and Other Public Collections in the Netherlands*, 3 vols. (The Hague: Martinus Nyhoff, 1967–70), I, 2.
7. H. Floris Cohen, *The Scientific Revolution: A Historiographical Inquiry* (Chicago, etc.: University of Chicago Press, 1994), 379.
8. Ann Kumar, *Java and Modern Europe: Ambiguous Encounters* (Richmond: Curzon, 1997), 181.

9. T. E. Behrend, "Manuscript Production in Nineteenth-Century Java; Codicology and the Writing of Javanese Literary History," *Bijdragen tot de Taal-, Land- en Volkenkunde* 149, 3 (1993): 12.
10. Ibid., 12.
11. Ibid., 23. Other manuscripts were produced on the orders of the colonial government under the guidance of the Dutch scholars. Serious copying of Javanese manuscripts for Dutch colonial library collections commenced in the second half of the century. It was associated with the names of A. B. Cohen Stuart, father and son C. F. and F. L. Winter, J. F. C. Gericke, J. A. Wilkens, and the great Javanese poet and copyist Ranggawarsita. Their interests were mainly literary and legal. Finally, some manuscripts were simply copied for cash, paid by private Dutch scholars or by Javanese lending libraries. Even Ranggawarsita copied manuscripts for money after 1870.
12. Pigeaud, *Literature of Java*.
13. Helaine Selin, ed., *Encyclopaedia of the History of Science, Technology, and Medicine in Non-Western Cultures* (Dordrecht etc: Kluwer Academic Publishers,1997), 500–20.
14. Ian Proudfoot, *Old Muslim Calendars of Southeast Asia*, Handbook of Oriental Studies, Section 3, South-East Asia, vol. 17 (Leiden etc.: Brill, 2006), 6.
15. Theodore G. Th. Pigeaud, "Een Stuk over Sterrenkunde Uit Het Anggastyaparwwa, Het Naksatrarupa," *Tijdschrift voor Indishe Taal-, Land-, en Volken-kunde* 65(1925). See also the comprehensive literature survey on the Javanese calendar in F. van den Bosch, "Der Javanische Mangsakalender," *Bijdragen tot de Taal-, Land- en Volkenkunde (BKI)* 136, 2/3 (1980).
16. "The Javans of the present day have no pretensions to astronomy as a science. The seasons are determined by reference to a system no longer perfectly understood, either in principle or application, but from the Hindu terms still in use for the days of the week, etc., and from the similarity of many of their superstitions to those of continental India [i.e., shouting during an eclipse to prevent the moon being eaten by a naga], it seems probable that if they ever possessed an astronomical system, it was derived from that quarter.... Some of the better informed have derived a few notions of astronomy from the Arabs, but their knowledge in this respect is at best extremely imperfect," Thomas Stamford Raffles, *The History of Java*, 2 vols., Oxford in Asia Historical Reprints (Kuala Lumpur etc.: Oxford University Press, 1978 [1817]), 473–74.
17. Amrit Gomperts, "Sanskrit Jyotisa Terms and Indian Astronomy in Old Javanese Inscriptions," in *Fruits of Inspiration: Studies in Honour of Prof. J.G. De Casparis, Retired Professor of the Early History and Archeology of South and Southeast Asia at the University of Leiden, the Netherlands on the Occasion of His 85th Birthday*, ed. Marijke J. Klokke and Karel R. van Kooij, Gonda Indological Studies, *vol.* 11 (Groningen: Egbert Forsten, 2001), 118.

18. Zoetmulder, *Kalangwan: A Survey of Old Javanese Literature*, 190.
19. Amrit Gomperts, "The Muhurtalaksana: A Brief Text on Time of Day, Gnomonic Shadow and Divination from Java, Compared to the Inscriptions and the Sanskrit Atmajyotisa," paper presented at the 13th World Sanskrit Conference, Edinburgh, Scotland, July 10–14, 2006. Ascension refers to the time it takes for *naksatra* (zodiacs) to cross the horizon. Due to the tilt of the earth, this time varies for different naksatras. The differences increase with latitude.
20. Kumar, *Java and Modern Europe: Ambiguous Encounters*, 147, 56–58; M. C. Ricklefs, *Modern Javanese Historical Tradition: A Study of an Original Kartasura Chronicle and Related Materials* (London: School of Oriental and African Studies, University of London, 1978), 226–28.
21. "If one would put a rope around the surface of the earth over the extremities of the north and the south poles, then its measure of circumference would be three times larger than its diameter" (Gomperts, "Sanskrit Jyotisa Terms and Indian Astronomy in Old Javanese Inscriptions," 95., retranslating Gonda's 1932 Dutch translation of Brahmandapurana). This in turn was based on Greek ideas about the sphericity of the earth.
22. Ricklefs, *Modern Javanese Historical Tradition*, 239–44.
23. Raffles, *The History of Java*, App. G.
24. M. C. Ricklefs, *Jogjakarta under Sultan Mangkubumi 1749–1792: A History of the Division of Java* (London: Oxford University Press, 1974), 176–226. Javanese tradition has the Majapahit empire falling with the death of the king in the Javanese year 1400 (1478 AD—in fact he had several Hindu-Javanese successors, but they were weak; see J. Noorduyn, "Majapahit in the Fifteenth Century," *Bijdragen tot de Taal-, Land- en Volkenkunde* 134, 2/3 (1978): 243–44. The vacuum was filled three years later by the rise of Demak. The next dynastic change—the end of Demak, the rise of Pajang—is supposed to have occurred in the Javanese years 1500–1503 (1578–1581 AD), though these dates are contrived to fit the '00-'03 pattern (dynastic collapse every hundred years and a new dynasty three years later). Certainly the Mataram kraton in Yogyakarta (Plered) did fall in the Javanese year 1600 (1677 AD), and a new kraton was established in Kartasura three years later. By the Javanese year 1700 (1774 AD) the Dutch were fully in control of Java, and dynastic wars had become impossible. Ricklefs speculates that the burst of courtly writing activity shortly after the Javanese year 1700 (1774 AD), in particular the manuscript Serat Surya Raja, was intended as prophecy to deal with the refractory historical reality of the divided Javanese kingdom (Yogyakarta and Surakarta) in an eschatological future.
25. C. Hooykaas, ed. *Cosmogony and Creation in Balinese Tradition*, Bibliotheca Indonesica, 9 (The Hague: Martinus Nijhoff,1974).

26. Theodore G. Th. Pigeaud, "Javaansche Wichelarij En Klassifikatie," in *Feestbundel: Koninklijk Bataviaasch Genootschap Van Kunsten En Wetenschappen, Bij Gelegenheid Van Zijn 150-Jarig Bestaan 1778–1928* (1929).
27. A. C. Crombie, *Styles of Scientific Thinking in the European Tradition: The History of Argument and Explanation Especially in the Mathematical and Biomedical Sciences and Arts*(London: Duckworth, 1994), vol.1.
28. Pigeaud, *Literature of Java*, I, 2–3.
29. Alfred North Whitehead, *Science and the Modern World* (London: Free Association Books, 1985 [1926]), 10.
30. James George Frazer, *The Golden Bough: A Study in Magic and Religion* (London etc.] Macmillan, 1987 [original 1922]).
31. Marcel Mauss, *A General Theory of Magic*, 2nd ed., Routledge Classics (London etc.: Routledge & Kegan Paul, 2001 [original French 1950]), 141–42.
32. B. A. J. Lovric, "Medical Semiology and the Semiotics of Dance," *Review of Indonesian and Malaysian Affairs* 24, 2 (1990).
33. Wolfgang Weck, *Heilkunde Und Volkstum Auf Bali* (Stuttgart: F. Enke, 1937).
34. Kumar, *Java and Modern Europe: Ambiguous Encounters*, 374.
35. W. P. Groeneveldt, *Historical Notes on Indonesia and Malaya: Compiled from Chinese Sources* (Djakarta: Bhratara, 1960 [1880]), 17.
36. M. C. Ricklefs, *War, Culture and Economy in Java, 1677–1726: Asian and European Imperialism in the Early Kartasura Period*, Southeast Asia Publications Series, vol. 24 (Sydney: Allen & Unwin, 1993), 131.
37. B. Schrieke, *Indonesian Sociological Studies: Selected Writings of B. Schrieke*, 2 vols., Selected Studies on Indonesia, vol. 2–3 (The Hague; Bandung: Van Hoeve, 1955–57), II, 122.
38. Gene Ammarell, "Sky Calendars of the Indo-Malay Archipelago: Regional Diversity/ Local Knowledge," *Indonesia* 45(1988); Gene Ammarell, "The Planetarium and the Plough: Interpreting Star Calendars of Rural Java," in *Songs from the Sky: Indigenous Astronomical and Cosmological Traditions of the World*, ed. Von Del Chamberlain, John B. Carlson, and M. Jane Young (College Park: Ocarina Books, 2005 [1996]).; Bosch, "Der Javanische Mangsakalender"; Alfred Maass, "Sternkunde Und Sterndeuterei Im Malaiischen Archipel," *Tijdschrift voor Indische Taal-, Land-, en Volken-kunde* 64(1924).
39. H. A. van Hien, *De Javaansche Geestenwereld, De Betrekking Die Tusschen De Geesten En De Zinnelijke Wereld Bestaat*, 5 vols. (Batavia: Kolff, 1922); L. Th. Mayer, *Een Blik in Het Javaansche Volksleven*, 2 vols. (Leiden: Brill, 1897), 135–75.
40. Hien, *De Javaansche Geestenwereld, De Betrekking Die Tusschen De Geesten En De Zinnelijke Wereld Bestaat.*

41. Joseph Needham and abridged by Colin A. Ronan, *The Shorter Science and Civilisation in China* (Cambridge etc.: Cambridge University Press, 1978);Simon Winchester, *The Man Who Loved China: The Fantastic Story of the Eccentric Scientist Who Unlocked the Mysteries of the Middle Kingdom* (New York: HarperCollins, 2008).
42. Ian Hacking, *The Social Construction of What?* (Cambridge, MA etc.: Harvard University Press, 1999).
43. Cohen, *The Scientific Revolution: A Historiographical Inquiry*, 480.
44. Ibid., 73–87.
45. Ibid., 475.
46. Ibid., 471.
47. Ibid., 456.
48. Ibid., 481.
49. Richard Herbert Howe, "Max Weber's Elective Affinities: Sociology within the Bounds of Pure Reason," *The American Journal of Sociology* 84, 2 (1978). "However, we can generalize about the degree of elective affinity between concrete structures of social action and concrete forms of economic organization; that means, we can state in general terms whether they further or impede or exclude one another—whether they are 'adequate' or 'inadequate' in relation to one another" (Max Weber, *Economy and Society* [New York: Bedminster Press, 1968], 341).
50. Max Weber, *The Protestant Ethic and the Spirit of Capitalism*, Routledge Classics (London etc.: Routledge, 2001 [original German 1920]).
51. H. H. Gerth and C. Wright Mills, eds., *From Max Weber: Essays in Sociology* (New York: Oxford University Press, 1967 [original 1946]), 62–63.
52. Kumar, *Java and Modern Europe: Ambiguous Encounters*, 435.
53. Ricklefs, *Jogjakarta under Sultan Mangkubumi 1749–1792: A History of the Division of Java*, 415.
54. Ibid., 362.
55. Peter Carey, *The Power of Prophecy: Prince Dipanagara and the End of an Old Order in Java, 1785–1855*, 2nd ed., Verhandelingen Van Het Koninklijk Instituut Voor Taal-, Land- En Volkenkunde, 249 (Leiden: KITLV Press, 2007), 331–53.
56. Ricklefs, *Jogjakarta under Sultan Mangkubumi 1749–1792: A History of the Division of Java*, 420.
57. Ricklefs, *Modern Javanese Historical Tradition: A Study of an Original Kartasura Chronicle and Related Materials*, 217–18.
58. M. C. Ricklefs, *Polarising Javanese Society: Islamic and Other Visions, C. 1830–1930* (Leiden: KITLV Press, 2007), 23.
59. Ricklefs, *Jogjakarta under Sultan Mangkubumi 1749–1792: A History of the Division of Java*, 422.
60. Ricklefs, *Polarising Javanese Society*, 50.
61. Kumar, *Java and Modern Europe*, 176.

62. Ibid., 73.
63. Ibid., 90.
64. To be sure, there had been remarkable exceptions. The Old Javanese text *Negarakertagama* describes the Majapahit court by means of many concrete observations of particular places. Similarly, in the area of historiography, which was elsewhere so often mythicizing, the *Babad ing Sangkala*, a unique document dated 1738 AD with links to Old Javanese, accurately describes the politics of the Kartasura court until about 1670. Ricklefs, *Modern Javanese Historical Tradition*, 202–3.
65. Pigeaud considered Cabolang a late nineteenth to early-twentieth-century document written mainly by Sumahatmaka of the Mangkunegaran House in Surakarta, but Kumar considers it part of the earlier Centhini cycle (Pigeaud, *Literature of Java*, I, 228–29; Kumar, *Java and Modern Europe: Ambiguous Encounters*).
66. Kumar, *Java and Modern Europe*, 411.
67. Ibid., 440.
68. J. A. C. Oudemans, *Ilmoe Alam of Wereldbeschrijving Voor De Inlandsche Scholen*, 5 vols. (Batavia: Landsdrukkerij, 1873–1885).
69. Ricklefs, *Polarising Javanese Society*, 158.
70. Harry A. Poeze, *In Het Land Van De Overheerser: Part I – Indonesiërs in Nederland 1600–1950*, 2nd ed., Verhandelingen Van Het Koninklijk Instituut Voor Taal-, Land- En Volkenkunde; 100: 1 (Dordrecht etc.: Foris, 1986), 37–39.
71. Elisabeth Quirine Hesselink, "Genezers Op De Koloniale Markt: Inheemse Dokters En Vroedvrouwen in Nederlandsch Oost-Indië, 1850–1915" (PhD, Universiteit van Amsterdam, 2009), 187–78.
72. Poeze, *In Het Land Van De Overheerser: Part I—Indonesiërs in Nederland 1600–1950*, 51.
73. Ibid., 32.
74. Liek Wilardjo, *Realita Dan Desiderata* (Yogyakarta: Duta Wacana University Press, 1990), 143–46.
75. Bambang Hidayat, "Indo-Malay Astronomy," in *Astronomy across Cultures: The History of Non-Western Astronomy*, ed. Helaine Selin (Dordrecht etc.: Kluwer Academic Publishers, 2000).
76. M. Sahari Besari, *Teknologi Di Nusantara: 40 Abad Hambatan Inovasi* (Jakarta: Salemba Teknika, 2008), 98, 218.

Bibliography

Ammarell, Gene. "Sky Calendars of the Indo-Malay Archipelago: Regional Diversity/Local Knowledge." *Indonesia* 45 (1988): 85–104.

———. "The Planetarium and the Plough: Interpreting Star Calendars of Rural Java." In *Songs from the Sky: Indigenous Astronomical and Cosmological Traditions of the World*, edited by Von Del Chamberlain, John B. Carlson, and M. Jane Young, 320–35. College Park: Ocarina Books, 2005 [1996].

Behrend, T. E. "Manuscript Production in Nineteenth-Century Java: Codicology and the Writing of Javanese Literary History." *Bijdragen tot de Taal-, Land- en Volkenkunde* 149, 3 (1993): 407–37.

Besari, M. Sahari. *Teknologi Di Nusantara: 40 Abad Hambatan Inovasi.* Jakarta: Salemba Teknika, 2008.

Bosch, F. van den. "Der Javanische Mangsakalender." *Bijdragen tot de Taal-, Land- en Volkenkunde (BKI)* 136, 2/3 (1980): 248–82.

Carey, Peter. *The Power of Prophecy: Prince Dipanagara and the End of an Old Order in Java, 1785–1855.* 2nd ed. Verhandelingen Van Het Koninklijk Instituut Voor Taal-, Land- En Volkenkunde, 249. Leiden: KITLV Press, 2007.

Cohen, H. Floris. *The Scientific Revolution: A Historiographical Inquiry.* Chicago: University of Chicago Press, 1994.

Crawfurd, John. *History of the Indian Archipelago: Containing an Account of the Manners, Arts, Languages, Religions, Institutions, and Commerce of Its Inhabitants.* 3 vols. Edinburgh: Constable, 1820.

Crombie, A. C. *Styles of Scientific Thinking in the European Tradition: The History of Argument and Explanation Especially in the Mathematical and Biomedical Sciences and Arts.* 3 vols. London: Duckworth, 1994.

Frazer, James George. *The Golden Bough: A Study in Magic and Religion.* London etc.: Macmillan, 1987 [original 1922].

Gerth, H. H., and C. Wright Mills, eds. *From Max Weber: Essays in Sociology.* New York: Oxford University Press, 1967 [original 1946].

Gomperts, Amrit. "Sanskrit Jyotisa Terms and Indian Astronomy in Old Javanese Inscriptions." In *Fruits of Inspiration: Studies in Honour of Prof. J.G. De Casparis, Retired Professor of the Early History and Archeology of South and Southeast Asia at the University of Leiden, the Netherlands on the Occasion of His 85th Birthday*, edited by Marijke J. Klokke and Karel R. van Kooij, 93–134. Groningen: Egbert Forsten, 2001.

———. "The Muhurtalaksana: A Brief Text on Time of Day, Gnomonic Shadow and Divination from Java, Compared to the Inscriptions and the Sanskrit Atmajyotisa." Paper presented at the 13th World Sanskrit Conference, Edinburgh, Scotland, July 10–14, 2006.

Groeneveldt, W. P. *Historical Notes on Indonesia and Malaya: Compiled from Chinese Sources.* Djakarta: Bhratara, 1960 [1880].

Groot, Hans. *Van Batavia Naar Weltevreden: Het Bataviaasch Genootschap Van Kunsten En Wetenschappen, 1778–1867.* Leiden KITLV: Uitgeverij, 2009.

Hacking, Ian. *The Social Construction of What?* Cambridge, MA. etc.: Harvard University Press, 1999.

Hageman Jcz, Johannes. *Handleiding Tot De Kennis Der Geschiedenis, Aardrijkskunde, Fabelleer En Tijdrekenkunde Van Java.* 2 vols. Batavia: Lange & Co, 1852.

Hesselink, Elisabeth Quirine. "Genezers Op De Koloniale Markt: Inheemse Dokters En Vroedvrouwen in Nederlandsch Oost-Indië, 1850–1915." PhD, Universiteit van Amsterdam, 2009.

Hidayat, Bambang. "Indo-Malay Astronomy." In *Astronomy across Cultures: The History of Non-Western Astronomy*, edited by Helaine Selin, 371–84. Dordrecht etc.: Kluwer Academic Publishers, 2000.

Hien, H. A. van. *De Javaansche Geestenwereld, De Betrekking Die Tusschen De Geesten En De Zinnelijke Wereld Bestaat.* 5 vols. Batavia: Kolff, 1922.

Hooykaas, C., ed. *Cosmogony and Creation in Balinese Tradition.* Bibliotheca Indonesica, vol. 9. The Hague: Martinus Nijhoff, 1974.

Howe, Richard Herbert. "Max Weber's Elective Affinities: Sociology within the Bounds of Pure Reason." *The American Journal of Sociology* 84, 2 (1978): 366–85.

Kumar, Ann. *Java and Modern Europe: Ambiguous Encounters.* Richmond: Curzon, 1997.

Lovric, B. A. J. "Medical Semiology and the Semiotics of Dance." *Review of Indonesian and Malaysian Affairs* 24, 2 (1990): 136–94.

Maass, Alfred. "Sternkunde Und Sterndeuterei Im Malaiischen Archipel." *Tijdschrift voor Indische Taal-, Land-, en Volken-kunde* 64 (1924): 1–172, 347–459.

Mauss, Marcel. *A General Theory of Magic.* 2 ed, Routledge Classics. London etc.: Routledge & Kegan Paul, 2001 [original French 1950].

Mayer, L. Th. *Een Blik in Het Javaansche Volksleven.* 2 vols. Leiden: Brill, 1897.

Needham, Joseph, and abridged by Colin A. Ronan. *The Shorter Science and Civilisation in China.* Cambridge etc.: Cambridge University Press, 1978.

Noorduyn, J. "Majapahit in the Fifteenth Century." *Bijdragen tot de Taal-, Land- en Volkenkunde* 134, 2/3 (1978): 207–74.

Oudemans, J. A. C. *Ilmoe Alam of Wereldbeschrijving Voor De Inlandsche Scholen.* 5 vols. Batavia: Landsdrukkerij, 1873–85.

Pigeaud, Theodore G. Th. "Een Stuk over Sterrenkunde Uit Het Anggastyaparwwa, Het Naksatrarupa." *Tijdschrift voor Indishe Taal-, Land-, en Volkenkunde* 65 (1925): 282–97.

———. "Javaansche Wichelarij En Klassifikatie." In *Feestbundel: Koninklijk Bataviaasch Genootschap Van Kunsten En Wetenschappen, Bij Gelegenheid Van Zijn 150-Jarig Bestaan 1778–1928*, 273–90. Weltevreden: Kolff, 1929.

———. *Literature of Java: Catalogue Raisonné of Javanese Manuscripts in the Library of the University of Leiden and Other Public Collections in the Netherlands.* 3 vols. The Hague: Martinus Nijhoff, 1967–70.

Poeze, Harry A. *In Het Land Van De Overheerser: Part I—Indonesiërs in Nederland 1600–1950.* 2 ed., Verhandelingen Van Het Koninklijk Instituut Voor Taal-, Land- En Volkenkunde; 100: 1. Dordrecht etc.: Foris, 1986.

Proudfoot, Ian. *Old Muslim Calendars of Southeast Asia.* Handbook of Oriental Studies, Section 3, South-East Asia, vol. 17. Leiden etc.: Brill, 2006.

Raffles, Thomas Stamford. *The History of Java.* 2 vols, Oxford in Asia Historical Reprints. Kuala Lumpur etc.: Oxford University Press, 1978 [1817].

Ricklefs, M. C. *Jogjakarta under Sultan Mangkubumi 1749–1792: A History of the Division of Java*. London: Oxford University Press, 1974.

Ricklefs, M. C. *Modern Javanese Historical Tradition: A Study of an Original Kartasura Chronicle and Related Materials*. London: School of Oriental and African Studies, University of London, 1978.

Ricklefs, M. C. *War, Culture and Economy in Java, 1677–1726: Asian and European Imperialism in the Early Kartasura Period*. Southeast Asia Publications Series 24. Sydney: Allen & Unwin, 1993.

Ricklefs, M. C. *Polarising Javanese Society: Islamic and Other Visions, C. 1830–1930*. Leiden: KITLV Press, 2007.

Schrieke, B. *Indonesian Sociological Studies: Selected Writings of B. Schrieke*. 2 vols. Selected Studies on Indonesia, vol. 2–3. The Hague; Bandung: Van Hoeve, 1955–57.

Selin, Helaine, ed. *Encyclopaedia of the History of Science, Technology, and Medicine in Non-Western Cultures*. Dordrecht etc: Kluwer Academic Publishers, 1997.

Weber, Max. *Economy and Society*. New York: Bedminster Press, 1968.

———. *The Protestant Ethic and the Spirit of Capitalism*. Routledge Classics. London etc.: Routledge, 2001 [original German 1920].

Weck, Wolfgang. *Heilkunde Und Volkstum Auf Bali*. Stuttgart: F. Enke, 1937.

Whitehead, Alfred North. *Science and the Modern World*. London: Free Association Books, 1985 [1926].

Wilardjo, Liek. *Realita Dan Desiderata*. Yogyakarta: Duta Wacana University Press, 1990.

Winchester, Simon. *The Man Who Loved China: The Fantastic Story of the Eccentric Scientist Who Unlocked the Mysteries of the Middle Kingdom*. New York: HarperCollins, 2008.

Zoetmulder, P. J. *Kalangwan: A Survey of Old Javanese Literature*. Translation Series, vol. 16. The Hague: Martinus Nijhoff, 1974.

Chapter 5

For the Common Good: Dutch Institutions and Western Scholarship on Indonesia around 1800

Peter Boomgaard

From the first arrival of Europeans at Indonesia's shores, they created and circulated knowledge. The Portuguese, trading with Java and the Moluccas from the early sixteenth century, were the first to publish their travel accounts, and much of our knowledge of the region around that time is based on their reports. After 1600, they no longer played an important role in the Indonesian Archipelago, and their days of knowledge creation concerning the region were over.[1]

The Dutch arrived just before 1600, and took up the baton. They would stay until the 1940s; they started to collect and publish information on the archipelago right away, and went on doing so ever after.[2] This chapter looks at the creation and circulation of knowledge regarding Indonesia by the Dutch and other Westerners during the last decades of the eighteenth century and the first of the nineteenth.

The Setting

The period around 1800 was not the finest hour of the Dutch state and its network of colonies and trading posts overseas. Between 1780 and 1815 the Netherlands often had to tolerate foreign armies on its territory, was involved in various wars, often in a subaltern capacity, and was cut off from its establishments in America, Africa, and Asia due to the same wars. This was to the detriment of its trade, and therefore of its economy in general. Many of its overseas settlements were taken over by the British and lost to the Dutch for ever. By 1815 the Netherlands miraculously was still in the possession of or had regained most of its former establishments in the Indonesian

Archipelago and the Caribbean (Suriname, Netherlands Antilles), while its near monopoly on the trade with Japan survived as well. But it had lost all other territories and trading posts.

The large trading companies that had connected the mother country and the colonies had disappeared as well—the Dutch East India Company (*Vereenigde Oostindische Compagnie*; VOC for short) and the West India Company (*West-Indische Compagnie*, WIC for short) were discontinued in 1795 and 1792 respectively. They had already been in the doldrums for a much longer period, and had left large debts.[3]

The (First) Golden Age

This was a far cry from the Dutch Golden Age, a period that lasted roughly from the 1580s to the 1680s, when the Dutch Republic went through a long phase of high economic growth rates, and arts, sciences, and technology flourished simultaneously. This was also a period of mercantile expansion, during which a network of colonies and trading posts was created in the Americas, Africa, and Asia by the WIC and the VOC.[4]

High rates of economic growth, mercantile expansion, and the flowering of Dutch science and technology went hand in hand. Perhaps the best example of this relationship can be seen in the development and role of Dutch botany. Botany was economically important because many plants had medicinal properties, and various plants played important roles in commerce, industry, and agriculture. During the seventeenth century, physicians-cum-botanists and amateur botanists followed the flag of the VOC and the WIC to the Americas (Brazil, Suriname), Africa (Cape of Good Hope), and Asia (Malabar, Ceylon, Java, Moluccas, and Japan).[5] There, they collected, dried, described, and made drawings of thousands of plants, shrubs, and trees. The descriptions and illustrations were sent to the Netherlands, where they were usually published in beautiful many-volume series. In the Netherlands, and in various Dutch establishments overseas, botanical or medicinal gardens were founded, and seeds, cuttings, tubers, etcetera, were exchanged and planted. The result was that in the early eighteenth century, the botanical gardens of Leiden and Amsterdam had the largest plant collections from America, Africa, and Asia in the world, plus the best collections of herbals and herbaria. So when the Swedish botanist Carl von Linné (Linnaeus) wanted to design and execute his grand project of a new system of all things natural, he came to the Netherlands.

That some of this botanical exchange stimulated economic growth in the colonies and in the mother country is proven by a crop like coffee. The Dutch brought it from Arabia to Java and to the Amsterdam botanical garden, from where the plant was sent on to Suriname. It became a huge commercial success in Java and Suriname, and therefore for Dutch merchants in the Republic as well. From Amsterdam and the Dutch colonies, coffee cultivation also spread to other Empires.[6]

Compared to the Golden Age, the period around 1800 was not only less illustrious, it was positively bleak in more than one sense. If we look at the decades from the 1760s to the 1820s, it cannot be denied that the scholarly harvest of the period, at least regarding some of the overseas settlements, was often disappointing. However, there were some bright spots.

Reverend Mohr and the Transit of Venus

Reverend Johan Maurits (or Moritz) Mohr (1716–75) was a German minister of the Dutch Reformed church in Batavia (modern-day Jakarta in western Java) from 1737, and rector of the Theological Seminar of the city from 1743 to 1753. He stands as an isolated figure at the beginning of the period we are dealing with, at a lonely height in Batavia's scholarly landscape. He himself was totally unimpressed with its intellectual climate, stating that one could only study how to make money in Batavia.[7] The story of his role in colonial science is fairly representative of Java around that time, and of the way in which at least some scientific activities were carried out on the initiative and with the support of what we might call "international scholarly networks."[8]

In the early morning of June 6, 1761, Mohr was present when the surveyor, Lieutenant-Captain Gerrit de Haan, head of the department of VOC mapmakers, came to Mohr's mansion in the environs of Batavia—possibly because the house offered an unimpeded view of the horizon—in order to observe the transit of Venus. This event had a long history behind it. The transit of Venus across the sun was important for the determination of the distance between the Earth and the Sun. Already in 1716, Edmund Halley had published the date of the next transit (June 1761), together with the method to be used for measuring it, in the *Philosophical Transactions* of the (British) Royal Society.[9] Halley mentioned Batavia as one of the spots where observations should take place. In 1759, the French astronomer Joseph-Nicolas Delisle sent a proposal regarding the places where the transit should be measured to the Académie Royale des Sciences. Since then, several expeditions were sent out to various places for this purpose, among which were French, British, Swedish, and Russian initiatives, often supported by the respective states.

Delisle also tried to get the Dutch involved in order to have someone in place in Batavia, and it seems likely that this request reached the governor-general in Batavia, who then ordered the observation of the transit, with the above-mentioned results. Mohr sent a report on the measurements to the Netherlands, where it was published in the 1763 issue of the *Verhandelingen* [Transactions] of the Hollandsche Maatschappij der Wetenschappen [Holland Society of Arts and Sciences]. In the same year he was elected a member of this society.

In the meantime Mohr, whose second marriage was to a rich woman, built a grander mansion than before, including an astronomical observatory. The building costs were a staggering 80,000 Rixdollars [*Rijksdaalders*], to which must be added 3,000 guilders [*guldens*] for the meteorological instruments he purchased in the Netherlands. Therefore, when the next transit of Venus occurred, in June 1769, Mohr had a brand new observatory at his disposal, equipped with state-of-the-art observational technology. He sent his observations, much more accurate than the ones dating from 1761, to the Netherlands. They were published in the *Verhandelingen* of the Hollandsche Maatschappij in 1770.

The case of Mohr shows how an amateur scientist could play a role in what was then cutting-edge astronomy. This kind of research had also practical implications, as it was used to calculate the longitude of places, thus being a handmaiden of geography, mapmaking, and navigation. Mohr, for instance, calculated the longitude of Batavia, and the differences in longitude between Greenwich, Paris, and Batavia.[10]

Voyages of Discovery

In the fall of 1770, the British explorer Captain James Cook (1728–1779), on his first voyage of discovery, weighed anchor at the Batavia roadstead, because his ship, the *Endeavour*, needed repairs. He paid a visit to Mohr, who gave him his report, translated into Latin, on the 1769 transit of Venus. This transit was the official motive for Cook's voyage (he would observe it on Tahiti), although the secret motive was something else—the search for *Terra Australis*, the fabled "Southland." Cook gave Mohr's report to the Royal Society, where it was read in November 1771, and in the same year published in the *Philosophical Transactions.*

Aboard Cook's ship was Sir Joseph Banks (1743–1820), rich British landowner, avid amateur botanist, and an important patron of the sciences. He would become the president of the Royal Society, and was appointed keeper of the Royal Botanical Gardens at Kew by King George III. It would be difficult to think of a more appropriate symbol of the link between science and empire in the later eighteenth century than Banks.

During his circumnavigation of the globe, another famous explorer, the French naturalist and admiral Louis-Antoine de Bougainville (1729–1811) called at Batavia in October 1768. Here, he visited Mohr, together with his astronomer, Pierre Antoine Véron (1736–1770), who was part of this French expedition in order to observe the 1769 transit of Venus as well. About Mohr, Bougainville wrote in his diary, published in 1771: "He has got the best instruments of all kinds from Europe, necessary for the nicest observations, and he is capable of making use of them."[11]

The voyages of discovery of Cook, Banks, and Bougainville—the first such maritime voyages after a period of more than a century—represented both the new spirit of scientific discovery—part of the second scientific revolution and symptom of the Enlightenment—and the imperial rivalry between England and France. It is a sign of the times that the Dutch no longer played a role regarding such voyages.[12] Neither the Dutch state nor the Princes of Orange—the so-called Stadholders [*Stadhouders*], who had a quasi-royal position in the Republic—were interested in financing such an undertaking, and there doesn't appear to have been a group of people such as the London Royal Society, which could and would put sufficient pressure on the Stadholder or the Estates-General (Dutch parliament and supreme governing body) to fund such a voyage.[13]

Reverend Mohr, admired by those who led the French and English voyages, mentioned and given space in their publications, was the best the Dutch in the Indonesian archipelago could do, and he was a German.

The Indies Enlightenment and the Batavia Society of Arts and Sciences

The period we are now about to discuss has been termed "the Indies Enlightenment" in Dutch colonial histories; it centered upon Jacobus Cornelis Mattheus Radermacher (1741–83).

When the Hollandsche Maatschappij der Wetenschappen, a learned society founded in 1752, wanted to establish an economic branch [*Oeconomische Tak*] in 1777, they sent word to the directors of the society—an honorary position—who resided in Bengal,

Ceylon [Sri Lanka], and Batavia, inviting them to join the new branch.[14] It was a wake-up call for the two directors living in Batavia to organize their own, local learned society in 1778. These two were the governor-general, Reinier de Klerk, and his son-in-law, Jacobus Radermacher. The new organization was called Bataviaasch Genootschap van Kunsten en Wetenschappen [Batavia Society of Arts and Sciences].

In its program of aims, the founders stated that the Batavia Society would attempt to advance all Arts [*Kunsten*] and Sciences [*Wetenschappen*]. It should be pointed out that the term "arts" in this expression did not refer to the fine arts but to practical knowledge, including technology. The Dutch word "*wetenschappen*" referred, and still refers, to all scholarly disciplines, and not only to the exact sciences. It looks, therefore, as if the founders thought that science and technology belonged together, a point that is still under discussion in the historiography on the eighteenth century.[15]

The Batavia Society, the program went on, would welcome all contributions regarding natural history, antiquities, and manners and customs of (local) peoples, in other words biology, history/archaeology, and (cultural) anthropology:

> However, its main aim is to pursue as objects of its research, such matters as might benefit agriculture, trade, and particularly the prosperity of this colony, and which will be deemed to be practically feasible in all respects.[16]

So the Batavia Society, though also interested in biology, ancient history, and anthropology, saw as it main purpose the economic development of the colony Batavia.

However, this statement, regarding the specific aims, was preceded in the program by a kind of preamble, in which the conviction was formulated that if the Protestant religion was to be successfully spread in the East, this could only be done by first giving the local population a taste of Western sciences, just as the Reformation had been preceded by the "reform of letters" (the Renaissance). Here, as was mentioned in the introductory chapter, Enlightenment science was not only seen as not antagonistic to Western religion, they were supposed to go together, the Book of Nature and Holy Scripture.[17]

Very much in line with other learned societies the world over, the society's motto was "For the Common Good" [*Tot Nut van het Gemeen*]. It is found on the medal struck on the occasion of its founding, an image of which adorns the frontispiece of its journal, not surprisingly called the *Verhandelingen [Transactions] van het Bataviaasch Genootschap van Kunsten en Wetenschappen.*

Offering prizes for essays was from the outset one of the main means to attain the stated aims. In the first volume of the *Verhandelingen*, some 50 questions were formulated as topics for these essays. About half the questions dealt with agricultural problems, including forestry, ten questions had to do with machines and implements, mainly regarding Batavia's water problems (dredging, sluices, etc.), eight with industrial (artisanal) production, and six with health and sickness. The overwhelming majority of the questions, therefore, were of a practical nature.

Only three questions had to do with the humanities—one about literature, one on languages, and one about religion, and even that last one was of a more or less practical nature, because the question was on how would the Muslims be able to convert the heathen inhabitants of these quarters! Muslims were encouraged explicitly to respond to this question. Of a more general nature was the exhortation to those who wanted to research one of these questions, to always ask the opinion of knowledgeable indigenous inhabitants.

The journal *Verhandelingen* was another vehicle for the society's aims. As was the case with the proposals for essays, the majority of the contributions during the first few years were of a practical nature, including readings of barometer and thermometer and of water levels in Batavia, and population statistics. Examples of topics dealt with in articles published in the *Verhandelingen* are given presently.

As will be shown shortly, the Batavia Society stood under the patronage of the governor-general and the Council of the Indies, the highest governing body of the VOC in Asia along with the governor-general, which, collectively, was called the High Government. The society, therefore, could easily request their support. This took the form, among other things, of directives, being forwarded by the High Government to all establishments of the VOC, ordering the local representatives to send information, for instance, regarding natural history, to the society's board. The society thus acquired information, manuscripts, and commodities from many places in Asia.

Finally, it should be mentioned that the Batavia Society had correspondents in the Dutch Republic, who were often well-known scholars, and elsewhere in Europe, like Sir Joseph Banks (Royal Academy), Marquis de Condorcet (Académie Royale), Carl Thunberg (Uppsala), and J. E. Euler (Russian Academy). Occasionally, the Batavia Society nominated celebrities as honorary members, as was the case with Goethe in 1826.

Jacobus Radermacher

Radermacher was the driving force behind the new society.[18] Thanks to the connections of his family, who were part of the patrician oligarchy of the Republic, the so-called regents [*regenten*], Radermacher had been able to leave for the Indies at the age of 16 in the rank of *onderkoopman* [junior merchant] of the VOC. He reached the rank of *opperkoopman* [senior merchant] very fast, in 1762, when he was only 21 years old. In 1761 he married the stepdaughter of the future governor-general, then already member of the Council of the Indies. After having returned to the Netherlands for three years, where he obtained a law degree, he resumed his career in Batavia, this time in the municipal government. In 1776 he was appointed a junior member of the Council of the Indies, and a senior member in 1781.

Radermacher's father had been an influential Freemason. In 1762, Radermacher himself founded the (short-lived) Freemasons' Lodge La Choisie in Batavia, and in 1764 he was appointed great-secretary of the Lodge of The Hague. He also became a member, and later president of the Lodge La Vertueuse, established in Batavia in 1769. He was a member or director of various learned societies in the Republic, and thus a fairly typical example of the link between these societies and freemasonry, both symptoms of the Enlightenment. During the early years of the Batavia Society, about one third of the members were Freemasons.[19]

Well-to-do and well-connected (he married twice into rich and influential families), a member of the administrative elite in Batavia, but not a professional scholar himself, Radermacher spent part of his fortune and his leisure time pursuing his scientific interests. He had a library, collected not only manuscripts, maps, drawings, scientific instruments, and "curiosities" for his curiosity cabinet, mainly *naturalia*, but also Javanese musical instruments. Later on, he would donate all of this to the Batavia Society.

He published extensively in the *Verhandelingen* of the Batavia Society. He coauthored the very first article in the first volume (1779), a description of the possessions of the VOC, which is mainly social and physical geography, population statistics, and a bit of anthropology. In the same volume he gave lists of words in Malay, Dutch, and Latin pertaining to "the three realms of nature" [plants, animals, minerals], and a comparative essay on Asian calendars. In the second volume (1780), he published on a number of equally diverse topics—a description of Borneo, an essay on people of different colors, one about the death penalty, a piece on the improvement of Dutch maritime maps, and a short article on a severe earthquake. In the volumes 3 (1781) and 4 (1786) he contributed altogether five articles describing countries and peoples of various parts of Asia.

The 'Directors' and the other Members of the Board

The governor-general and all the members of the Council of the Indies became "directors" of the Batavia Society, honorary positions that ensured that the highest VOC functionaries in Asia would support the new organization.

Radermacher became the president of the board, which consisted of ten other members ("directing members") in addition to him. Four of the ten were VOC employees with the rank of merchant or junior merchant, two were clergymen, one practiced law, one was part of the municipal administration, one was a physician, and one a notary. None of these people was, therefore, a professional scholar, but a number of them did have university training. About a few of them we have some information, mainly because they published in the *Verhandelingen* of the Batavia Society.

Far from famous or even well known, but probably the best known of the whole lot, is Willem van Hogendorp. Like Radermacher, he came from a well-connected family; he studied law and was a Freemason. In 1778 he was in the employ of the VOC in the rank of *koopman* [merchant], being also a trader on his own account. He was, moreover, a man of letters, who wrote theater plays and poems.[20] In the *Verhandelingen* he was the coauthor of the above-mentioned first article in the first volume. In this and the next issues he also wrote about medical matters (variolation against smallpox), the land and people of Timor, and Timorese words, while in another article he discussed Voltaire's interpretation of a certain passage in the Bible.

All the other members of the board are even more obscure. Johannes Hooyman, minister of the Lutheran church, and owner of an estate in the environs of Batavia, wrote various detailed and well-informed articles about agriculture, particularly estate agriculture (sugar, peanuts), around Batavia, in addition to essays on such varied topics as the Danish mission at the Coromandel Coast of India, spinning and weaving cotton in Java, and Java's edible birds' nests. He came to a sticky end, killed by one of his own slaves.

Josua van Iperen, also a minister, but of the Dutch-Reformed Church in Batavia, became one of the two secretaries of the Batavia Society. He was already a member of three learned societies in the Netherlands before he came to the Indies. In Batavia he gave public lectures during the short time span allotted to him (he died one year after he arrived, possibly of malaria). He also made various contributions to the *Verhandelingen*, such as a series, three years running, on Javanese history according to Javanese sources, an essay on historical knowledge in general, and, as coauthor, on a species of ape to be found in Java, in addition to two poems.

Jacobus van der Steege was a physician who made his fortune as regent of the outer hospitals of Batavia. He was actively engaged in the battle against smallpox, a topic on which he published as well, in addition to other medical topics and magnetism.

Finally a few words about Karl Friedrich Baron von Wurmb, a German, who came to Batavia in or shortly before 1775 and died there in 1781. He was the other secretary to the society. We know somewhat more about him, as his letters to his brother and his sister-in-law were published.[21] In 1776, he was made bookkeeper of the main hospital, his first job in the Indies; he was promoted in 1780, being appointed administrator of a number of VOC warehouses. He was interested and possibly trained in natural history, about which he wrote often in his letters to Germany and published many contributions in the *Verhandelingen* (volumes 1–4, on plants, animals, minerals, and mining). He collected butterflies, shells, and crustaceans (and, one assumes, dried plants), some of which he sent home to Germany, as well as edible birds' nests, for his brother's curiosity cabinet. He was apparently interested in music, was rather negative regarding the level of musical performances among the Dutch in Batavia, and also about Javanese music and musical instruments. He was also rather negative about the intellectual climate in Batavia, although he mentioned some exceptions, like Mohr, and lamented the lack of schools there.[22] He kept abreast of developments in Europe by reading the learned and political journals his brother sent him.

The founding of the Batavia Society appears to have been a pleasant surprise to Von Wurmb. Who would have imagined that possibility in such an outpost, and a Dutch establishment to boot? He was asked to become secretary, with the specific tasks to administer the collections, including the *naturalia*, and the small botanical garden that had been donated to the society, to give demonstrations in the field of physics, for instance, with the electrification machine, and to take care of the meteorological observations, probably with some of the instruments of Mohr that the Batavia Society had acquired.

We will now look at some of the topics that the members of the Batavia Society and other Europeans in the Indonesian Archipelago studied.

Statistics

Statistics was a new science in the eighteenth century, pioneered by German scholars such as Gottfried Achenwall, who appears to have been the person who used the term *Statistik* for the first time (in 1749), and August Ludwig Schlözer, both of whom taught at the university of Göttingen. At that time, the term "statistics" had a slightly different meaning from that of today—all facts relevant for the power of the state, and not only—as is the case today—numerical data.[23]

In the first volumes of the *Verhandelingen*, we find quite an amount of data that both we and the scholars in the eighteenth century would call statistics, among which population data hold pride of place. However, the quality of these figures was often poor, to put it mildly. One example may suffice.

On p. 7 of volume 1 of the *Verhandelingen* (1779), the figure for the entire population of Java was given as 915,588. If one looks at the details of this calculation, it becomes clear that some figures were quoted from the description of the East by François Valentyn, published in 1724–26, a given that does not really inspire much confidence in the reliability of these figures in general. In volume 3 (1781), revised figures for Java's population were published, which now was supposed to consist of 2,029, 915 "souls" (p. 274), or more than twice as much as the estimate dated 1779.

The first serious attempt by the VOC in Java to collect reliable population statistics dates back to 1795. It was not a real census, whereby all people are actually counted, during a short period of time (a month, for instance). In all probability the local Dutch civil servants asked the village heads for data. The number of people that were registered as living in the areas under the sway of the VOC in Java turned out to be 2.5 million, but that left the regions where the Dutch had no authority, with probably some two-fifths of the total population of the island of Java, unaccounted for. Therefore, compared to 1781, the population figure would have more or less doubled if the part of Java still under its own rulers had also been counted in 1795. Population counts of the areas under Dutch authority, held in 1802 and 1808, showed a small increase, but no more revolutionary improvements of the figures occurred.

Between 1811 and 1815, Java was in British hands, and the lieutenant-governor, Thomas Stamford Raffles, for the first time organized a population count for the whole island in 1815, when Java's registered population turned out to be 4.5 million. In hindsight, we must assume that Java's population was perhaps as high as 8 million in

1815, or almost double the number of people registered in that year. Therefore, in 1779, when the island's population must have been between 6.5 and 7 million, the figure published by the Batavia Society was less than 15 percent of the real value!

To my knowledge, the details of the 1795 and 1808 enumerations were never published during the colonial period, and the 1802 count only in 1866. These figures may have been regarded by the government as strategic information that should not be freely available.[24]

The wish to have accurate statistics,of which the enumeration of 1795 was the result, may have been prompted by the need for more revenue from the possessions of the bankrupt VOC, but it was also a sign of the times. States wanted to be better informed about their revenue-producing potential, and that kind of information could only be produced by the state itself, not by a learned society.

When in 1778 the Gentlemen XVII, the governing body of the VOC in the Dutch Republic, endorsed the founding of the Batavia Society, they expressed their expectation that the society would mainly concern itself with matters of importance to the VOC, but they also urged the society not to publish any business secrets. Regarding population figures of their possessions, they need not have worried.

Völker- und Länderkunde

Between 1781 and 1793, the well-known Johann Reinhold Forster (1729–1798), botanist of Cook's second expedition, and his son, the even more famous Georg Forster (1754–1794), who published a detailed account of Cook's voyage based on his father's diaries, were the editors of a journal published in Leipzig entitled *Beiträge zur Völker- und Länderkunde*, in 27 volumes, which was read in the Netherlands and elsewhere in Europe, and in the United States. This journal dealt with what we would now call "cultural anthropology" and geography, both disciplines in which German scholars were pioneers. The University of Göttingen, mentioned earlier, then one of the most progressive German universities, and a center of the European Enlightenment, must be regarded as one of the most important nuclei of these disciplines. Here, the name of Schlözer should be mentioned again, as one of the first scholars to use the word "ethnography" (in 1771), an older term for cultural anthropology. A few years earlier (1767), one of his Göttingen colleagues, J. F. Schöpperlin, had also used the term ethnography, while a third Göttingen professor, Johann Christoph Gatterer, did so a few years later (1775). As a result of their classes, and the books they published, the notion of ethnography or

Völkerkunde, was adopted amazingly fast by other European scholars, including those in the Netherlands.

Material for the new discipline came from Northern and Eastern Europe, Siberia, Central Asia, and the Middle East. Much material had been produced by Russian imperial expeditions in their Asian possessions, in which German scholars were overrepresented. Therefore, the birth of the discipline was, somewhat surprisingly perhaps, linked to Russian empire building, not to that of the Spaniards, the Portuguese, the Dutch, the English, or the French.[25]

It is likely that word of these developments had reached Java around the time the Batavia Society was founded. As was noted earlier, Von Wurmb had his brother send him learned and political journals from Germany in the late 1770s and early 1780s.

Another German baron, Von Wollzogen, journeyed in the Indonesian Archipelago from 1790 to 1792 as captain-lieutenant of the Würtemberg regiment, a professional soldier, therefore, not a scholar. A friend sent him what he called the *Lichtenbergische Magazine*, of which the official title was *Göttingisches Magazin der Wissenschaften und Literatur*, with G. C. Lichtenberg and Georg Forster (whose *Voyage Round the World* was so rich in ethnographic detail) as editors, a journal, therefore, right from the heartland of German ethnography. Wollzogen, obviously a man with scholarly inclinations—in his spare time he collected *naturalia* from flora and fauna, mainly in southern Sulawesi—was strongly influenced by Forster's *Voyage*; he wanted to try to write down everything following Forster's research methods.

Von Wollzogen refers to the many private libraries in Batavia; when their owners died, the books often could be bought very cheap. We also know that Radermacher regularly ordered books from Europe. From 1780 onward, the Batavia Society had a library as well.

We know that Germans took notice of the articles in the *Verhandelingen* of the Batavia Society—around 1790, Wollzogen mentioned a contribution on mousedeer in the *Göttingisches Magazin*, an article that must have been based on Von Wurmb's essay on this animal in the *Verhandelingen*, volume 3 (1780).[26]

The first volumes of the *Verhandelingen* contain many essays that can be regarded as contributions to the new disciplines of geography and ethnography. In volumes 1 and 2, Radermacher and Van Hogendorp published descriptions of lands and people in the Indonesian Archipelago, including some about which very little was known, such as Borneo and Timor. Radermacher continued to do so in the volumes 3 and 4 (Sumatra, Sulawesi, Flores, Lombok, Bali). He also wrote about Japan and India, while in volume 4 anonymous contributions appeared about Japan, China, and the parts of eastern Asia then recently under Russian authority. The last contribution, therefore, referred to the same region German scholars were interested in.

Radermacher and Van Hogendorp were rather coy about their sources, and one assumes they used the VOC archives and library in Batavia. Thus, they presented new material that would have been difficult to find elsewhere. However, the accounts they published are rather short and superficial, and in hindsight we can say that they could have published much more detailed and interesting accounts, based on the same VOC archives. Were they afraid to give away too many trade secrets? Or did their daytime jobs keep them from spending more time on these publications?

It seems fair to say that the first well-documented publications in the field of geography and anthropology of the Indonesian Archipelago did not come from Dutch scholars or even from foreign scholars supported by the Dutch authorities, but from three scholars from the British Empire. I am referring to William Marsden, Thomas Stamford Raffles, and John Crawfurd, who published their pathbreaking monographs between 1783 and 1820.[27]

Granted that Raffles and Marsden were members of the Batavia Society and published in the volumes 7 (1814) and 8 (1816) of the *Verhandelingen*, but it would be surely wrong to argue that this should be seen as support given by the Batavia Society to these two high EIC officials. It was the other way around—these two British officials supported the Batavia Society and gave them a boost after a difficult period. In the 1790s and during the first decade of the nineteenth century, the society had almost ceased to exist, and between 1792 and 1814 no volumes of the *Verhandelingen* were published. Of course, Raffles and Marsden used the Dutch archives and libraries in Batavia, but much in their books was the fruit of new research, undertaken or commissioned by them, and the kind of books they produced overshadow anything that in the field of geography and anthropology had been published by the Dutch.

However, when the first volumes of the *Verhandelingen* were published, these books had not yet been printed, and many scholars may have had their first taste of the geography and ethnography of the Indonesian Archipelago from the pages of this journal.

Plants and Animals

The late eighteenth and early nineteenth centuries appear to have been a good time for scholars specialized in natural history to spend some time in Java, sometimes a considerable amount of time. Daniel Carlsson Solander, Carl Peter Thunberg, Karl Friedrich Baron von Wurmb, Clas Fredrik Hornstedt, Francisco Noroña, Louis Auguste Deschamps, Thomas Horsfield, and Jean Baptiste Louis Claude Theodore Leschenault de la Tour, all of them non-Dutch scholars trained in botany and/or zoology, all stayed for some months or even years in Java, between 1770 and 1815.[28]

The Swedish botanist Solander, pupil of Linnaeus, came with Cook and Banks to Java, where they stayed less than three months in 1770, waiting for their ship to be repaired. However, he fell ill with malaria, and of the one hundred new families and one thousand new species of plants with which Banks and Solander returned—mainly from Australia (Botany Bay!) and the Pacific—only some three hundred hailed from Java. Moreover, most of this material was never published, and the small part that was did not come off the press until 1900. However, all this material was kept in Banks's library, and later in the British Museum. Solander spent the rest of his rather short life in London, where he became Banks's secretary and librarian.[29]

More successful, longer lived, and better known was Thunberg, another Swedish botanist and pupil of Linné. He studied in the botanical gardens of Amsterdam and Leiden before he went on a tour of the VOC settlements, where he was to collect specimens for these gardens. He stayed for a long time at the Cape and in Japan, and for shorter periods in Java and Ceylon, doing botanical research. In Java, his research—in 1775 and 1777—was on occasion carried out on collecting trips together with Radermacher and Von Wurmb. Thunberg tells us in his book about this journey that

> Radermacher gave me as a companion and a guide for my botanical trips [across Java] a good Javanese, rather well-informed about the trees and the plants of his country, and who even knew their names in Malay [in addition to his native Javanese, one assumes]. At the same time he carefully pointed out to me the medicinal use that the people of his country made of the various plants.[30]

Before returning to the country of his birth, Thunberg, back in Amsterdam, made a short trip to London, where he met Sir Joseph Banks, and Johann Reinhold Forster, botanist of Cook's second voyage. After his return to Sweden, he would be appointed to Linné's chair in Uppsala. His botanical publications were mainly about Japan and the Cape, but he wrote about his travels, including to Java, in a less specialized publication for a broader public. From all the places where he did research, he had sent seeds and cuttings to the botanical gardens of Leiden and Amsterdam, in addition to dried specimens of plants. The latter were used by Maarten Houttuyn, for his 37-volume *Natuurlijke Historie*, published between 1761 and 1785.

Thunberg left the artificial system of classifying plant families designed by Linné in favor of a more natural system, pioneered by Michel Adanson and Antoine Laurent de Jussieu. As a result of his publications and his specimens, many new genera and species were added to the corpus of scientifically identified and named plants. He published extensively in the many *Transactions* of various learned societies, being a member of 66 of them.[31]

After the untimely death of Von Wurmb, Radermacher wrote to Thunberg, then back in Uppsala, asking him to send someone who could take over Von Wurmb's tasks regarding the classification of *naturalia*. Thunberg sent his pupil Hornstedt, who had just defended his dissertation, in 1782. He arrived in Java in 1783, where he received a monthly stipend not only from the governor-general, but also from the Batavia Society, and from the Royal Swedish Academy of Science. Hornstedt was housed in the building of the Batavia Society where he catalogued the significant collection of natural specimens sent to Batavia from all over Asia, and oversaw its botanical gardens. He also did some collecting of his own, close to Batavia, and his diary contains a list of medicinal plants identified with Chinese characters, which suggests that Chinese pharmacists were his informants. On account of ill health, he returned to Europe in 1784, accompanied among others by an orangutan, which, however, died at sea. He came home with a large collection of natural specimens, which were partly used for his dissertation on the edible fruits of Java (University of Greifswald, 1786). He made his career—less distinguished than that of Thunberg—in Sweden in natural history and medicine. The book he wanted to write on the zoology of Java never materialized.[32]

The next four scholars had no relationship to the Swedish "dynasty." Spanish botanist Francisco Noroña studied medicine in his native Seville, and among other disciplines natural history in London and Paris. In the early 1780s, he began a scientific journey that would take him to Ceylon, southern India, and the Philippines, where he stayed from 1784 to 1786, writing about the cultivation of cinnamon in Ceylon, and about the raising of silkworms.[33] After a conflict with the Spanish governor-general in Manila, he traveled to Batavia, where he met Reverend Hooyman, one of the board members of the Batavia Society. Hooyman became Noroña's patron, introducing him to the governor-general, who asked Noroña to lead an expedition through the mountains of western Java, an area hitherto not much explored by botanists. Other members of the expedition were a painter and four Malay (?) herbalists.[34]

When Noroña discovered that Hooyman had been embezzling funds for the expedition, he left Java on the first outbound vessel. He died on the way back to Europe. Three contributions were posthumously published in the fifth volume (1790) of the *Verhandelingen* of the Batavia Society—two descriptions of trees, and a list of the 800 plants from Java that he collected (*Relatio Plantarum Javanensium*). Like Thunberg, Noroña broke with Linné's artificial classification, and applied the natural plant classification of Adanson. Most of his work was never published, and a few hundred drawings of Javanese plants (and some animals) can only be found in manuscript form in archives and libraries in Paris, London, and Berlin.[35]

French surgeon-naturalist Deschamps formed part of the Entrecasteaux expedition that got stranded at Surabaya, eastern Java, in 1793. The governor of Java's northeast coast offered him to stay and to carry out natural history investigations, which he accepted. From 1794 to 1798, he and his [indigenous?] collectors went on several long trips to both the coastal and the inland areas of West, Central, and East Java, to my knowledge the first naturalist to have covered the entire island from west to east and from north to south, and also the first one to collect specimens in many of the mountainous areas. During a part of these trips he was accompanied by some young assistants, possibly from the Semarang Naval Academy, who helped him with descriptions and drawings. Afterward, he settled in Batavia as a physician until 1802, the year he went back to France. Deschamps published very little, perhaps because on his voyage home the British robbed him of his collections, parts of which ended up in the British Museum. His claim to fame is that he was the first white man to find the genus *Rafflesia*, later named after Raffles by Dr. Joseph Arnold.

The American physician and naturalist Thomas Horsfield came to Java in 1801 and remained there until 1819. In 1802, the High Government appointed him chief surgeon, charged with researching indigenous herbs and plants for medical purposes. Horsfield became a member of the Batavia Society, which gave the government positive advice regarding his plans. In 1804, he obtained permission for a long research trip through Java, which would last until 1812. In 1811, Horsfield met Raffles, who employed him forthwith. During his trips, he sent double specimens to the Batavia Society. During his botanical journey, he also came across the French naturalist Leschenault. Many of his findings would be published in the volumes 7 (1814) and 8 (1816) of the *Transactions.*

After his stay in Java, he found employment in England, based on the glowing references Raffles had given him. He played an important role in the Zoological Society of London and the Royal Entomological Society of London. Here he wrote his *Zoological Researches in Java* (1821–24), and his *Plantae Javanicae Rariores* (with two coauthors, 1838–52). He was the first one to publish extensively on Java's wild animals.

It was Horsfield's bad luck that he placed his very extensive collection of plants from Java, 2,196 specimens, in the hands of Robert Brown, who had undertaken the task of publishing the descriptions in 1821, but did not produce the first part of it until 1838. When the fourth and last part was published, in 1852, only 50 species had been described. By that time, botanist C. L. Blume, director of the National Herbarium [*'s Rijks Herbarium*] in Leiden, had published his major works on the flora of Java in Batavia and Leiden, thus stealing Horsfield's thunder.[36]

Finally, botanist and ornithologist Leschenault de la Tour must be mentioned briefly. He did his botanical research in Java between1804 and 1806, possibly mainly in Eastern Java, including the island of Madura, while almost all other researchers had focused on Western Java. He also collected plants on the island of Timor in 1803, after having investigated Australian plants for a number of years. He became a member of the Batavia Society, and assisted the governor of Java's northeast Coast, Nicolaus Engelhard, with his private collection of plants.[37] Leschenault did not publish much, but his collections could be studied in the Museum of Natural History in Paris. His data on Javanese birds was published by Cuvier and Vieillot, and he sent skins and notes to C. J. Temminck, director of the National Museum of Natural History [*'s Rijks Museum van Natuurlijke Historie*] in Leiden. As the only one of the eight naturalists, he also visited Suriname, in the Dutch West Indies.[38]

In historiography, the period 1770–1815 does not stand out as important with regard to knowledge about Java's plants and animals. Generally speaking, the eighteenth century is so much overshadowed by the Golden Age of naturalists like Bontius, Hendrik Adriaan van Reede tot Drakestein, Paul Hermann, and, most of all, Rumphius, that the former has often been eclipsed. One easily forgets, however, that of these four, only Bontius did all his research in Java, and that the number of species he described was very small. Van Reede did Malabar (in India), Hermann did Ceylon, and Rumphius mainly Ambon. Of the ca. 1,700 plant forms Rumphius published, only 125 were from Java.[39] Therefore, when the eight scholars undertook their investigations, they did not have much to go on.

A few features of this episode are worth emphasizing. All naturalists mentioned here were non-Dutch nationals, but all of them got permission to collects plants, animals, and minerals. All but one (Solander) were either entirely or partly financed by Dutch institutions or private persons, in the Netherlands or in the Indies. Almost all researchers were supported in one way or another by the Batavia Society, and several of them published part of their findings in the *Verhandelingen*, or included excerpts from the *Verhandelingen* in other publications.

Solander, Thunberg, and Hornstedt were all Swedes, belonging to the Linnaean "mafia." The Swedish connection had been established early in the eighteenth century, when Linné came to the Netherlands to study the plants, the herbaria, and the herbals found in the botanical gardens and libraries, particularly looking for plants from Asia, Africa, and the Americas. To this Leiden/Amsterdam-Uppsala connection, we have to add London, where many of the naturalists dealt with here, or their collections, ended up gravitating toward the new hub of long-distance commerce and science (or center of calculation, as Latour would call it) and the inspiring figure of Sir Joseph Banks.

Finally, it should be mentioned that many of the eight biologists did come to Java as part of a much longer voyage, beginning with the voyage of Captain Cook, which brought Solander to the shores of Java. They were not part of any Dutch (imperial) project.

Information, therefore, regarding botany, zoology, and perhaps to a lesser extent geology (minerals), was circulating rather easily between countries and empires, even though in the period studied there were many wars being fought, in which the English, the French, and the Dutch were implicated. The Batavia Society played a positive role, both in terms of support for research and as an outlet for publications.

The "New" Dutch State

However, it is also clear that all these efforts were not the result a centrally organized endeavor (by a "center of calculation"), and did not lead to a "cycle of accumulation" in the terminology of Latour. Neither was there a collective, unified, integrated product at the end of all these investigations. The Batavia Society did not have the wherewithal for such a project, the VOC was severely weakened or defunct, and there was no institution in the Dutch Republic to take up the slack.

After Napoleon's defeat, the country became a centralized kingdom under King Willem I, son of the last Stadholder. King Willem had the will, the power, and the means to launch such a project, and he founded the Committee for Natural History of the Netherlands Indies in 1822. The 1820s also witnessed the founding of two important institutions in the Netherlands—the National Museum of Natural History and the National Herbarium—both in Leiden, and both with the support of the king. Between them, the members of the committee and the staffs of the two museums produced the authoritative—and often sumptuous—publications on Java's flora and fauna that had been so sadly missing during the period discussed here.[40]

Thus Leiden—or one should probably say the Netherlands—became an important "center of calculation" for tropical/colonial botany and zoology in the nineteenth century, getting a "cycle of accumulation" going, and helping to usher in a new golden age in Dutch science. The scientific knowledge thus generated by the new Dutch Empire could and would also be used to create the means needed to support and perpetuate that same Empire.

The eight naturalists who came to Java around 1800 produced new information, but they lacked the power and the means to turn their findings into something that could be called scientific knowledge. The (imperial/colonial) institution—or institutions—needed for such a transformation was just not there, the "Empire" being in the doldrums. Scientific knowledge production in botany and zoology, therefore, had to wait for the New Dutch State and the New Empire.

Concluding Remarks

During the period of the late-eighteenth-century voyages of discovery and the second Scientific Revolution, the Dutch did not play as important a role as they did during the voyages of discovery in the seventeenth century and the first Scientific Revolution, or as they would do after 1815. However, there was an interest in scholarly endeavor, both in the Dutch Republic and in its colonies, and "amateurs" and scholars employed by the VOC took part in the scientific developments of the period.

In this contribution, I emphasized the activities of the Batavia Society of Arts and Sciences and its luminaries. Most of them were amateurs with some academic schooling, but not necessarily in the fields they were interested in. A person like Radermacher was not a scholar as we now understand the term, but rather a wealthy facilitator with excellent government connections, and the will to make a difference, perhaps comparable, although on a more modest scale, with someone like Joseph Banks. His own writings, mainly on statistics, geography, and ethnography, were not of high quality. It is a moot point whether he was unable or unwilling to provide the readership of the *Verhandelingen*, the journal of the Batavia Society, with better-informed and more detailed contributions.

However, the Batavia Society and it board members were also patrons of botany and zoology, and in that capacity they quite supported some new research in Java between 1770 and 1815. The fact that all the biological researchers in this period were foreigners does not appear to have mattered. They were permitted to enter the usually carefully guarded inland areas, were often paid by the government or its officials, some of their findings were published in the *Verhandelingen*, and they could take their collections with them when they left the island. Three of the researchers represented the Linnaean connection Uppsala-Leiden/Amsterdam. But in many cases, the researchers or their collections ended up in England, where Joseph Banks was the central figure in all things botanical, and perhaps we should think of this botanical network as the Uppsala-Leiden/Amsterdam-London triangle.

Nevertheless, the most important studies on Javanese botany would not be published until much later, and not in Batavia but in the Netherlands, by scholars paid by the Dutch government. Although, at least in this field, the collection and the distribution of information was not hindered by considerations of national identity, it would still take a long time for publication in a more systematic fashion to occur, and information to be turned into scientific knowledge. In this sense, Indonesian botany and zoology were clearly a colonial science in the making during the period under consideration.

People in Batavia were not working entirely in a vacuum—Mohr, the Batavia Society, and many of its individual members bought state-of-the-art instruments, books, and learned journals in the Dutch Republic and other European countries. So, although Batavia was not exactly a center of vigorous intellectual life, given the absence of even secondary education, motivated people could and did keep abreast of scientific development in Europe, and in 1778 there was apparently sufficient critical mass for the establishment of its own learned society, the first of its kind in Asia.

We saw that Bougainville and Cook reported on Reverend Mohr's activities regarding the transit of Venus in their books, thus informing the European public about what went on in Batavia in this respect. This can be generalized: scholars in Europe could remain informed of scientific activities in the Indonesian Archipelago, as copies of the *Verhandelingen* of the Batavia Society were sent out to sister societies in the Netherlands and elsewhere. At the same time, a number of the scholars in the Indies were members of other learned societies in Europe, or at least published in their journals, as was the case with Mohr and Radermacher.

According to Drayton, Joseph Banks thought that imperial botany in the 1780s and 90s could play a role in the "improvement" of agriculture in Europe and in the British colonies. Radermacher and the other board members during the early years of the Batavia Society also saw the prosperity of the colony as their main aim, and many articles in the *Verhandelingen* testify to this. The society's motto ,"for the common good," was there to constantly remind the members of this goal. However, which "commonality" did it refer to? Was the indigenous population included? If we look at the articles in the early issues of the *Verhandelingen* on agriculture, for instance, the emphasis appears to be on estate agriculture, not subsistence production.

There is no record of indigenous Indonesians who during these years played a role in all this as scholars in their own right. During these early years, the Batavia Society did not have Indonesian members, the possibility of which was not even discussed until 1825. Indonesians, it would seem, only played a role as informants, as plant collectors, and possibly as draftsmen. As Jean Taylor put it: the Batavia Society was a debating club for gentlemen born into prosperous families in the Netherlands and educated there. It was formed by immigrants, basically for immigrants, and the most active members were outsiders too.[41]

Joseph Banks "organized a network of officially sponsored surveys and scientific investigations through the Royal Society, the Admiralty, and the East India Company." As we have seen, the Batavia Society used the official VOC lines of command for the same purpose, but there was a vast difference of scale (and of quality?) between Banks's network and that of the VOC. The VOC headquarters in Batavia had started to systematically gather information at a much earlier date, as witness an instruction to its merchants, dated December 1, 1670, which required detailed information on its establishments and the people who lived there. In addition to the usual commercially interesting facts (commodities traded, natural resources), they wanted to be informed about not only population figures, physical geography, law, type of state and ruler, but also anthropological data on, for instance, religion, ceremonies, food, clothing, and marriages. But this information, if it ever came in, was not meant for publication.[42]

Finally, I was surprised to find that the Germans, not having an Empire of their own, played such an important role as regards gathering, processing, and distributing scholarly "colonial" information, including theory formation, partly no doubt because they participated in the data gathering of the Russian Empire. Apparently, for the creation and circulation of "colonial" science, it was not necessary to be a colonial metropole.

Notes

1. Luís Filipe F. R. Thomaz, "O Malogrado Establecimento Oficial dos Portugueses em Sunda," in *Aquém e Além da Taprobana: Estudos Luso-Orientais à Memória de Jean Aubin e Denys Lombard*, ed. Luís Filipe F. R. Thomaz (Lisboa: Universidade Nova de Lisboa, 2002); Geoffrey C. Gunn, *First Globalization: The Eurasian Exchange, 1500–1800* (Lanham, etc.: Rowman and Littlefield, 2003), 23–25. This does not apply to Eastern Timor.
2. For a short overview of Dutch scholarly research regarding Indonesia during the last four centuries, see Peter Boomgaard, "The Making and Unmaking of Tropical Science: Dutch Research on Indonesia, 1600–2000," *Bijdragen tot de Taal-, Land- en Volkenkunde* 162, 2/3 (2006). The first data collection regarding Indonesia by a Dutch person, Jan Huygen van Linschoten, who sailed with the Portuguese to Asia, was already published shortly before 1600.
3. On the VOC, see Femme S. Gaastra, *The Dutch East India Company: Expansion and Decline* (Zutphen: Walburg Pers, 2003); on the WIC, see Henk den Heijer, "The Dutch West India Company, 1621–1791," in *Riches from Atlantic Commerce: Dutch Transatlantic Trade and Shipping, 1585–1817,* ed. Johannes Postma and Victor Enthoven (Leiden: Brill, 2003).
4. On the Dutch Golden Age in general, with emphasis on economic aspects, see Jan de Vries and Ad van der Woude, *The First Modern Economy: Success, Failure, and Perseverance of the Dutch Economy, 1500–1815* (Cambridge, etc.: Cambridge University Press, 1997); Jonathan I. Israel, *The Dutch Republic: Its Rise, Greatness, and Fall, 1477–1806* (Oxford: Clarendon Press, 1998). On science and technology in the seventeenth-century Dutch Republic, see E. J. Dijksterhuis, *Simon Stevin: Science in the Netherlands around 1600* (The Hague: Martinus Nijhoff, 1970); Klaas van Berkel, Albert van Helden, and L. C. Palm, *A History of Science in the Netherlands: Survey, Themes and References* (Leiden: Brill, 1999); Eric Jorink, *Reading the Book of Nature in the Dutch Golden Age, 1575–1715* (Leiden: Brill, 2010).

5. On Suriname, see Oostindie, chapter 10, this volume; on the Cape, see Huigen, chapter 9, this volume; on Japan, see Rietbergen, chapter 8, this volume.
6. Lucille H. Brockway, *Science and Colonial Expansion: The Role of the British Botanical Gardens* (New Haven/London: Yale University Press, 1979); Richard Drayton, *Nature's Government: Science, Imperial Britain, and the "Improvement" of the World* (New Haven/London: Yale University Press, 2000); Londa Schiebinger, *Plants and Empire: Colonial Bioprospecting in the Atlantic World* (Cambridge MA, etc.: Harvard University Press, 2004), 10–11; Harold J. Cook, *Matters of Exchange: Commerce, Medicine, and Science in the Dutch Golden Age* (New Haven/London: Yale University Press, 2007).
7. Letter Johan Mauritz Mohr to Jan Jacob Schultens, Batavia, April 28, 1759, Leiden University Library, B.PL. 245 XII. I owe this reference to Michael Laffan.
8. Most of this section is based on Huib J. Zuidervaart and Rob H. van Gent, "'A Bare Outpost of Learned European Culture on the Edge of the Jungles of Java': Johan Maurits Mohr (1716–1775) and the Emergence of Instrumental and Institutional Science in Dutch Colonial Indonesia," *Isis* 95 (2004), and J. P. M. Groot, *Van Batavia naar Weltevreden: Het Bataviaasch Genootschap van Kunsten en Wetenschappen 1778–1867* (Leiden: KITLV Press, 2009).
9. Well-known because of the eponymous comet.
10. Boomgaard, "Dutch Research on Indonesia," 197.
11. Quoted from Zuidervaart and Van Gent, "A Bare Outpost," 17.
12. Perhaps the attempt at circumnavigation by Jacob Roggeveen, in 1722, could be seen as the last Dutch voyage of discovery. The voyage yielded information on Easter Island.
13. On the voyages of discovery in the late eighteenth century, see for example, Daniel A. Baugh, "Seapower and Science: The Motives for Pacific Exploration," in *Scientific Aspects of European Expansion,* ed. William K. Storey (Aldershot: Variorum, 1996); George Forster, A *Voyage Round the World* (Honolulu: University of Hawai'i Press, 2000); Nicholas Thomas, *Discoveries: The Voyages of Captain Cook* (London: Penguin, 2003). On Joseph Banks, see Drayton, *Nature's Government*; Richard Holmes, *The Age of Wonder: How the Romantic Generation Discovered the Beauty and Terror of Science* (London: Harper Press, 2008). See also Boomgaard, introduction, this volume, and Reyes, chapter 2, this volume.
14. This section is largely based on Jean Gelman Taylor, *The Social World of Batavia: European and Eurasian in Dutch Asia* (Madison: University of Wisconsin Press, 1983), 78–96; Groot, *Van Batavia naar Weltevreden*, 45–101; and on the early volumes of the journal *Verhandelingen van het Bataviaasch Genootschap.*
15. See Boomgaard, introduction, this volume.

16. *Verhandelingen van het Bataviaasch Genootschap* 1 (1779), 9 (my translation).
17. See Boomgaard, introduction, this volume.
18. Much of the following two sections is based on Groot, *Van Batavia naar Weltevreden*, 46–60, and the first six volumes of the *Verhandelingen van het Bataviaasch Genootschap* (1779, 1780, 1781, 1786, 1790, 1792). On Radermacher, see also Taylor, *The Social World of Batavia*, 85–87, 91–94; Kees Zandvliet, *The Dutch Encounter with Asia 1600–1950* (Amsterdam: Rijksmuseum/Zwolle: Waanders, 2002), 149–51.
19. On Enlightenment and Freemasonry, see for example, Margaret C. Jacob, *Living the Enlightenment: Freemasonry and Politics in Eighteenth-Century Europe* (New York/Oxford: Oxford University Press, 1991); on Freemasonry and the Batavia Society, see Groot, *Van Batavia naar Weltevreden*, 36–37.
20. On this aspect, see E. du Perron, *Een Lettré uit de 18e Eeuw: Willem van Hogendorp* (Den Haag: Leopolds Uitgevers-Maatschappij, 1940).
21. See *Briefe des Herrn von Wurmb und des Herrn Baron von Wollzogen auf ihren Reisen nach Afrika und Ostindien in den Jahren 1774 bis 1792* (Gotha: Ettinger, 1794).
22. Batavia had, indeed, not much "higher learning" to offer around this time. The Theological Seminar and the Naval Academy, both founded in 1743, had both been closed down in 1755, due to cutbacks. There would be no Naval Academy until 1785, when one was founded in Semarang, to the East of Batavia (I. J. Brugmans, *Geschiedenis van het Onderwijs in Nederlandsch-Indië* [Groningen/Batavia: J. B. Wolters, 1938], 46–53; Zandvliet, *The Dutch Encounter*, 269–78).
23. On the term "statistics" and the scholars who taught it in the eighteenth century, see A. Th. van Deursen, *Geschiedenis en Toekomstverwachting: Het Onderwijs in de Statistiek aan de Universiteiten van de Achttiende Eeuw* (Kampen: Kok, 1971) [Inaugural Lecture]; Mohammed Rassem and Justin Stagl, eds., *Statistik und Staatsbeschreibung in der Neuzeit, vornehmlich im 16.–18. Jahrhundert* (Paderborn, etc.: Ferdinand Schöningh, 1980).
24. For details on Java's population enumerations around 1800, see Peter Boomgaard and Hans Gooszen, *Population Trends 1795–1942*, Changing Economy in Indonesia, vol. 11 (Amsterdam: Royal Tropical Institute, 1991).
25. Han F. Vermeulen, "Early History of Ethnography and Ethnology in the German Enlightenment: Anthropological Discourse in Europe and Asia, 1710–1808" (PhD diss., Leiden University, 2008).
26. *Briefe*, 123, 199, 351, 403, 413–14; Groot, *Van Batavia naar Weltevreden*, 84, 89.
27. William Marsden, *The History of Sumatra, Containing an Account of the Government, Laws, Customs, and Manners of the Native Inhabitants,*

with a Description of the Natural Productions, and a Relation of the Ancient Political State of That Island (London: Payne, 1783); Thomas Stamford Raffles, *The History of Java* (London: Black, Parbury and Allen/London: Murray, 1817); John Crawfurd, *History of the Indian Archipelago: Containing an Account of the Manners, Arts, Languages, Religions, Institutions, and Commerce of its Inhabitants* (Edinburgh: Constable/London: Hurst, Robinson and Co., 1820).

28. Much of this section is based on M. J. van Steenis-Kruseman and C. G. G. J. van Steenis, *Malaysian Plant Collectors and Collections, Being a Cyclopaedia of Botanical Exploration in Malaysia (…)*, Flora Malesiana, series I, vol. 1 (Djakarta: Noordhoff-Kolff, 1950); additional references are given for each individual researcher below.
29. On Solander, see Roy Anthony Rauschenberg, *Daniel Carl Solander, Naturalist on the "Endeavour"* (Philadelphia: American Philosophical Society, 1968). The Natural History Museum in London has a manuscript list, drawn up by Solander, entitled *Plantae Javanenses*, enumerating 338 species.
30. C. P. Thunberg, *Voyages de–au Japon, par le Cap de Bonne-Espérance, les Isles de la Sonde, etc.* (Paris: Benoit Dandré, 1796), I, 386 (my translation).
31. On Thunberg, see Thunberg, *Voyages*; D. O. Wijnands, E. J. A. Zevenhuizen, and J. Heninger, *Een Sieraad voor de Stad; De Amsterdamse Hortus Botanicus 1638–1993* (Amsterdam: Amsterdam University Press, 1994), 120–23; on the shift from an artificial to a natural system of plant families, see Drayton, *Nature's Government*, 18–19.
32. On Hornstedt, see Ann Kumar, "A Swedish View of Batavia in 1783–84: Hornstedt's Letters," *Archipel* 37 (1989); Christina Skott, "'Ask about Everything': Clas Fredrik Hornstedt in Java, 1784," in *Intercultural Exchange in Southeast Asia: History and Society in the Early Modern World*, ed. Tara Alberts and D. R. M. Irving (London: Tauris, 2013). On the role of the Swedes in Java's natural history, see also Christina Skott, "The VOC and Swedish Natural History: The Transmission of Scientific Knowledge in the Eighteenth Century," in *Dutch Trading Companies as Knowledge Networks*, ed. Siegfried Huigen et al. (Leiden: Brill, 2010).
33. On collecting in the Philippines around this time, see Reyes, chapter 2, this volume.
34. Malay herbalists is possible, but Malay-speaking Sundanese herbalists would be more logical.
35. On Noroña, see Susana Pinar, "The Scientific Voyages of Francisco Noroña (1748–1788) in Southeast Asia and the Indian Ocean," *Itinerario* 19, 2 (1995).
36. On Horsefield, see M. J. Sirks, *Indisch Natuuronderzoek: Een Beknopte Geschiedenis van de Beoefening der Natuurwetenschappen in de Nederlandsche Koloniën* (Amsterdam: Koloniaal Instituut,

1915), 81–82; Thomas Horsfield, *Zoological Researches in Java, and the Neighbouring Islands* (Singapore, etc.: Oxford University Press, 1990); Groot, *Van Batavia naar Weltevreden*, 109–11.

37. On Engelhard's collecting activities, see Zandvliet, *The Dutch Encounter*, 269–78.
38. On Leschenault, see Sirks, *Indisch Natuuronderzoek*, 80; Groot, *Van Batavia naar Weltevreden*, 110. On Suriname, see Oostindie, chapter 10, this volume.
39. On the Golden Age of Dutch botany in Asia, see Klaas van Berkel, "Een Onwillige Mecenas? De Rol van de VOC bij het Natuurwetenschappelijk Onderzoek in de Zeventiende Eeuw," in *VOC en Cultuur: Wetenschappelijke en Culturele Relaties tussen Europa en Azië ten Tijde van de Verenigde Oostindische Compagnie*, ed. J. Bethlehem and A. C. Meijer (Amsterdam: Schiphouwer and Brinkman, 1993); Peter Boomgaard, "Rumphius (1627–1702): A Botanist in the Indonesian Archipelago, his Network, and His Plants" (paper presented at the Environmental Studies Seminar, Center for International Studies, University of Chicago, 2007); Cook, *Matters of Exchange*.
40. Andreas Weber, "Hybrid Ambitions: Science, Governance, and Empire in the Career of Caspar G. C. Reinwardt (1773–1854)" (PhD diss., Leiden University, 2012).
41. Taylor, *The Social World of Batavia*, 86, 88; Drayton, *Nature's Government*, 80.
42. *Nederlandsch-Indisch Plakaatboek 1602–1811* (Batavia: Landsdrukkerij, 1885–1900), vol. 2 530–33. The quote is from C. A. Bayly, *The Birth of the Modern World 1780–1914* (Oxford, etc.: Blackwell, 2004), 110.

Bibliography

Baugh, Daniel A. "Seapower and Science: The Motives for Pacific Exploration." In *Scientific Aspects of European Expansion*, edited by William K. Storey, 85–139. Aldershot: Variorum, 1996.

Bayly, C. A. *The Birth of the Modern World 1780–1914*. Oxford, etc.: Blackwell, 2004.

Berkel, Klaas van. "Een Onwillige Mecenas? De Rol van de VOC bij het Natuurwetenschappelijk Onderzoek in de Zeventiende Eeuw." In *VOC en Cultuur: Wetenschappelijke en Culturele Relaties tussen Europa en Azië ten Tijde van de Verenigde Oostindische Compagnie*, edited by J. Bethlehem and A. C. Meijer, 39–58. Amsterdam: Schiphouwer and Brinkman, 1993.

Berkel, Klaas van, Albert van Helden, and L. C. Palm. *A History of Science in the Netherlands: Survey, Themes and References*. Leiden: Brill, 1999.

Boomgaard, Peter. "The Making and Unmaking of Tropical Science: Dutch Research on Indonesia, 1600–2000." *Bijdragen tot de Taal-, Land- en Volkenkunde* 162, 2/3 (2006): 191–217.

Boomgaard, Peter. "Rumphius (1627–1702): A Botanist in the Indonesian Archipelago, His Network, and His Plants." Paper presented at the Environmental Studies Seminar, Center for International Studies, University of Chicago, 2007.

Boomgaard, Peter, and Hans Gooszen. *Population Trends 1795–1942.* Changing Economy in Indonesia, volume 11. Amsterdam: Royal Tropical Institute, 1991.

Briefe des Herrn von Wurmb und des Herrn Baron von Wollzogen auf ihren Reisen nach Afrika und Ostindien in den Jahren 1774 bis 1792. Gotha: Ettinger, 1794.

Brockway, Lucille H. *Science and Colonial Expansion: The Role of the British Botanical Gardens.* New Haven/London: Yale University Press, 1979.

Brugmans, I. J. *Geschiedenis van het Onderwijs in Nederlandsch-Indië.* Groningen/Batavia: J. B. Wolters, 1938.

Cook, Harold J. *Matters of Exchange: Commerce, Medicine, and Science in the Dutch Golden Age.* New Haven/London: Yale University Press, 2007.

Crawfurd, John. *History of the Indian Archipelago: Containing an Account of the Manners, Arts, Languages, Religions, Institutions, and Commerce of Its Inhabitants.* Edinburgh: Constable/London: Hurst, Robinson and Co., 1820.

Chijs, J. A. van der, ed. *Nederlandsch-Indisch Plakaatboek 1602–1811.* 16 vols. Batavia: Landsdrukkerij, 1885–1900.

Deursen, A. Th. van. *Geschiedenis en Toekomstverwachting: Het Onderwijs in de Statistiek aan de Universiteiten van de Achttiende Eeuw.* Kampen: Kok, 1971 [Inaugural Lecture].

Dijksterhuis, E. J. *Simon Stevin: Science in the Netherlands around 1600.* The Hague: Martinus Nijhoff, 1970.

Drayton, Richard. *Nature's Government: Science, Imperial Britain, and the "Improvement" of the World.* New Haven/London: Yale University Press, 2000.

Forster, George. *A Voyage Round the World.* 2 vols. Honolulu: University of Hawai'i Press, 2000.

Gaastra, Femme S. *The Dutch East India Company: Expansion and Decline.* Zutphen: Walburg Pers, 2003.

Groot, J. P. M. *Van Batavia naar Weltevreden: Het Bataviaasch Genootschap van Kunsten en Wetenschappen 1778–1867.* Leiden: KITLV Press, 2009.

Gunn, Geoffrey C. *First Globalization: The Eurasian Exchange, 1500–1800.* Lanham, etc.: Rowman and Littlefield, 2003.

Heijer, Henk den. "The Dutch West India Company, 1621–1791." In *Riches from Atlantic Commerce: Dutch Transatlantic Trade and Shipping, 1585–1817,* edited by Johannes Postma and Victor Enthoven, 77–112. Leiden: Brill, 2003.

Holmes, Richard. *The Age of Wonder: How the Romantic Generation Discovered the Beauty and Terror of Science.* London: Harper Press, 2008.

Horsfield, Thomas. *Zoological Researches in Java, and the Neighbouring Islands: With a Memoir by John Bastin.* Singapore, etc.: Oxford University Press, 1990.

Israel, Jonathan I. *The Dutch Republic: Its Rise, Greatness, and Fall, 1477–1806*. Oxford: Clarendon Press, 1998.

Jacob, Margaret C. *Living the Enlightenment: Freemasonry and Politics in Eighteenth-Century Europe*. New York/Oxford: Oxford University Press, 1991.

Jorink, Eric. *Reading the Book of Nature in the Dutch Golden Age, 1575–1715*. Leiden: Brill, 2010.

Kumar, Ann. "A Swedish View of Batavia in 1783–84: Hornstedt's Letters." *Archipel* 37 (1989): 247–62.

Marsden, William. *The History of Sumatra, Containing an Account of the Government, Laws, Customs, and Manners of the Native Inhabitants, with a Description of the Natural Productions, and a Relation of the Ancient Political State of That Island*. London: Payne, 1783.

Perron, E. du. *Een Lettré uit de 18e Eeuw: Willem van Hogendorp*. Den Haag: Leopolds Uitgevers-Maatschappij, 1940.

Pinar, Susana. "The Scientific Voyages of Francisco Noroña (1748–1788) in Southeast Asia and the Indian Ocean." *Itinerario* 19, 2 (1995): 161–63.

Raffles, Thomas Stamford. *The History of Java*. 2 vols. London: Black, Parbury and Allen/London: Murray, 1817.

Rassem, Mohammed, and Justin Stagl, eds. *Statistik und Staatsbeschreibung in der Neuzeit, vornehmlich im 16.—18. Jahrhundert*. Paderborn, etc.: Ferdinand Schöningh, 1980.

Rauschenberg, Roy Anthony. *Daniel Carl Solander, Naturalist on the "Endeavour."* Transactions of the American Philosophical Society, new series, vol. 58, pt. 8. Philadelphia: American Philosophical Society, 1968

Schiebinger, Londa. *Plants and Empire: Colonial Bioprospecting in the Atlantic World*. Cambridge MA, etc.: Harvard University Press, 2004.

Sirks, M. J. *Indisch Natuuronderzoek: Een Beknopte geschiedenis van de Beoefening der Natuurwetenschappen in de Nederlandsche Koloniën*. Amsterdam: Koloniaal Instituut, 1915.

Skott, Christina. "The VOC and Swedish Natural History: The Transmission of Scientific Knowledge in the Eighteenth Century." In *Dutch Trading Companies as Knowledge Networks*, edited by Siegfried Huigen et al., 361–92. Intersections: Yearbook for Early Modern Studies, vol. 13. Leiden: Brill, 2010.

———. "'Ask about Everything': Clas Fredrik Hornstedt in Java, 1784." In *Intercultural Exchange in Southeast Asia: History and Society in the Early Modern World*, edited by Tara Alberts and D. R. M. Irving, 161–202. London: Tauris, 2013.

Steenis-Kruseman, M. J. van, and C. G. G. J. van Steenis. *Malaysian Plant Collectors and Collections, Being a Cyclopaedia of Botanical Exploration in Malaysia (…)*. Flora Malesiana, series I, vol. 1. Djakarta: Noordhoff-Kolff, 1950.

Taylor, Jean Gelman. *The Social World of Batavia: European and Eurasian in Dutch Asia*. Madison: University of Wisconsin Press, 1983.

Thomas, Nicholas. *Discoveries: The Voyages of Captain Cook*. London: Penguin, 2003.

Thomaz, Luís Filipe F.R. "O Malogrado Establecimento Oficial dos Portugueses em Sunda," in *Aquém e Além da Taprobana: Estudos Luso-Orientais à Memória de Jean Aubin e Denys Lombard*, edited by Luís Filipe F. R. Thomaz, 381–607. Lisboa: Universidade Nova de Lisboa, 2002.

Thunberg, C. P. *Voyages de—au Japon, par le Cap de Bonne-Espérance, les Isles de la Sonde, etc.* Paris: Benoit Dandré, 1796.

Vermeulen, Han F. "Early History of Ethnography and Ethnology in the German Enlightenment: Anthropological Discourse in Europe and Asia, 1710–1808." PhD diss., Leiden University, 2008.

Vries, Jan de, and Ad van der Woude. *The First Modern Economy: Success, Failure, and Perseverance of the Dutch Economy, 1500–1815.* Cambridge, etc.: Cambridge University Press, 1997.

Weber, Andreas. "Hybrid Ambitions: Science, Governance, and Empire in the Career of Caspar G. C. Reinwardt (1773–1854)." PhD diss., Leiden University, 2012.

Wijnands, D. O., E. J. A. Zevenhuizen, and J. Heninger. *Een Sieraad voor de Stad; De Amsterdamse Hortus Botanicus 1638–1993.* Amsterdam: Amsterdam University Press, 1994.

Zandvliet, Kees. *The Dutch Encounter with Asia 1600–1950.* Amsterdam: Rijksmuseum/Zwolle: Waanders, 2002.

Zuidervaart, Huib J., and Rob H. van Gent. "'A Bare Outpost of Learned European Culture on the Edge of the Jungles of Java': Johan Maurits Mohr (1716–1775) and the Emergence of Instrumental and Institutional Science in Dutch Colonial Indonesia." *Isis* 95 (2004): 1–33.

Chapter 6

"A Religion That Is Extremely Easy and Unusually Light to Take On": Dutch and English Knowledge of Islam in Southeast Asia, ca. 1595–1811

Michael Laffan

Leyden in the Library

In late August 1811, a member of Lord Minto's expedition to Java, the giddy 35-year-old Scottish folklorist and Orientalist John Leyden set foot in the Dutch colonial capital of Batavia with his eyes firmly set on locating its collections of Malay manuscripts. What awaited him, however, was not merely two centuries of accumulated manuscripts, for there was apparently something sinister in the musty air of an unventilated library, perhaps even that of the Batavia Society of Arts and Sciences, and he contracted an infection that was to take his life within a few short days.

Leyden's death presented quite a stumbling block to one man's plans for a work chronicling Java's past, present, and (hopefully Britannic) future. The disappointed individual in question was Leyden's patron, Thomas Stamford Raffles (1781–1826), who would take on the post of lieutenant governor until 1816, publishing his landmark work on the subject in 1817, and in emulation of his patron William Marsden (1754–1836), the first secretary of the Admiralty who had done the same for Sumatra over three decades beforehand.[1] Yet unlike the work of Marsden, whom he certainly acknowledged, Raffles's text on the history, fauna, flora, and peoples of the island relied to a significant degree on both Dutch knowledge and Dutch works in the making.[2] For Raffles had arrived within living memory of an active push on the part of the local Europeans to document the world around them, both for their use and, indeed, for that of the general public,

Christian and Muslim. After all, this had been the very raison d'être of the Batavia Society.

While we know much of Java and its neighboring isles as a consequence, there is little in Raffles's work that specifically treats Islam in a historical sense, or that goes beyond acknowledging its majoritarian position in the archipelago. As a consequence, one might gather that the Dutch had had little interest in, or yet knowledge of, the dominant faith of the region. Such a point is underlined by the fact that a more sustained (if not any more positive) treatment of the faith and its adherents would be offered within the pages of John Crawfurd's 1820 *History of the Indian Archipelago*, which rested on a much closer reading of indigenous sources.[3] This difference, however, brings us to the question at hand. Namely: What did the Dutch know of Islam prior to Lord Minto's expedition?

Netherlandic Forebears

It has long been a commonplace to declare that the Dutch knew little about, and cared even less for, Islam in their expanding colonies of Southeast Asia before the nineteenth century, being animated purely by the prospect of commercial gain. Yet this is too easy a claim to make, based purely on the surviving printed record. I would rather contend that various servants of the VOC knew as much as they felt that they needed to know about a religion that was by no means new to them, and that one should differentiate between their scholarly and practical concerns. For when we do speak more precisely, we can see that while (often metropolitan) scholarly knowledge was deployed with the quixotic hope of the ultimate conversion of Muslims to Calvinism, on the ground the increasing spread of VOC rule necessitated a more pragmatic understanding of Muslim law—if only to the point where it did not clash with their own. In both cases, there was little broader sense of public utility in seeing such information disseminated.

From the very beginning, Europeans knew full well that they would encounter Muslims, and certainly Arabic speakers, en route to the much desired Spice Islands. For such reasons, Columbus had taken a converso interpreter, Luis de Torres, with him on his westward voyages to the East Indies. The ubiquity of Islam was very much confirmed by the account of the eastern way thither provided by Jan Huygen van Linschoten (1562–1611) in his *Itinerario* of 1595–96, and then by the first Dutch fleet under Cornelis de Houtman (1565–99), which returned in 1597. In the meantime, in 1596, the newly appointed professor of Hebrew at Leiden University, Frans van Ravelingen

(Raphelengius, 1539–97), was tasked with producing Arabic passes requesting safe passage for the increasing numbers of Dutch ships headed for Southeast Asia.[4]

What the recipients of such documents would have made of them is moot, however, given the heavy-handedness of men like the Houtman brothers, whom Steenbrink argues were representative of the sort of Dutchmen already rendered immune to Islam by virtue of an existing familiarity with its practitioners.[5] In any case, the account of the second fleet of Jacob Van Neck (d. 1638), which made it all the way to the Moluccas, provided some of the first printed glimpses of Muslim life in the distant Indies, with an account of the mosques (shown in an engraving of a royal procession) and the ritual of prayer. As the accompanying text explained:

> The inhabitants are in general heathens, having the Mahometish belief, to which they have very great devotion. They shall neither come or go from the market without first making their prayers in their Temple, which they call Musquita in their language.... And once they have done their cleansing and washing, they go into their Church and make their prayers with calls and cries. It is so loud that one can hear it from over 20 houses away and with the following words, which they generally say two or three times: Stofferolla, Stofferolla, Ascehad an la, Ascehed an la, Yll la, Ascehad an la, Yll lol la, Yll lol la, Yll lol la, Machumed die rossulla. And as they say the last word they pass their hands over their faces, in which they evince great devotion. Moreover they say some other prayers, which they do quietly, and mostly in murmurs, which go pretty much as follows: They spread a small rug on the ground, upon which they stand casting their eyes heavenwards two or three times. They then fall firmly upon their knees, placing their head on the ground two or three times. And they do this often together and in their houses, and even in public, in their outrigger vessels, on the roads, and on the beach.[6]

This passage was moreover published together with excerpts of a much older text on Islam that had been available in Europe since the twelfth century. Known as *The Book of a Thousand Questions*, this compendium of questions and answers, the questions supposedly put to Muhammad by a Jewish leader and answered by Muhammad, had been composed to aid new converts, or even to stimulate conversion to Islam in the first place.[7] Whereas Steenbrink sees this juxtaposition as yet more evidence of a notion of Islam as a known quantity to be described only when it deviated from Middle Eastern Standards, one might also adjudge its reprinting as having been a matter of happy convenience, given that the Arabic copies of precisely that text were found in the Moluccas by such visitors as Francois Valentijn (see next) a century later.

But we are getting ahead of ourselves. Even if original texts had been brought back to the West as well, it would sometimes take decades, if not centuries, for them to be properly identified. Notable exports included fragments of the Qur'an, the anonymous *Idah fi l-fiqh*, and a catechism composed by `Umar b. Muhammad al-Nasafi (d. 1142).[8] In the case of one tract, apparently brought from the Moluccas by a member of Van Neck's fleet, the wait was a mere decade, and we have a brief mention of its contents provided by a North African visitor, Ahmad b. Qasim al-Hajari, who went to Holland in 1613. Apparently it was a surprise to both al-Hajari and his learned host, the Orientalist Erpenius (Thomas van Erpe, 1584–1624), that what they had before them was a tract concerned with mysticism.[9]

Grounded Visions

Whereas Erpenius is celebrated today as one of the first lights of Dutch Orientalism, his concerns were firmly directed toward Arabic and the Middle East, as were those of his successor Golius, whom al-Hajari may also have met during his Dutch sojourn. As it happens, too, Erpenius was not the first Dutch speaker of Arabic that al-Hajari had ever met. This honor seems to belong to Pieter Maertensz. Coy, whom the States General had dispatched to Marrakesh in 1607 by virtue of his having learned enough Arabic when he was in the Moluccas (perhaps as a member of Van Neck's expedition).[10]

Not all those who ventured eastward began their careers as soldiers. In the Indies, one could occasionally meet with humanists of the same stamp as Erpenius. Being generally convinced of the superiority of their own faith, such savants could engage with Muslims and recognize that they possessed significant learning. An early case in point is the physician Jacobus Bontius (1592–1631), who profited gratefully from the help of local Javanese men and women whose knowledge of herbs entitled them to claim a place in his memory as respected teachers.[11]

Still, the subject of their discussions revolved more about nonsectarian issues, and, as Hal Cook has argued, their being commemorated in print rested on the perceived public utility of their information.[12] For such reasons then, we often have little information on the agents of the company who did manage to acquire a significant level of knowledge of the languages and customs of the peoples that they dealt with.

Among such men were two soldiers; the commandant of the Batavian garrison, Isaac de Saint Martin (1629–96), whose extensive library was said to have included dozens of Malay and Javanese manuscripts, and the much celebrated German-born Georg Rumph, also known as Rumphius (1627–1702).[13] While we know little of Saint Martin himself, Rumphius's publications—largely released posthumously—have generated a significant degree of scholarly interest, if mainly for their data on natural history. Still there is even something to be said for his attitude toward Islam here and there. For having also depended upon and admired Muslim informants on some questions, a note of derision could be found in some of his observations about the practices of some. For example, he mocked the late Lord of Giri—whose family had been famous as saints and Islamizers to much of the eastern part of the archipelago—as a "false holyman" engaged in the deceit of Muslims, Chinese, and Christians alike.[14] Similarly, he dismissed the indigenous practice of withdrawal and contemplation (*tapa*) as contravening the avowed faith of the islanders of Bima, especially as it involved communion with the Jinn:

> There is a mountain called Tolocco on Bima, which is their refuge, about a day's journey from the King's fort on the beach, which is flat and wide on top, covered with sown land and fruit trees; behind the Village, on a higher hill, is a hole with fresh water, where the King goes to perform *Batappa*, and where a *Djing* or Devil appears to him in the water with large buffalo horns on his forehead, and who points out these stones to him on rocks in shallow water. *Batappa* is a Godless relique of their Heathendom, which the Moors perform against their law, and, therefore, in secret; when they desire something from a *Djing*, that is, *Daemon* (which they distinguish from Satan or the Devil) or when they want to learn a new trick, or want riches, or how to be lucky and invulnerable in warfare, how to rob, steal, or commit thievery, gamble, or love, etc., they go to such distant places and high mountains, stay for a while, day and night, and bring some offerings to the *Djing*, firmly resolved not to be scared by its appearance nor to let themselves be chased off, and so the *Djing* finally gives them a small piece of wood or a little stone, which they are supposed to wear in order to get the things they prayed for, and so they even think that they are Religious in their fashion, and that is why they call these crystals *Batu Djing*.[15]

Then again, similar criticisms can also be found in Muslim writings against local practices deemed to contravene the law, and one might wonder whether Rumphius's reformed sensibilities would have been approved of by later Islamic literati of a similar bent, such as the

Malay scholars `Abd al-Samad al-Falimbani (ca. 1704–ca. 1789) and Muhammad Arshad al-Banjari (1710–1812).[16]

Christian Intentions

Another VOC servant with more than passing knowledge of Islam was Herbert de Jager (active at Batavia 1682–97), whose manuscripts passed to Saint Martin at his death in 1696.[17] De Jager, who was regarded highly by the latter, and who corresponded with Rumphius, was reputed to know Sanskrit, Persian, and Singhalese, and was furthermore employed in 1683 to provide training in Arabic and Malay "as a means of furthering the Malay religion"—by which the Dutch meant Protestant Christianity.[18] For we should not forget that among the pragmatic reasons for studying Malay was the equipping of ministers with a linguistic arsenal to explain their faith in local terms, and indeed to repel the charges of "Papists" and Muslim alike.

While it is true that, seen with two centuries of hindsight, the VOC was not especially interested in spreading Christianity (beyond annexing the Catholics stranded by their Portuguese rivals), there had long been enough print-minded enthusiasts for Christ in the Netherlands who believed that the Company should be engaged in precisely such activities. One such lobbyist was Justus Heurnius (1587–1652), who had declared that God had "laid bare the riches of India for us in order that the Kingdom of Christ...shall be spread *en route* to and in the immense lands of the East."[19] Doubtless he would have been overjoyed to hear of the early successes claimed by a party of missionaries who visited the tiny island of Rozengain, in the Banda Archipelago, in August of 1622. There, with the aid of a local aristocrat who could write in Arabic script, they had laid out the core elements of their faith. And on being questioned about their knowledge of Islam, they had also declared that they knew more than enough about that religion, even lecturing them on the alleged crimes of Muhammad, "the leader of a number of thieves, robbers and murderers," and what they believed to have the stipulations of the Qur'an.[20]

In the wake of the 1618–19 Synod of Dort, which had succeeded in revising the preamble to the formal charter of the VOC to include a statement that it would henceforth protect "the public faith," Heurnius became an advocate for the establishment at Leiden of a training school for Asia-bound ministers or *predikanten*.[21] Thus, it was that the Collegium Indicum opened its doors in Leiden in 1622 under his friend Antonius Walaeus (1573–1639), though he deemed an education in what he vaguely called "the most usual language of these lands," as secondary to the provision of sound moral training in Leiden itself for their labors against Papists as the real enemy of the faithful. At any rate Walaeus proposed that a knowledge of the languages of the East could be provided by some "orthodox" man with some experience in that tongue in the Indies.[22]

A likely candidate for this task might ultimately have been Heurnius himself, who ventured out as a preacher between 1622 and 1638. However the Collegium was closed in 1632, and it seems in any case that by his return he had become rather disenchanted with Christianity's chances among the inhabitants of Muslim-dominated areas; or those ruled by the Company for that matter. The inhabitants of one island even tried to poison him, and the VOC imprisoned him in the face of his persistant demands for church independence.[23]

Whereas Heurnius is remembered today as the author of the first Sino-Dutch dictionary, finished at Batavia in 1628, he was also engaged in the expansion of the Malay-Dutch lexicon written by another early proselytizer, Sebastiaan Danckaerts, whose work had been printed by order of the Amsterdam Municipal Chamber in 1623.[24] While the evidence is coincidental, Heurnius is a likely candidate for another early Malay-Dutch dictionary now held at the Bodleian Library. Having crossed the English Channel with the sale of Golius's private library in 1696, this is the sole Malay document in the Arabist's collection, and the numerous mistaken transliterations and misrecognitions of Arabic terms, along with the fact of its transcription in roman script, make it unlikely that it was a product of his own hand.

Whatever the provenance, there is evidence within its pages of attempts to recall discussions of Islamic terms (which are relatively restricted, it must be said) or examples of straightforward misinterpretation. A prime example of this is the entry on the "Aijaan thabida" as being things that were "visible and demonstrable."[25] Yet this can only refer to the advanced notion of the *a`yan thabita*. Often found in Neoplatonic mystical treatises referring to God's "eternal essences," these are actually deemed imperceptible to any but the most experienced gnostic.[26]

Such misinterpretation is to be expected, for it should be said on balance that missionaries like Heurnius were far less concerned with the heights of Islamic philosophy than the vocabulary required to render their own scriptures in Malay. This was a constant desideratum for the missionaries who always sought to surpass the efforts of their predecessors. A part of Matthew's gospel, originally completed at Ambon by Albert Cornelisz. Ruyl in 1612, would appear on the metropolitan press in 1629. A full edition of his version of the Four Gospels and the

Acts of the Apostles would be brought forward in 1651 with the help of Heurnius after his return from the Indies in 1639. Laboring much to the same end, metropolitan academicians like Gisbertus Voetius (1589–1676), a professor at the University of Utrecht, could rehash older Spanish polemics on Islam and issue guides for outgoing *predikanten* without paying any heed to primary sources.[27]

Yet it was in the primary sources in the archipelago itself that the real keys were believed to lie, with respect to the most easterly of all potential believers. And it would be there that the Bible scholars would set to work in earnest to make the linguistic breakthrough, given that the various tracts produced by Heurnius and his ilk had made scarcely an impression. In 1691, Melchior Leijdecker (1645–1701) would be engaged at Batavia to produce a complete Bible and revised dictionary making use of the manuscripts of the VOC's General Secretariat.

François Valentijn

The scholarly and the ecclesiastic strands of Dutch writing on Islam come together in the person of Leijdecker's famously avaricious rival François Valentijn (1666–1727), who would proffer a Bible of his own during his first tour of duty as a Malay-speaking *predikant* at Ambon from 1686 to 1694. Going beyond the vagaries of Walaeus, Valentijn had a rather different take on the linguistic needs of the church, remarking that the "High" Malay traditionally used by Dutch preachers was effectively a clone of an Islamicate culture whose conceptual language and phraseology implicitly excluded most non-Muslims, and was therefore incomprehensible to them. Valentijn instead urged that a localized "Low" Malay, which he used in his own work both as a preacher and (now) Bible translator, would be far more effective in reaching out to the inhabitants of the Indies.[28]

Valentijn, too, would not find success in his claims, which were probably weakened by an increasingly bad reputation among his parishioners for an overriding interest in making money, among other charges, including making off with pages from the church register. He was, moreover, seen as a linguistic parvenu. Even if his teacher had been none other than Rumphius, it was considered somewhat suspicious that he had been able to offer a Malay Bible for publication in so short a time, and with so little experience (see next).

Even if he was a less than successful minister, Valentijn was undoubtedly a scholar who learned much more than Malay from Rumphius and his own expanding network of informants. This was made abundantly clear with the publication of his monumental *Oud en Nieuw Oost-Indien*, published after a second ignominious return from the Indies. Certainly his is perhaps one of the most important Dutch works from which a broad history of Islam in Southeast Asia can be extracted. Yet, much as he made extensive use of the unpublished work of his mentor Rumphius, he seemed similarly disinclined to pursue an extended discussion of the religion itself.

When he did recount something of Islam at Ambon, for example, he compared Portuguese and Malay sources regarding the spread of that religion by Pati Tuban of Java (Sunan Bonang) and knew enough to differentiate between the prevailing Sunni rather than Shi`i form of belief brought to Java, he believed, by the Arabs. Based on contacts with learned Muslims, such as a "priest" from Hila called Hasan Sulayman, he cast most Ambonese as ignorant of their own doctrines, but nevertheless more pious than the Dutch of the region.[29] In his section on "Macassarese Matters," by comparison, he merely added that the religion was "very widespread among these blind Heathens," alleging that it was "a Religion that is extremely easy and unusually light for them to take on."[30] And as he reflected on the widespread popularity of the Qur'an, he begrudgingly noted:

> One must add here that the Mohammedans have not so far spread themselves by use of violence, but that all of their believers have (with the exception of some, who have carelessly married a Mohammedan female, or carelessly uttered the words of their creed...) voluntarily declared themselves Mohammedan and taken this religion.[31]

According to Steenbrink, what little Valentijn said elsewhere in his compendium was based on Voetius's guidebook.[32] On the whole, then, Islam was presented as the regrettable (and inherently foreign) faith of a simple people. This did not mean, however, that Valentijn had not planned to say more, being well aware of the call of another of Voetius's students, the Utrecht professor Hadrianus Relandus (1676–1718), who had urged the Dutch to make an investment in learning Arabic and about Islam in 1705.[33] Both men were clearly in contact with each other, given that some of Relandus's manuscripts had come to him by way of Valentijn.[34]

It is also clear that Valentijn had planned to produce some sort of work of his own on the subject. For while Leijdecker had devoted some attention to a translation of the *Idah*,[35] of which copies were in wide circulation in the Indies, Valentijn turned his hand to completing notes on an elementary compendium, the *Ma`rifat al-islam*, of which he possessed at least two copies, alongside Muslim accounts of the Prophets Moses and Joseph.[36] According to the later Dutch Islamologist Snouck Hurgronje (1857–1936), Valentijn was ultimately persuaded by a friend that the "useful" notes he kept would not find much interest among the Dutch public, and he determined to drop the project.[37] This indicates that Snouck saw, or knew someone who saw, them in the nineteenth century, though their location is unknown today, and we can only point to their existence through the sale catalogues for his library.[38] Certainly one wonders what use Snouck, the towering Orientalist of the later nineteenth century, would have had for them.

A Swiss Grammarian

One also wonders if these notes on Islam were even Valentijn's in the first place. According to George Henrik Werndly (1694–1744), a Swiss *predikant* sent to Batavia from 1723 to assist Pieter Worm (also known as Petrus van der Vorm, 1664–1731) with the final production of Leijdecker's Bible in 1724, the text offered by Valentijn at the end of the previous century had more than likely been written by an earlier churchman, Simon De Larges, who had died at Ambon in 1677.[39]

Either way, it is to Werndly's *Maleische Spraakkunst* (a summary grammar and catalogue of Malay works issued in 1736), that we can finally turn for some more solid indication of what these churchmen-humanists could access of Islam, together with the added bonus that this work, much like Valentijn's magnum opus, would be printed en masse and repeatedly into the nineteenth century. In his text Werndly spent the bulk of his time recounting his fellow Europeans' efforts in coming to grips with the language (and culture) of the peoples of the Indies, before turning to what was, in effect, a catalogue of the holdings of the Archives of the Court of Chancery in Batavia, which was largely created as a result of the deaths of past bibliophiles like St. Martin. For it is in this section that Werndly listed the 69 books that he felt provided an acceptable representation of an ideal Malay library. It was also, as we shall see, the same listing that would whet John Leyden's appetite the following century.

Based on the sometimes unfamiliar titles—for these differed from place to place—Werndly's list included many romances and royal epics alongside works of Islamic law. The latter ranged from the elementary texts like the *Usul Agama Islam* and the *Ma`rifat al-Islam*, to advanced treatises by Hamza al-Fansuri (d. 1527) and

Shams al-Din of Pasai (d.1630), whose *Mir'at al-mu'min* had been rendered into Dutch (though never published) by Van der Vorm.[40] As the late Ian Proudfoot has pointed out, Werndly was himself well aware of the limitations of his survey, noting the likely existence of many more books.[41] His *Spraakkunst* even included an appendix listing an additional eight titles, including collections of the acts and sayings of the Prophet and the catechism of Abu l-Layth al-Samarqandi (d. 983 or 993).

To collect, catalogue, and even translate was one thing, but to appreciate was another. And on this point Werndly was ambiguous. Whereas he both praised and damned the contents of certain books, he seemed far more interested in the utility of the texts for providing language fit for a Malay Bible than relating how Islam was understood locally. In specific relation to the *Mir'at al-mu'min*, he outlined its organization and pointed to its usefulness for "obtaining a knowledge of the religious technical terms of the Mohammedan clergy themselves."[42]

Of "Mohammedan Priests" and Their Successes

The so-called Mohammedan clergy (*geestelijkheid*) or body of "priests," as the Dutch so often misnamed the *'ulama'*, were long an evident, if officially unwelcome, presence, even at Batavia. But it was to them that the Dutch turned when it was decreed by Governor-General Jacob Mossel (in office 1750–61) that a compendium was required to properly administer (and thus divide) the native and Chinese communities on Java. The resultant text dealing, it is said, with "the most important Mahomedan laws and customs concerning inheritance, marriage and divorce," was composed after consultations between the (unnamed) "Delegate for Native Affairs" and "some Machometan priests and *kampong* officials."[43] A committee then produced a draft in February of 1756, and it was further scrutinized by Dutch and the local "priests" before finding its way into the official edicts issued in 1760.[44]

Beyond the practicalities of law and inheritance, the proscriptions regarding marriage and divorce must have been useful too for dealing with the various "renegades," or at least those who, in Valentijn's words, had so "carelessly" uttered the words of the Islamic creed. For it was to them, or at least their children, that the Dutch also turned when they sought to deal with local authorities. And the expansion of Dutch rule over the course of the eighteenth century would have made the employment of properly trained officials all the more desirable. Perhaps it was for this reason that edicts of 1757 and 1759 recommended the training of officials in both Arabic and Malay as the languages of administration, and then their examination with regard to questions of Muslim law and practices.[45]

Whereas the VOC seemed almost happier to deal with Muslim rulers than with the Hindu Balinese, effectively encouraging the Islamization of the eastern end of Java at the expense of the latter in the 1760s, questions would begin to be asked about how was it that the Muslims had been so successful in the first place (and indeed as to why Christianization had been such an abject failure). Thus it was that, in a meeting held at Batavia in 1782, a prize was offered by the recently established Society for the Arts and Sciences to whoever could explain how "Muhamed, the Imams, and the following teachers and missionaries of the Muselmans" had managed to convert the "heathens" of the various regions and islands of the archipelago. It was also recommended that Muslims too could answer the question, or at the very least that they should be consulted.[46]

Unfortunately, nobody ever stepped forward to claim the prize, and in any case officials who were observed asking a few too many questions were discouraged. Or at least that is what one might garner from the dismissal of Andries Hartsinck from his post as resident of Surakarta in 1790 when it was discovered that he had been dressing as a Javanese and attending lessons with Sultan Pakubuwana IV.[47] In any event, it is highly unlikely that he was doing so to claim the 100 ducats on offer, and the company for which he worked was by now in dire straits, being ultimately wound up by the government of the Batavian Republic in 1795 [the new name of the Dutch Republic].

Even if new administrators would be dispatched to the Indies, attempting their own interventions in coming to grips with the Muslims under their rule, this was very little and very late, and would be effectively wiped out by the arrival of the British in 1811. For example, Marshall Daendels, the governor-general to the puppet French regime in Amsterdam from 1808 to 1811, ordered that an investigation be conducted into the state of learning of the indigenous peoples of Java in preparation for supplying something of a universal education.[48] He is also said to have commissioned a translation of the widely used Muslim juridical manual of Ibn Hajar al-Haytami (d. 1565), a task taken on by the governor of Java's Northeast Coast, Nicolaus Engelhard (1761–1831).[49] The manuscript never saw the (published) light of day unlike other aspects of his work, which did appear, though in a different language. Sent to Raffles, these were smoothly incorporated into his *History of Java*.

Other Eyes

This has been thus far a "Dutch" story, setting aside the fact that many who wrote in that language were from neighboring Germanic lands, and I have offered little in the way of discussion of how other Europeans might have garnered information on Islam in Southeast Asia from Dutch publications. But as will be clear by now, there was little to be gained beyond the pages of Valentijn and Werndly. And even then their works were likely to be either ignored or mined for very different purposes.

A case in point is the Anglo-Irish Orientalist William Marsden, who was stationed at Bengkulu (Bencoolen) on Sumatra between 1771 and 1779. In the first edition of his *History of Sumatra* (1783), Marsden expressed his astonishment at the apparent lack of Portuguese and Dutch writing on that island, though he corrected himself in later editions, praising Valentijn's work and apologizing for not having consulted the proceedings of the Batavia Society and a work published in German by another Dutchman.[50] In any case, much like the Dutch scholars before (and not so far to the side of) him, he was not specifically interested in Islam. Despite collecting significant amounts of Islamic material, including Arabic and Malay texts, which may be largely sourced to the Sultanate of Palembang, he dismissed the religion as a foreign "Malay" accretion that had done much to remove the Sumatrans from their genuine cultural genius.[51]

In later life, Marsden, who was well aware of earlier Dutch work on linguistics in the archipelago—compiling a dictionary that drew on the work of Danckaerts, Heurnius, Werndley, and others—became a friend to Raffles, who supported Leyden in turn. Certainly Leyden depended on earlier Dutch scholars. For when one examines his notes of 1802 on the works "written in Malays," one finds a copy of Werndly's listing of 1736, though without the Arabic script or his appendix.[52] It is no small irony, to my mind, that Leyden met his end perhaps by virtue of inhaling too much of the air surrounding these very works, or what was left of them when the British invaded.

It was only subsequent to the arrival of the new missionary bodies, such as the London Mission Society (founded 1795) and the Nederlansche zendelinggenootschap (founded in Rotterdam in 1797) and then, more spectacularly, with the outbreak of the Java and Padri Wars of the 1820s and 1830s, that one can speak of an active attempt to comprehend the mechanisms that made Islam such a force in the region. But that is very much another story, and of other times, in which information about Islam in Southeast Asia, while not always accurate, would begin to pour onto the presses to be sifted, debated, and, largely, bemoaned by scholar-officials to come.[53]

Conclusion

Seen over the course of over two centuries of interaction, the relative lack of published information on Islam in Southeast Asia would appear to betoken a general lack of interest in the faith of the rivals of the Dutch and the English alike. While there is some truth to this, the rare examples of curious humanists like St. Martin and Rumphius, as well as the missionaries they engaged with and even educated, from Heurnius to Valentijn, show that there were figures with knowledge of the rival local faith, but they felt but little impelled to make their learning more widely known. A knowledge of Islam, it seems, was not seen as being of utility to the Common Good, or at least the Common Good of the West and those parts of the East they sought to dominate.

Notes

1. T. S. Raffles, *The History of Java*, 2 vols. (London: Black, Parbury, and Allen; and John Murray, 1817); William Marsden, *The History of Sumatra: Containing an Account of the Government, Laws, Customs, and Manners of the Native Inhabitants, with a Description of the Natural Productions, and a Relation of the Ancient Political State of That Island* (London: Thomas Payne and Son, 1783).
2. D. E. Weatherbee, "Raffles' Sources for Traditional Javanese Historiography and the Mackenzie collections," *Indonesia* 26 (October 1978): 63–95.
3. John Crawfurd, *History of the Indian Archipelago*, 3 vols. (Edinburgh: Constable, 1820).
4. The pass is reproduced in Alastair Hamilton, *Arab Culture and Ottoman Magnificence in Antwerp's Golden Age* (The Arcadian Library in Association with Oxford University Press, 2001), 86.
5. Karel Steenbrink, *Dutch Colonialism and Indonesian Islam: Contacts and Conflicts, 1596–1950* (Amsterdam and Atlanta: Rodopi, 1993), 16.
6. Jacob Cornelisz van Neck, *Het tvveede boeck, Iournael oft dagh-register, inhoudende een warachtich verhael... vande reyse, gedaen... onder't beleydt vanden admirael Iacob Cornelisz. Neck* (Middelburch: Barent Langhenes, 1601), 26b.
7. Steenbrink, *Dutch Colonialism*, 33–35; G. F. Pijper, *Het boek der duizend vragen* (Leiden: Brill, 1924); G. W. J. Drewes, "Javanese Versions of the 'Questions of `Abdallah b. Salam,'" *BKI* 142, 2/3 (1986): 325–27; and Ronit Ricci, *Islam Translated: Literature, Conversion,*

and the Arabic Cosmopolis of South and Southeast Asia (Chicago and London: University of Chicago Press, 2011).

8. Peter G. Riddell, "Rotterdam MS 96 D 16: The Oldest Surviving Qur'an from the Malay World," *Indonesia and the Malay World* 30/86 (2002): 9–20; G. W. J. Drewes, *Een 16de Eeuwse Maleise Vertaling van de Burda van al-Bûṣîrî (Arabisch Lofdicht op Mohammad)* (The Hague: Nijhoff, 1955); Muhammad Naguib al-Attas, *The Oldest Known Malay Manuscript: A 16th Century Translation of the Aqai'd of al-Nasafi* (Kuala Lumpur: University of Malaya Press, 1988).
9. For an account of this incident, see Aḥmad ibn Qāsim al-Ḥajarī, *Kitāb Nāṣir al-dīn `alā 'l-qawm al-kāfirīn* [The Supporter of Religion against the Infidels], trans. and ed. P. S. van Koningsveld, Q. al-Samarrai, and G. A. Wiegers (Madrid: Consejo Superior de Investigaciones Científicas, Agencia Española de Cooperación Internacional, 1997), 198–99.
10. al-Ḥajar ī, *Kitāb Nāṣir al-dīn*, 179–82.
11. Harold J. Cook, *Matters of Exchange: Commerce, Medicine, and Science in the Dutch Golden Age* (New Haven and London: Yale, 2007), 204–6.
12. Cook, *Matters of Exchange*, 2007.
13. F. de Haan, "Uit oude notarispapieren I," *Tijdschrift voor Indische Taal-, Land- en Volkenkunde* 42 (1900): 297–308; Georgius Everhardus Rumphius, *The Ambonese Curiosity Cabinet*, trans., ed., and annot. E. M. Beekman (New Haven and London: Yale, 1999). On the linguistic proficiencies of the Company's servants in general, see C. van Dijk, "De VOC en de kennis van de taal- en volkenkunde van insulair Zuidoost-Azië," in *VOC en Cultuur: Wetenschappelijke en culturele relaties tussen Europa en Azië ten tijde van de Vereenigde Oostindische Compagnie*, ed. J. Bethlehem and A. C. Meijer (Amsterdam: Schiphouwer en Brinkman, 1993), 59–76, esp. 67–71.
14. Rumphius, *Curiosity Cabinet*, 238.
15. Ibid., 267.
16. The "reformist" activities of these men are described in Azyumardi Azra, *The Origins of Islamic Reformism in Southeast Asia: Networks of Malay-Indonesian and Middle Eastern "Ulama" in the Seventeenth and Eighteenth Centuries* (Leiden: KITLV Press, 2004).
17. De Haan, "Uit oude notarispapieren," 306–7.
18. Ibid., 307.
19. Justus Heurnius, *De legatione evangelica ad Indos capassenda, admonito* (Leiden: Elsevier, 1618), as quoted in J. A. Grothe, "Het Seminarie van Walaeus," *Berigten der Utrechtsche Zendings-Vereeniging* 23 (1882): 17–27, 33–44, 49–57, esp. 22.
20. "Corte beschrijvinghe hoe ende op wat maniere de Rosengeynders (volgens haer versoeck) om Christen te worden gehandelt is," in

Archief voor de geschiedenis der oude Hollandsche zending, ed. J. A. Grothe, 6 vols. (Utrecht: Van Bentum (1884–91), vol. 5, 152–67.
21. J. A. van der Chijs, ed., *Nederlandsch-Indisch Plakaatboek, 1602–1811*, 17 vols. (Batavia and the Hague, Landsdrukkerij and Nijhoff, 1885–1900), vol. 1, 108; Grothe, "Het Seminarie van Walaeus."
22. Walaeus stipulated that just as local students should be well armed against Papists, they should also be prepared for the "Jews, Mohammedans and Heathens, so that they will possess the most powerful arguments on hand to fend off their intrigues and superstitions and to defend the truth of our belief." See Grothe, "Het Seminarie van Walaeus," 37, 42.
23. On the poisoning by the "Uliassers" in 1635, see Grothe, *Archief*, vol. 6, 330.
24. Koos Kuiper, "The Earliest Monument of Dutch Sinological Studies: Justus Heurnius' Manuscript Dutch-Chinese Dictionary and Chinese-Latin Compendium Doctrinae Christianae (Batavia, 1628)," *Quaerendo* 35 (2005): 95–186; Jacobus Richardus Callenbach, *Justus Heurnius: Eene bijdrage tot de geschiedenis des Christendoms in Nederlandsch Oost-Indie* (Nijkerk: C. C. Callenbach, 1897), 36–37.
25. Bodleian Library, MS Marsh 712, 3b.
26. S. van den Bergh, "'Ayn," in *Encyclopaedia of Islam*, ed. P. Bearman, et al., 12 vols., 2nd ed. (Leiden etc.: Brill, 1954–2004), vol. 1, 784.
27. Steenbrink, *Dutch Colonialism*, 49–52; J. van Amersfoort, and W. J. van Asselt, *Liever Turks dan Paaps? De visies van Johannes Coccejus, Gisbertus Voetius en Adrianus Relandus op de islam* (Zoetermeer: Boekencentrum, 1997), 19–23.
28. R. R. F. Habiboe, *Tot verheffing van mijne natie: Het leven en werk van François Valentijn (1666–1727)* (Franeker: Van Wijnen, 2004), 53–63.
29. See, for example, F. Valentyn, *Oude en Nieuw Oost-Indiën, vervattende een naauwkeurige en uitvoerige verhandeling van Nederlants Mogentheyd in die Gewesten, enz.met meer dan 1050 prentverbeeldingen verrykt…en met…kaarten opgeheldert*, 5 vols. (Dordrecht etc.: Van Braam, 1724–26), vol. 3,i, 19–27.
30. Valentyn, *Oude en Nieuw Oost-Indiën*, vol. 3, ii, 233.
31. Ibid., vol. 4, ii, 4.
32. Steenbrink, *Dutch Colonialism*, 41.
33. See the preface to his *Religione Mohammedica* as translated in Van Amersfoort and Van Asselt, *Liever Turks dan Paaps?*, 124–25.
34. See, for examples, Leiden University Library, Or. 1692 and Or. 1945 as described in Jan Just Witkam, *Inventory of the Oriental Manuscripts of the Royal Netherlands Academy of Arts and Sciences in Amsterdam* (Leiden: Ter Lugt, 2006), vol. 2, 219, 291.
35. Witkam, *Inventory*, vol. 1, 18–19.

36. *Catalogus Exquisitissimorum & Excellentissimorum Librorum, In omni fere materia, facultate & lingua, praecipué in linguis Orientalibus, quibus Simul Elegentissima nec minus curiosissima Manuscripta eminent. Viri Reverendi Fr. Valentyn, Verbi divini Ministri (dum viveret) fidelissimi ad Amboinam & ad Bandam in India Orientali, Quorum publica fiet auctio in Aula magna (vulgo) De Groote Zaal van 't Hof, ad diem 23. Febr. Seqq. 1728* (Hagae Comitum: Apud Alberts & van der Kloof, 1728), 5–6. Cf. Habiboe, *Tot verheffing van mijne natie*, 121–22.
37. C. Snouck Hurgonje, "Een rector der Mekkaansche universiteit (met aanhangsel)," in *Verspreide geschriften van C. Snouck Hurgronje*, ed. A. J. Wensick, 6 vols. (Bonn: K. Schroeder, 1923–27), vol. 3, 65–122, esp. 77.
38. "Kerknieuws," *Boekzaal der Geleerde Werelt* (November 1727): 610–32, 619. See also Habiboe, *Tot verheffing van mijne natie*, 118.
39. Habiboe, *Tot verheffing van mijne natie*, 64.
40. I had a chance to briefly view a copy of Worm's lengthy translation of Shams al-Din's *Mir'at al-mu'min* subsequent to writing this paper. While space precludes detailed consideration here, its existence solely in manuscript form points once more to the individual cleric's knowledge of the religion of his adversaries, and to the collective disinterest in having that knowledge reprinted. See "Verklaaringe vande Muhhamedaansche Godts-geleertheijt... uijt de arabische in de Nederduijtsche tale overgebraagt," Leiden University Library, BPL 310; LOr. 1700 and LOr. 14.383.
41. Ian Proudfoot, "An Expedition into the Politics of Malay Philology," *Journal of the Malaysian Branch of the Royal Asiatic Society* 76 (2003): 1–53, esp. 20; George Henrik Werndly, *Maleische Spraakkunst, Uit de eige schriften der Maleiers opgemaakt* (Amsterdam: Wetstein, 1736), 356–57.
42. Werndly, *Maleische Spraakkunst*, 354–55.
43. Van der Chijs, *Plakaatboek*, vol. 7, 392. A *kampong* refers in Malay to a settlement, usually a village, though with the spread of Batavia it took on the meaning of an internal ward within the city.
44. Van der Chijs, *Plakaatboek*, vol. 7, 393–407. While the original manuscript is yet to surface, it may prove to be a manual that Van der Lith claimed was (re)produced at Batavia under the auspices of Pieter Haksteen, then Secretaris van Schepenen, in 1760. P. A. van der Lith, "De koloniale wetgeving tegenover de Europeesche en de Inlandsche Rechtsbegippen," *De Gids* 3 (1882): 193–231, esp. 221. Haksteen had collaborated on a compendium of Chinese Laws in 1754. P. Haksteen and Reinier de Klerk, eds., *Compendium der Chinese Wetten*, KITLV MS H 458.
45. British Library, MSS Eur/Mack Private 32, "Realia Livre A," vol. 3," 308–9.

46. "Voorbericht," *Verhandelingen van het Bataviaasch Genootschap van Kunsten en Wetenschappen* I (1779): 30–31, esp. 29–30.
47. M. C. Ricklefs, *Mystic Synthesis in Java: A History of Islamization from the Fourteenth to the Early Nineteenth Centuries* (Norwalk: EastBridge, 2006), 175.
48. J. A. van der Chijs, "Bijdragen tot de geschiedenis van het inlandsch onderwijs in Nederlandsch-Indië: aan officiële bronnen ontleend, " *Tijdschrift voor Indische taal-, land-, en Volkenkunde* 34, 5 (1864): 212–323, esp. 212.
49. C. Snouck Hurgronje to director of Education, Batavia, February 18, 1892, in *Ambtelijke adviezen van C. Snouck Hurgronje 1889–1936*, ed. E. Gobée and C. Adriaanse, 3 vols. (The Hague: Nijhoff, 1957–65), vol. 3, 1844–45. See notes regarding Engelhard's manuscript in Taco Roorda, *Kitab Toehpah: Een Javaansch handboek voor het Mohammedaansch regt* (Leiden: Brill, 1874), v–vi.
50. Compare p. 4 of Marsden's first edition, cited above, with the third edition published by McCreery of London in 1811.
51. Some of Marsden's gatherings may be found today in the library of the School of Oriental and African Studies, London.
52. British Library, IOL Add. 26568, 117b–18a.
53. I have tried to go some way to discuss what came next elsewhere. See Michael Laffan, *The Makings of Indonesian Islam: Orientalism and the Narration of a Sufi Past* (Princeton: Princeton University Press, 2011).

Bibliography

Amersfoort, J. van, and W. J. van Asselt. *Liever Turks dan Paaps? De visies van Johannes Coccejus, Gisbertus Voetius en Adrianus Relandus op de islam.* Zoetermeer: Boekencentrum, 1997.

Attas, Muhammad Naguib al-. *The Oldest Known Malay Manuscript: A 16th Century Translation of the Aqai'd of al-Nasafi.* Kuala Lumpur: University of Malaya Press, 1988.

Azra, Azyumardi. *The Origins of Islamic Reformism in Southeast Asia: Networks of Malay-Indonesian and Middle Eastern "Ulama" in the Seventeenth and Eighteenth Centuries.* Leiden: KITLV Press, 2004.

Bearman, P. et al., ed. *Encyclopaedia of Islam.* 12 vols. 2nd ed. Leiden etc.: Brill, 1954–2004.

Bodleian Library, MS Marsh 712.

British Library, IOL Add. 26568, 117b–18a.

British Library, MSS Eur/Mack Private 32.

Callenbach, Jacobus Richardus. *Justus Heurnius: Eene bijdrage tot de geschiedenis des Christendoms in Nederlandsch Oost-Indie.* Nijkerk: C. C. Callenbach, 1897.

Catalogus Exquisitissimorum & Excellentissimorum Librorum, In omni fere materia, facultate & lingua, praecipué in linguis Orientalibus, quibus

Simul Elegentissima nec minus curiosissima Manuscripta eminent. Viri Reverendi Fr. Valentyn, Verbi divini Ministri (dum viveret) fidelissimi ad Amboinam & ad Bandam in India Orientali, Quorum publica fiet auctio in Aula magna (vulgo) De Groote Zaal van 't Hof, ad diem 23. Febr. Seqq. 1728. Hagae Comitum: Apud Alberts & van der Kloof, 1728.

Chijs, J. A. van der. "Bijdragen tot de geschiedenis van het inlandsch onderwijs in Nederlandsch-Indië: aan officiële bronnen ontleend." *Tijdschrift voor Indische taal-, land-, en volkenkunde* 34, 5 (1864): 212–323.

———, ed. *Nederlandsch-Indisch Plakaatboek, 1602–1811.* 17 vols. Batavia and the Hague, Landsdrukkerij and Nijhoff, 1885–1900.

Cook, Harold J. *Matters of Exchange: Commerce, Medicine, and Science in the Dutch Golden Age.* New Haven and London: Yale, 2007.

Dijk, C. van. "De VOC en de kennis van de taal- en volkenkunde van insulair Zuidoost-Azië." In *VOC en Cultuur: Wetenschappelijke en culturele relaties tussen Europa en Azië ten tijde van de Vereenigde Oostindische Compagnie*, edited by J. Bethlehem and A. C. Meijer, 59–76. Amsterdam: Schiphouwer en Brinkman, 1993.

Drewes, G. W. J. *Een 16de Eeuwse Maleise Vertaling van de Burda van al-Bûṣîrî (Arabisch Lofdicht op Mohammad).* The Hague: Nijhoff, 1955.

———. "Javanese Versions of the 'Questions of `Abdallah b. Salam." *BKI* 142, 2/3 (1986): 325–27

Gobée, E. and C. Adriaanse, eds. *Ambtelijke adviezen van C. Snouck Hurgronje 1889–1936.* 3 vols. The Hague: Nijhoff, 1957–65.

Grothe, J. A. "Het Seminarie van Walaeus." *Berigten der Utrechtsche Zendings-Vereeniging* 23 (1882): 17–27, 33–44, 49–57.

———. ed. *Archief voor de geschiedenis der oude Hollandsche zending.* 6 vols. Utrecht: Van Bentum, 1884–91.

Haan, F. de."Uit oude notarispapieren I." *Tijdschrift voor Indische Taal-, Land- en Volkenkunde* 42 (1900): 297–308.

Habiboe, R.R.F. *Tot verheffing van mijne natie: Het leven en werk van François Valentijn (1666–1727).* Franeker: Van Wijnen, 2004.

Ḥajarī, Aḥmad ibn Qāsim al-. *Kitāb Nāṣir al-dīn `alā 'l-qawm al-kāfirīn* [The Supporter of Religion against the Infidels]. Translated and edited by P. S. van Koningsveld, Q. al-Samarrai, and G. A. Wiegers. Madrid: Consejo Superior de Investigaciones Científicas, Agencia Española de Cooperación Internacional, 1997.

Hamilton, Alastair. *Arab Culture and Ottoman Magnificence in Antwerp's Golden Age.* The Arcadian Library in Association with Oxford University Press, 2001.

"Kerknieuws." *Boekzaal der Geleerde Werelt* (November 1727): 610–32.

KITLV MS H 458

Kuiper, Koos. "The Earliest Monument of Dutch Sinological studies: Justus Heurnius' Manuscript Dutch-Chinese Dictionary and Chinese-Latin Compendium Doctrinae Christianae (Batavia, 1628)." *Quaerendo* 35 (2005): 95–186.

Leiden University Library, Or. 1692 and Or. 1945

Lith, P.A. van der. "De koloniale wetgeving tegenover de Europeesche en de Inlandsche Rechtsbegippen." *De Gids* 3 (1882): 193–231.

Marsden, William. *The History of Sumatra: Containing an Account of the Government, Laws, Customs, and Manners of the Native Inhabitants, with a Description of the Natural Productions, and a Relation of the Ancient Political State of That Island.* London: Thomas Payne and Son, 1783.

———. *The History of Sumatra.* 3rd ed. London: M'Creery, 1811.

Neck, Jacob Cornelisz van. *Het tvveede boeck, Iournael oft dagh-register, inhoudende een warachtich verhael... vande reyse, gedaen... onder't beleydt vanden admirael Iacob Cornelisz. Neck.* Middelburch: Barent Langhenes, 1601.

Pijper, G. F. *Het boek der duizend vragen.* Leiden: Brill, 1924.

Proudfoot, Ian. "An Expedition into the Politics of Malay Philology." *Journal of the Malaysian Branch of the Royal Asiatic Society* 76 (2003): 1–53

Raffles, T. S. *The History of Java.* 2 vols. London: Black, Parbury, and Allen; and John Murray, 1817.

Ricci, Ronit. *Islam Translated: Literature, Conversion, and the Arabic Cosmopolis of South and Southeast Asia.* Chicago and London: University of Chicago Press, 2011.

Ricklefs, M. C. *Mystic Synthesis in Java: A History of Islamization from the Fourteenth to the Early Nineteenth Centuries.* Norwalk: EastBridge, 2006.

Riddell, Peter G. "Rotterdam MS 96 D 16: The Oldest Surviving Qur'an from the Malay World." *Indonesia and the Malay World* 30/86 (2002): 9–20

Roorda, Taco. *Kitab Toehpah: Een Javaansch handboek voor het Mohammedaansch regt.* Leiden: Brill, 1874.

Rumphius, Georgius Everhardus. *The Ambonese Curiosity Cabinet.* Translated, edited, and annotated by E. M. Beekman. New Haven and London: Yale, 1999.

Steenbrink, Karel. *Dutch Colonialism and Indonesian Islam: Contacts and conflicts, 1596–1950.* Amsterdam and Atlanta: Rodopi, 1993.

Valentyn, F. *Oude en Nieuw Oost-Indiën, vervattende een naauwkeurige en uitvoerige verhandeling van Nederlants Mogentheyd in die Gewesten, enz. met meer dan 1050 prentverbeeldingen verrykt... en met... kaarten opgeheldert.* 5 vols. Dordrecht etc.: Van Braam, 1724–26.

"Voorbericht." *Verhandelingen van het Bataviaasch Genootschap van Kunsten en Wetenschappen* I (1779): 30–31.

Weatherbee, D. E. "Raffles' Sources for Traditional Javanese Historiography and the Mackenzie Collections." *Indonesia* 26 (October 1978): 63–95.

Wensick, A.J., ed. *Verspreide geschriften van C. Snouck Hurgronje.* 6 vols. Bonn: K. Schroeder, 1923–27.

Werndly, George Henrik. *Maleische Spraakkunst, Uit de eige schriften der Maleiers opgemaakt.* Amsterdam: Wetstein, 1736.

Witkam, Jan Just. *Inventory of the Oriental Manuscripts of the Royal Netherlands Academy of Arts and Sciences in Amsterdam.* Leiden: Ter Lugt, 2006.

Chapter 7

A Moral Obligation of the Nation-State: Archaeology and Regime Change in Java and the Netherlands in the Early Nineteenth Century

Marieke Bloembergen and Martijn Eickhoff

In the (post-)modern Western world, "archaeological" sites, especially sites that contain ruinous buildings, are often regarded as places of great importance; they can give people a sense of belonging to certain ethnic groups, religious worldviews, or civilizing missions. Since the rise of nation-states, these sites time and again came to symbolize a combination of national, cultural, and political origins. Sometimes a ruin was also regarded as a prefiguration of future prosperity, or even as a place that symbolizes a parallel or even lost world. But the spectrum of meanings connected to these places is much greater than this. This can become clear if we look at the ways ruins were depicted and explained in the early modern time in Europe, and in the context of Western colonial expansion.[1] During the seventeenth and eighteenth centuries in Europe, ruins—"real" ones or so-called follies in landscape gardens—often figured in allegorical explanations that contained strong moral or religious connotations. They functioned as "memento mori"-signs or as warnings against decay and decadence. Western travelers in the non-Western world in early modern and modern times, who encountered "archaeological" sites with ruined buildings seem to have given them more or less the same meaning as they used to do with ruins closer to home. In an Oriental context, it might have been "the other's" past that was at stake, but we still see that people connected the ruins to notions of decay and decadence. This description of 1891 of the ruined kraton (palace) Ratu Boku near Yogyakarta by the Dutch engineer and self-made archaeologist Jan Willen IJzerman can serve as a telling example:

> Nowadays this is a lonely and deserted place, whereas in the past it was full of life and movement. It was a place where powerful sovereigns were protected against hostile attacks by nature, and where they could look down full of contempt on their subjects, who, on their order, in the plains north- and southwards were building temples that expressed their personal pride, as these temples were considered by them as imperishable witnesses of their own glory and grandeur.[2]

This description contains a central topos of the European perception of ruinous sites in Asia. This topos connected ruined temples or palaces to the despotism of the former political leaders in the region. Especially during the heydays of European colonialism in the nineteenth and early twentieth centuries, this topos spread through Asia and Europe by different media: site descriptions, novels, drawings, photographs, or performances. The final act of the ballet *La Bayadère*, composed by the Austrian composer Ludwig Minkus and staged for the first time in 1877 at the Imperial Bolshoi Kamenny Theatre in St. Petersburg, can serve as a revealing example. The ballet tells a story about decadent and despotic behavior at a court in India. In the last "shaadi"—or marriage—scene, the gods punish the people for their immoralities by destroying the temple with thunder and lightning. This leads to the death of the people who are attending the nuptial celebration and to the transformation of the temple into a ruin. It is tempting to project this powerful late-nineteenth-century "Orientalist" image of a temple ruin as a symbol of supranatural, divine punishment for immoral behavior onto the general Western view on ruined sites in Asia. But this is an all too easy explanation. It is worthwhile to analyze the appropriation and appreciation of these sites, rooted in local circumstances and connected to global developments, in more detail.

In this article we analyze how at the beginning of the nineteenth century, Dutch and British civil servants and military officials made sense of the Hindu-Buddhist temple ruins on the island of Java, in the context of regime changes in both Asia and Europe. According to the cultural historian Göran Blix, who investigated European ways of seeing ruined sites, the period around 1800 was special because during that time a "reconstructive gaze" developed in the new trend to reconstruct ruins on paper. Through this gaze, so Blix reasons, vanished pasts could be recalled, and thereby also appropriated.[3] The first drawings of Javanese ruined temples, that reached a wider European audience, were printed in Thomas Stamford Raffles's *The History of Java* from 1817. The book contained thirteen drawings of ruined temples and, indeed, three reconstructions. In the book, Raffles wrote that these antiquities of Java had not drawn much attention until that moment. He spoke of numerous remains of former art and grandeur that existed in the ruins of temples that either lay completely buried under rubbish or that were only partially examined. As an explanation he referred to the Dutch devotion to commerce that was too exclusive to allow any interest in this subject. At the same time there was, according to Raffles, a narrow-minded policy of the Dutch, which denied antiquarians from other nations to conduct research in Java.[4]

This last remark of Raffles, clearly a self-confirming observation, seems meaningful if we wish to know more about the actual practice of archaeological explorations in Java in the early nineteenth century, and their political implications—and how they relate to comparable activities in the Netherlands. For, to what extent were the parallel regime changes in Java and the Netherlands of the early nineteenth century an incentive for archaeological activities with regard to the past of Southeast Asia?[5] Why were military expeditions and governmental reforms flanked by activities on archaeological sites on Java? Can it be concluded—following Raffles's critical statements on the narrow-minded policy of the Dutch—that the depictions of temple ruins on Java served as a means for social critique directed toward the Dutch colonial power itself? Can we discern, at the start of the nineteenth century, in the interactions between Java, Great Britain and the Netherlands, the origin of a discourse in which it was self-evident that the archaeological sites of Java, after their "discovery," had to be safeguarded by the colonial state? And did these sites, which hosted an ongoing and multiple process of signification, already at that time give prestige to the colonial state in development? Benedict Anderson has pointed to the connection between an "archaeological push" around 1900 in Southeast Asia and (post-)colonial nationalism. Against that background, the Hindu-Buddhist temples that had been cleaned, investigated, and restored in the late nineteenth century and early twentieth century, could become "state regalia" of the late colonial state. In this article we extend Anderson's time frame by focusing on the first half of the nineteenth century; maybe we can find an explanation why these Hindu-Buddhist temple sites—rediscovered and investigated in a period when the majority of the Javanese population had turned to (various forms of) Islamic faith—and not, for example, old palaces (kratons) or mosques, gained this status.[6]

This perspective on archaeological sites, in which we investigate how early archaeological interventions at the sites and the following dissemination of site-related objects, documentation, and images stirred early heritage formation, differs from the traditional way of writing the history of colonial archaeology. Although we depend on the same sources as the more traditional historiographers, we are not focusing, like for example Bruce Trigger, on the abuse of science by colonial archaeologists exclusively.[7] We acknowledge the existence of colonial hierarchies and the role knowledge played in creating these hierarchies,[8] but on the basis of our approach we hope to be able to understand how archaeological activities were an integral part of the complex interactions between the European nations and the world they were colonizing in general.

Archaeology and Regime Change in the Netherlands and Java

With regard to Europe, the late eighteenth and early nineteenth centuries can be regarded as the period in which many central political and cultural concepts received their fixed and modern meaning. The then recent experiences with historical discontinuity and destruction, in particular during the French Revolution, evoked a new focus on the past, on change, and future developments. This phenomenon made "archaeology," with its multiple material links to the past, into an extremely meaningful activity. In the context of the non-Western world, where since the start of the colonial expansion the confrontation of Europeans with many different cultural traditions had sharpened European awareness of difference, change, and development, archaeological sites and activities gained the potential of becoming markers of a new era.[9] As a result, archaeological activities could become public events, in particular since Napoleon's reign; Napoleon's Egyptian campaign of 1798–1801 was partly a scholarly-archaeological enterprise, whereas the subsequent publication of the *Voyage dans la basse et la haute Egypte* by Vivant Denon in 1802, and the opening of Musée Napoléon in 1803 made a great impression on contemporaries.[10]

In the following decades, in many European countries archaeology became institutionalized in a national context by museums and universities. At the same time, long-established interpretative frameworks, notably the Bible and the classical tradition, gradually lost their pivotal meaning. As an inevitable result, national perspectives made their entry in archaeological interpretations. In the different European countries, the developments were however not identical. During the

Napoleonic period, in Great Britain and its imperial territories there was, for example, no strong concept of cultural patrimony. It was the competition with continental European states with their tradition of strong central government that inspired the British participation in the international race for antiquities around the Mediterranean, and evoked the discussions about the protection of antiquities in, for example, India. Imperial officials and military officers, connected by networks of knowledge exchange, took the initiative, whereas formal institutions of governance played a minor role.[11]

At first sight there was no significant increase in archaeological activities in the Netherlands during the late eighteenth and early nineteenth centuries. On a political level there was an immense discontinuity: after the revolutionary Batavian period (1795–1806) and so-called French period (1806–1813), the Dutch Republic became a kingdom (1813), and the family of Stadtholders (who had governed until 1795) became the royal family. During this period of political change, there were no institutional structures that were able to offer any kind of scholarly continuity with regard to archaeology. But at the same time new institutional structures came into being that in later decades would create this same continuity. The Koninklijk Instituut van Wetenschappen, Letterkunde en Schoone Kunsten, (Royal Institute of Sciences, Literature and Fine Arts, hereafter Royal institute) which was founded by Louis Bonaparte in 1808, would later, when the Netherlands had become a kingdom, play a small but initiating role in this field. Moreover in 1818, the newly established king, Willem I, would appoint at Leiden University, the Netherlands, the first professor in Archaeology, Casper J. C. Reuvens (1793–1835); that same year, Reuvens also became the director of the newly founded Museum of Antiquities in Leiden. Thanks to the financial support of the new state, Reuvens could start collecting antiquities on a scale the Netherlands had not witnessed before. This included the acquisition of statues from the classical world, Egypt, and the Dutch East Indies.[12]

The regime changes in the Netherlands had a great impact on the political status of the Dutch overseas territories. In 1798, after the liquidation of the Vereenigde Oost-Indische Compagnie (Dutch East Indies Company, VOC), the territories in Asia became part of the Batavian Republic. In 1807, Louis Bonaparte appointed Herman W. Daendels for the post of governor-general of the Dutch East Indies. Daendels became responsible, among others, for huge infrastructural activities in Java. When the kingdom of Holland was incorporated into France in 1810, Daendels returned to Holland. A year later, in

1811, after Lord Minto had defeated the Franco-Dutch troops and conquered Java, Thomas Stamfort Raffles was appointed lieutenant-governor.[13] Raffles stayed in Java until 1815. He left for England after the island was returned to the control of the Netherlands under the terms of the Anglo-Dutch Treaty of 1814.

In Java, at the start of this period of regime changes there was no institutional landscape that was able to offer any kind of scholarly continuity with regard to the archaeology of the island. The year 1813 was however a turning point as it witnessed the "reanimation" (as Raffles's biographer Boulger would write in 1897) of the Bataviaasch Genootschap voor Kunsten en Wetenschappen (Batavia Society of Arts and Sciences), the learned society that was founded in Batavia in 1778.[14] Two years earlier, in 1811, Lieutenant Colonel Colin Mackenzie (1754–1821), who had joined the expedition to seize Java from the Dutch, had undertaken an important archaeological initiative. As "surveyor-general" of Java, he had among other things been given the task of compiling statistics about Java on which Raffles could base his governmental reforms.[15] He had to register the records of the Dutch colonial administration and direct a land tenure commission as well as conduct fiscal-administrative surveys. But he immediately also started collecting archaeological information on the island, like he had done between 1796 and 1811 in India. In the years 1811–1813, Mackenzie traveled through Java accompanied by a group of Dutch engineers, among whom were H. C. Cornelius and a team of draughtsmen from Java and India. Some of these draughtsmen had been trained earlier in India. Whereas Mackenzie made detailed notes, his team was responsible for the measuring and drawing of the temples. It was not easy for him to find reliable informants and translators. He made use of an archaeological questionnaire (in English, French, and Dutch) that was sent out to potential informants on the island, and he was successful in making contact with the Javanese aristocracy, being interested in their history, customs, and literature.[16] Mackenzie did not keep the collected information for himself. In 1814, he published his Javanese "archaeological" travel diary of January 1812 in the *Verhandelingen van het Bataviaasch Genootschap der Kunsten en Wetenschappen [Proceedings of the Batavia Society of Arts and Sciences].* In it he described remains of ancient cities and temples as part of the "historical" landscape of Java.[17]

Whereas Raffles in 1817 criticized the Dutch's narrow-minded policy toward Javanese antiquities in his *The History of Java*, he had been much more polite, or even optimistic, in 1813 in an address to the members of the Batavia Society. Being lieutenant-governor of the island, he praised the history of its society and explained its recent decline as a result of the war in Europe that "has desolated the finest countries." During the tenure of his own government—he spoke of the "present promising circumstances"—he however foresaw a "revival of the institution." In his speech he also praised the work on the antiquities and native history of the island by Mackenzie, "whose researches into antiquities of Western India so eminently qualify him for similar pursuits in this quarter."[18] Apart from Mackenzie, Raffles had recruited several specialists who collected archaeological information from all around the island: Colonel Adams, Captain G. Baker, Major M. Johnson, J. C. Lawrence, T. Horsfield, H. C. Cornelis, and J. W. B. Wardenaar.[19]

Archaeology thus became a part of the political reform program; supporting archaeology was understood as a way of supporting Raffles's governmental reorganizations and opposing the former Dutch misrule.[20] The sending of an archaeological questionnaire by Mackenzie should therefore not only be regarded as a scholarly activity, but as an act of legitimating the British government on the island as well. Answering it meant showing loyalty to the new government, and as such the questionnaire created a supporting network for this same government. Javanese sites with temple ruins were in this context understood as places that marked an obligation; their decay was connected to both the "degeneration" of Javanese society and the misrule of the VOC period. Raffles would later write about this situation: "The indifference of the natives has been as great as that of their conquerors."[21] Against this background, the conservation of temples—removing the vegetation was the first step—symbolized a new government that recognized it had more duties then just commercial ones. At the same time British nationalist imperial objectives were never far away: Mackenzie pleaded in 1813 in a letter to Raffles for the "preservation" of archaeological sites: "It might at some future day call to remembrance an event that will be always deemed interesting to the Nation at large, the incorporation of Java in the British empire."[22]

Raffles and Mackenzie thus ignored the ways in which the local Javanese, notwithstanding their Islamic faith, still regarded (elements of) these sites as sacred. The two men also appeared ignorant of the fact that during the previous Dutch regime, some archaeological activities at Javanese temple sites were initiated; however, the individual status that could be obtained by dealing with archaeological sites and statues seemed to have been central in these activities, whereas "the nation" was nonexistent. In 1802, Nicolaus Engelhard, who was the governor of Java's northeastern coast between 1801 and 1808, visited the courts of Solo and Yogyakarta to discuss problems—so-called *differentien*—with the native rulers who had arisen after the installation of the Batavian Republic in the Netherlands in 1795 and the failed British invasion of Java of 1800.[23] During his stay in Yogyakarta he made a trip to three sites with ruined buildings—the Prambanan temples (build around 850 AD), Kota Gede (a kraton and royal graveyard dating from around 1700), and the royal *pasanggrahan* (place of retreat) Gimbirowati.[24] It is not known precisely when and how Engelhard's interest in archaeological sites was triggered. In spite of that, it is noticeable that in the years following his visit he continued his archaeological initiatives. In 1805 he ordered lieutenant-engineer H. C. Cornelius, who at that time was building a fortress in Klaten, to clean the temples of Prambanan and to make drawings of them. During this project, which lasted till 1807, Cornelius was assisted by the ensigns J. W. B. Wardenaar, A. F. Van der Geugten, and J. A. Dubois.[25] Around the same time, Engelhard started collecting statues from Javanese antiquity for his residency's garden "De Vrijheid" in Semarang. Six of them were taken from the Singasari temple, a thirteenth-century Buddhist temple near Malang (Eastern Java) that surveying Dutch officials discovered in 1803.[26]

It is also reported that Engelhard, after his visit to the *pasanggrahan* Gimbirowati—a place of retreat and meditation for the sultan, with two twin statues of giants like those found at Prambanan—chiselled his name on the building, thus marking his presence at the site with graffiti.[27] Following Blix, this act of graffiti can be explained not only as marking a visit, but also as a way to appropriate a great but vanished past. In general Engelhard considered the Javanese antiquities to be of no importance for the Javanese people. In his elaborate answer to the questionnaire of Mackenzie, he wrote that the people who inhabit the areas near these antiquities "do not honour them although they bring offers to some of these idols, without being able to explain why." He stated, moreover, that the Javanese people did not have any memory regarding the people who once built these temples and who, so he believed, later disappeared.[28] However, here Engelhard seems to have been prejudiced; not knowing why they bring offerings does not imply that these offerings are meaningless. The same can be concluded about the archaeological intervention at the sites involved: in 1814, Colonel Adams would report that the local people living near the Singasari temple—after the removal of the six statues—had transferred other, similar objects into the jungle to prevent further removals. [29]

In the later Dutch overviews of the history of archaeology in the Dutch East Indies, it is a topos that archaeological activities of Engelhard have been plagiarized by Raffles: he used his insights in his *The History of Java* of 1817 without giving credits to the author.[30] Raffles mentioned Engelhard's Semarang collection and spoke of "several very beautiful subjects in stone," but stressed that Engelhard, by his collecting practices, ruined the temple site of Singasari.[31] The work of the Dutch archaeologist N. J. Krom, who, from the start of his career, combined archaeological and historical research, can serve as a revealing example here. In his inaugural lecture of 1919, he summarized the archaeological developments in Java of the period around 1800. He argued that 1819 was the year that, what he called "Indische" (Indies), archaeology in *The Netherlands* ("our Fatherland") began. The year was marked by a visit of Reuvens to the East India Company Museum in London, the arrival in the Netherlands of three Singasari statues, and a subsequent exposé written by Reuvens on these statues.[32] The statues had been part of the Engelhard collection in Semarang (drawings of two of them were printed in Raffles's *The History of Java*)[33] and were then placed in the garden of the Royal Institute in Amsterdam.[34] Krom also mentioned that Raffles in his *The History of Java* had used information gathered by Engelhard without mentioning his name.[35] At the same time he acknowledged that Raffles brought together and continued what others had started, and thus succeeded in creating a wider perspective on the Javanese past.[36] Here we see clearly how in the context of the strong Dutch nation-state of the early twentieth century, archaeology in the period of regime change in Java of one century earlier was received with mixed emotions. On the one hand, it was in this period in which Krom situated the beginning of a national Dutch East Indies archaeological tradition, while on the other hand he could not ignore the initiatives of the British. These initiatives were however judged with ambivalence. Thus, Krom emphasized earlier Dutch activities, and he complicated the idea of "discovering" temple sites, which Raffles supposedly had done in 1814 with Borobudur, the giant eighth-century Buddhist shrine near Magelang in Central Java, today a world heritage site. Krom argued that Raffles "found" the temple, only after he "was told" about Borobudur in 1814, charging moreover a Dutchman, Cornelius, with the task of cleaning up and drawing this temple.[37] He also added that the Borobudur never had been a forgotten site. There were some Javanese sources from the late eighteenth century that mentioned the site. One of them was about a Javanese prince who visited the site in 1757—just before his death—to see the "thousands of statues." Moreover, Krom continued,

Cornelius in 1814 not only made the first drawing of Borobudur, but also described it in his work "Beschrijving van de ruïnes, bij den inlander bekend als onder de benaming Borro-Boedoer" [Description of the ruins, known to the natives under the name Borro-Boedoer].[38] In other words, the fact that Javanese people had a special name for the site proved that Borobudur was not a discovery of Raffles.

A Dutch Concern?

In the Netherlands, Reuvens was the first archaeologist with an institutional backing who published about the archaeology of Java. Reuvens, who as a young man—from 1811 till 1814—lived in Paris when his father represented the Netherlands there, became very impressed by the Napoleonic archaeological initiatives. During his stay in Paris he, for example, frequented the Musée Napoléon. But later in his academic career, he was first inspired by the German archaeological tradition of his time. In his work he managed to combine the Dutch antiquarian and humanistic tradition with universalism of the Enlightenment era.[39] In his research on Javanese antiquities, he primarily focused on the relation between mythology and objects, and on supposed "classical" influences on Java. As a member of the Royal Academy, he wrote in 1824 an essay—*Verhandeling over drie groote steenen beelden* [Exposition on three large stone statues]—on the Singasari statues that had arrived in Amsterdam in 1819.[40] He also took a stand in the Dutch-English archaeological rivalry with regard to the Javanese past in which he focused on the unjust British depreciation of the Dutch initiatives in this field. He wrote, for example, that since he left his work unmentioned, Raffles apparently had a very low opinion of Engelhard. Reuvens explicitly referred to the passage in *The History of Java* in which Raffles described how Engelhard "ruined" the temple of Singasari:

> In one of these niches we observed an image lying flat on the ground, with its head of; in another, the pedestal of an image, which we were informed had been taken away by Mr. Engelhard; and where the traces of a third niche appeared, the stones had been removed, and a deep hole dug, so as to disfigure, and in a great measure demolish, this part of the building. This was also attributed to Mr. Engelhard's agents.[41]

To defend Engelhard, Reuvens stated that these were hasty observations and he brought to mind that it was Engelhard who had started the cleaning of the Prambanan temple and who had commissioned

Cornelius with making drawings of the temple.[42] At the same time Reuvens acknowledged that Raffles himself had contributed to the knowledge of Javanese antiquity as well. He concluded: "We must and want, being members of a commonwealth of science, not judge according to ethnic lines [Reuvens used the word "volk" M.B & M.E.], and show these writers our gratitude."[43] Thus, Reuvens was well aware of the contemporary nationalist appropriation of Javanese archaeology, whereas he at the same time tried to maintain his enlightened universalistic ideals.

A few years later, in 1827, Reuvens received a personal letter from Engelhard who at that time lived in Buitenzorg (Bogor). Engelhard thanked Reuvens for sending his study of the three Singasari statues. He recalled that these statues had been his personal gift to King Willem I of the Netherlands. In 1819, so he wrote, he had actually planned to send six statues, but three of them remained in Buitenzorg. He now conceived the plan to send them to the Netherlands in the near future. In addition, he also considered offering the king the drawings of the Prambanan temple made by Cornelius. In his letter Engelhard confirmed that the explorers of the British Interregnum in Java, like Raffles, Crawfurd, Horsfield, and Mackenzie had used the information he had collected earlier, but projected it as their own work. He even stated that the population statistics with regard to Java and Madura on pages 243 to 288 in Part II of *The History of Java* were actually his work; he had sent them to the government in 1803.[44] Apparently, Engelhard did not realize that his elaborate answer to Mackenzie's questionnaire had not only been a scholarly activity but all the same an act of loyalty to the new British government. This loyalty was not without reason. A few years earlier, during Daendels's government, his position—being a so-called *oudgast* (old guest)—had been quite awkward; in 1808 Daendels finally fired him.[45] Actually, in his letter to Reuvens, Engelhard performed a similar act of loyalty, again with the help of archaeology, but now directed toward the new king of the Netherlands.

In 1832, two years after the Java War of 1825–1830 ended in a Dutch victory and, from the Dutch perspective, the colonization of Java seemed to have been completed, Reuvens again showed his commitment to Javanese antiquities. In that year he sent a memorandum to J.C. Baud, who was to become governor-general in Batavia, in which he pleaded for measures to safeguard archaeological sites and objects from vandalism, looting, and neglect.[46] He wrote: "There seems to be a feeling on Java that these monuments are communal property and that everyone, especially the higher civil servants, can take away what they like." He then stated: "To stop this conduct completely by declaring the monument property of the government, from which nothing can be taken, is probably impossible and could have bad side effects."[47] He therefore suggested asking the owners of the areas involved to make drawings of objects and to measure the shape of the temples in detail. A society—the *Maatschappij tot Uitgave van Javaansche Monumenten* (Society for the Publication of the Javanese Monuments)—would, according to Reuvens, be the best way to solve this problem. To promote his plan, Reuvens spoke about how the interest of the fatherland was at stake and posed the rhetoric question: "Why should Dutch' purses and Dutch' love of one's country not be able to put this idea into practice?"[48]

Reuvens's recommendation of 1832 that taking care of archaeological sites and objects should and could be primarily a state-regulated civil obligation did not have any effect. His motivation was very much in line with the humanistic social critique he had formulated in the Netherlands, which was likewise without much success. With regard to the Netherlands he had repeatedly spoken of the decline of cultural life, and the dominance of a mercantile spirit, which could only be stopped by a good classical education and state investments in archaeological research and collections.[49] By expanding his social critique to the Dutch East Indies, Reuvens ironically partly recycled Raffles's statements of 1817 about the Dutch narrow-minded policy toward Javanese antiquities. Subsequently, in 1840, Governor-General C.S.W. van Hogendorp commissioned a new archaeological regulation. However, it was a letter from Baud, who at that time was minister for the colonies in The Hague, that initiated this regulation. In this letter Baud announced the visit of the French archaeologist Ernest De Sancigny to Java. He then realized that he urgently needed to take measures. The new regulation of 1840 officially prohibited the export of antiquities from Java without the permission of the governor-general, and it obliged local authorities to make lists of the antiquities in their region. In case of archaeological discoveries they had to report them to the governor-general.[50]

Following this regulation, in 1842, Governor-General Pieter Merkus authorized the Batavia Society to ask the regional authorities in Java to send, on the basis of these lists made in 1840, archaeological objects to Batavia. There was one restriction: the transportation of the objects should not interfere with indigenous appropriations of these objects. Once in Batavia, the objects would become part of the museum of the Batavia Society and as such they were described as "national property."[51] This sudden use of the category "national property" implies that the concern for archaeology at that moment was not seen anymore as a state-regulated civil responsibility, like Reuvens had suggested, but as a national obligation of the—internally expanding—colonial state.

Concluding Remarks

In the first half of the nineteenth century, there was a strong interconnection between the regime changes in Java and the Netherlands, the rivalries between the colonizing powers and the archaeological activities in Java and the Netherlands. In the context of military expeditions, governmental reforms and processes of state formation, the tasks of the preserving of archaeological sites and the creation of archaeological collections increasingly came to be regarded as an obligation of the colonial state, and this would turn out to be an ideal to stay. In the early nineteenth century, in the interactions between Java, Great Britain, and the Netherlands, we see the start of a process in which archaeological sites and objects in Java became part of a state-centered heritage discourse. As a result it became self-evident that the archaeological sites of Java, after their "discovery," had to be safeguarded and restored by the colonial state, while objects from these sites were to be collected by this same state as "national property."

However, the different colonial officials would never have a unified "heritage"-monopoly. Not only did they encounter local engagements with these sites, the officials themselves were not univocal in the ways they appreciated and interpreted sites and objects. Raffles used the drawings of these temples in his *The History of Java* as a metaphor for the "degeneration" of the Javanese society and as a means of critique directed toward the former Dutch misrule. Engelhard on the other hand regarded archaeological activities primarily as a means to connect to the great but disappeared people who, so he presumed, once—before the contemporary Javanese people—inhabited the island. Against this background it is revealing that the engravings of the Prambanan that Raffles used in 1817 were based on drawings made by Cornelius in 1807 when he still worked for Engelhard. On one of the original drawings, Cornelius had depicted how Wardenaar, Van der Geugten, and Dubois measured the temple with the help of Javanese workers who were removing vegetation and stones. The engraving in Raffles's book does neither show the Dutch officials nor the Javanese workmen. It depicts the ruined temple full of vegetation with—as staffage—a few Javanese people standing and sitting next to it.[52] As such it represents primarily a romantic, exotic place filled with passivity and decay.[53]

In the course of the nineteenth century and early twentieth century, the Dutch colonial government made herself known to the outside world as a benevolent ethical ruler that took good care of Javanese Hindu and Buddhist remains of the past. Notwithstanding the Orientalist perceptions of the Javanese culture and the past, keeping objects at its original location and connecting archaeological research with the interest of local people, elites, and rulers became important topics. But at the same time the colonial state also created new ruins, especially during its attacks on the cities and palaces of so-called rebellious, despotic principalities in the Indonesian archipelago. There is an intrinsic connection between the taking care of archaeological sites with Hindu-Buddhist temples by the colonial state, and the destruction of the contemporary sites of indigenous power, which often had a strong religious—Islamic—connotation. It might explain why the Hindu-Buddhist archaeological sites and not the kratons of the contemporary principalities (often connected to Islamic sites like holy graves or mosques) became the "regalia" of the colonial state.[54]

Notes

1. For some general historical overviews of depictions of ruins and their cultural meaning (mostly concentrating on the Western world), see among others: Marcello Barbanera, "Metamorfosi delle rovine e identià cultural," in *Relitti rilitti. Metamorfosi delle rovine e identià culturale,* ed. Marcello Barbanera (Torino: Bollati Boringhieri, 2009), 15–88; Jeannot Simmen, *Ruinen Faszination in der Graphik vom 16. Jahrhunderd bis in die Gegenwart* (Dortmund: Harenberg, 1980); Carolyn Springer, *The Marble Wilderness: Ruins and Representation in Italian Romanticism* (Cambridge: Cambridge University Press, 1987); Paul Zucker, *Fascination of Decay. Ruins: Relic—Symbo—Ornament* (Ridgewood: Gregg Press, 1968).
2. "Eenzaam en verlaten is het thans hier, waar vroeger zoveel leven en beweging heerschte; waar machtige vorsten tegen iederen vijandelijken aanval door de natuur beveiligd, met verachting konden neerzien op hunnen onderdanen, die in de vlakten noord- en zuidwaards op hun bevel trotsche tempelgebouwen deden verrijzen, naar zij meenden onvergankelijke getuigen van hun roem en grootsheid." See Jan W. IJzerman, *Beschrijving der Oudheden nabij de grens der residenties Soerakarta en Djogdjakarta* (Batavia: Landsdrukkerij, 1891), 112; cf. Maarten Kuitenbrouwer, *Tussen oriëntalisme en wetenschap. Het Koninklijk Instituut voor Taal-, Land- en Volkenkunde in historisch verband 1851–2002* (Leiden: KITLV Uitgeverij, 2001), 152.

3. Goran Blix, *From Paris to Pompeii: French Romanticism and the Cultural Politics of Archaeology* (Philadelphia: University of Pennsylvania Press, 2009), 159–60.
4. Thomas S. Raffles, *The History of Java,* vol. 2 (London: Black, Parbury and Allen, 1817), 5–6.
5. For an analysis of the archaeological activities in Java between 1800 and 1850 with a stronger focus on indigenous appropriations, see: Marieke Bloembergen and Martijn Eickhoff, "A Wind of Change on Java's Ruined Temples: Archaeological Activities, Imperial Circuits and Heritage Awareness in Java and the Netherlands (1800–1850)," *BMGN/Low Countries Historical Review* 128,1 (2013): 81–104; Pauline Lunsingh Scheurleer, "Collecting Javanese Antiquities: The Appropriation of a Newly Discovered Hindu-Buddhist Civilisation," in *Colonial Collections Revisited,* ed. Pieter ter Keurs (Leiden: Research School CNWS, 2007), 71–114.
6. Benedict Anderson, *Imagined Communities: Reflections on the Origin and Spread of Nationalism,* rev. ed. (London, New York: Verso, 1991),184. See also Marieke Bloembergen and Martijn Eickhoff, "Conserving the Past, Mobilising the Indonesian Future: Archaeological Sites, Regime Change and Heritage Politics in Indonesia in the 1950s," *Bijdragen tot de Taal-, Land- en Volkenkunde* 167, 4 (2011): 405–36.
7. Bruce C. Trigger, "Alternative Archaeologies: Nationalist, Colonialist, Imperialist," *Man. The Journal of the Royal Anthropological Institute* 19, 3 (1984): 360–63.
8. See for example, Marieke Bloembergen and Martijn Eickhoff, "Colonial Archaeological Heroes Revisited: Post-colonial Perspectives on the 'Discovery' of the Prehistoric Past of the Dutch East Indies," *eTopoi. Journal for Ancient Studies,* http://journal.topoi.org/index.php/etopoi, in print.
9. Bloembergen and Eickhoff, "A Wind of Change."
10. Peter Rietbergen, "Oriëntalisme: een theorie van ficties—de fictie van een theorie? Een poging tot contextualisering en herinterpretatie," *Tijdschrift voor Geschiedenis* 111 (1998): 577.
11. Holger Hoock, *Empires of the Imagination: Politics, War and the Arts in the British World, 1750–1850* (London: Profile Books, 2010), 380–85; cf. Brenda Deen Schildgen, *Heritage or Heresy: Preservation and Destruction of Religious Art and Architecture in Europe* (New York: Palgrave Macmillan, 2008), 121–32.
12. For Reuvens, see Johannes A. Brongers, *Een vroeg begin van de moderne archeologie: Leven en werken van Cas Reuvens (1793–1835). Documentatie van een geleerden leven*, Nederlandse archeologische rapporten 23 (Amersfoort: ROB, 2002); Erich H. P. Cordfunke et al., eds, *'Loffelijke verdiensten van de archeologie': C.J.C. Reuvens als grondlegger van de moderne Nederlandse archeologie* (Hilversum: Verloren, 2007); Ruurd B. Halbertsma, *Scholars, Travellers and Trade: The Pioneer Years of the National Museum of Antiquities in*

Leiden, 1818–1840 (London: Routledge, 2003); Mirjam Hoijtink, *Caspar Reuvens and the Museums of Antiquities in Europe, 1800–1840* (Turnhout: Brepols, 2012).

13. Jan A. Somers, *Nederlandsch-Indië: Staatkundige ontwikkelingen binnen een koloniale relatie* (Zutphen, Walburg Pers, 2005), 81–106.
14. Demetrius C. Boulger, *The Life of Sir Stamford Raffles* (London: Knight, 1897), 177. For the "reanimation," see Hans Groot, *Van Batavia naar Weltevreden: Het Bataviaasch Genootschap van Kunsten en Wetenschappen, 1778–1867* (Leiden: KITLV Uitgevers, 2009), 157–84.
15. For the Javanese policies of Raffles, see John S. Bastin, *The Development of Raffles' Ideas on the Land Rent System In Java and the Work of the Mackenzie Land Tenure Commission.* PhD diss., University of Leiden, 1954; John Bastin, *The Native Policies of Sir Stamford Raffles in Java and Sumatra: An Economic Interpretation* (Oxford: Clarendon Press, 1957).
16. For this questionnaire, see Nicolaas J. Krom, "Engelhard over de Javaansche Oudheden," *Bijdragen tot de Taal-, Land- en Volkenkunde van Nederlandsch Indië* 76 (1929): 435–48. For Mackenzie's collecting activities in Java, see Seda Kouznetsova, "Colin Mackenzie as Collector of Javanese Manuscripts and Manuscript BL MSS JAV. 29," *Indonesia and the Malay World* 36–106 (2008): 375–94; Hoock, *Empires of the Imagination,* 324–25; Groot, *Van Batavia naar Weltevreden,* 168.
17. Colin Mackenzie, "Narrative of a journey to examine the remains of an ancient city and temples at Brambana in Java," *Verhandelingen van het Bataviaasch Genootschap der Kunsten en Wetenschappen* 7 (1814): ix, 1–53.
18. Thomas S. Raffles, "A Discourse delivered at a Meeting of the Society of Arts and Sciences in Batavia, on the Twenty-fourth day of April 1813, being the Anniversary of the Institute," *Verhandelingen van het Bataviaasch Genootschap der Kunsten en Wetenschappen* 7 (1814): 3, 14, 34. See also: Groot, *Van Batavia naar Weltevreden,* 173–74.
19. Nicolaas J. Krom, *Inleiding tot de Hindoe-Javaansche Kunst I* (The Hague: Martinus Nijhof, 1923), 6.
20. For Raffles, the "degenerate Javan" and the Dutch "tyranny" on Java, see Sarah Tiffin, "Raffles and the Barometer of Civilisation: Images and Descriptions of Ruined Candis in The History of Java," *JRAS* 3–18–3 (2008): 345, 355, 359. For Raffles's ideas on Hinduism/Buddhism and Islam, see Syed M. K. Aljunied, "Sir Thomas Stamford Raffles' Discourse on the Malay World: A Revisionist Perspective," *Sojourn* 20, 1 (2005): 1–22.
21. Raffles, *The History of Java,* vol. 2, 6.
22. MacKenzie to Raffles, April 14, 1813, British Library, London, Colonial Office, Mss Eur F 148/47, 1.

23. Dagregister van den reis van de Gouverneur Nicolaus Engelhard naar Sourakarta en Djogjakarta, 1802—Arsip Karesidenan Yogyakarta 1724–1891 nr. 348, Arsip Nasional Republik Indonesia, Jakarta. For Engelhard, see Frederik de Haan, "De Historie van een Oudgast," *Tijdschrift voor Indische Taal-, Land- en Volkenkunde* 18 (1901): 195–225. For his visit to the courts, see Jean H. M. Kommers, *Besturen in een onbekende wereld: Het Europese binnenlandse bestuur in Nederlands-Indië: 1800–1830. Een Antropologische studie*. Vol. I. PhD diss., Katholieke Universiteit Nijmegen, 1979, 70–74.
24. The authors thank Sri Margana for identifying these locations.
25. John Bastin and Pauline Rohatgi, *Prints of Southeast Asia in the India Office Library. The East India Compagny in Malaysia and Indonesia 1786–1824* (London: Her Majesty's Stationery Office, 1979), 168; Krom, *Inleiding tot de Hindoe-Javaansche Kunst,* 5.
26. For the discovery of the Singasari temple, see Nandana Chutiwongs, "Çandi Singasari—A Recent Study," in *Interpreting Southeast Asia's Past: Monument, Image and Text,* ed. Elisabeth A. Bacus, Ian C. Glover, and Peter D. Sharrock (Singapore: NUS Press, 2008), 101; cf.Jessy Blom, *The Antiquities of Singasari* (Leiden: Burgersdijk & Niermans 1939), 7–11; Jan L. A. Brandes, Hendrik L. Leydie Melville, and J. Knebel eds., *Beschrijving van Tjandi Singasari: en De Wolkentoneelen van Panataram*. Archeologisch Onderzoek op Java en Madura II. The Hague: Nijhoff, 1909, 1–3, 53–56.
27. 'Zondag den 29e Augustus', page 268, Dagregister van den reis van de Gouverneur Nicolaus Engelhard naar Sourakarta en Djogjakarta, 1802—Arsip Karesidenan Yogyakarta 1724–1891 nr. 348, Arsip Nasional Republik Indonesia, Jakarta.
28. Krom, "Engelhard over de Javaansche Oudheden," 440.
29. Tiffin, "Raffles and the Barometer of Civilisation," 357.
30. Engelhard was not the only "victim." In 1816 Raffles left Java with a collection of manuscripts to be published in volume 9 of the *Verhandelingen van het Bataviaasch Genootschap der Kunsten en Wetenschappen* that was to be printed in London. When they received the volume, the members of the Batavia Society were, according to Groot, "bitter," as many of the manuscripts were not printed. Although there is no proof, as vital archival documents are missing, it seems clear that parts of the manuscripts were printed in *The History of Java*. See: Groot, *Van Batavia naar Weltevreden,* 175–77.
31. Raffles, *The History of Java,* vol. 2, 41, 55.
32. Nicolaas J. Krom, *De Sumatraansche Periode der Javaansche Geschiedenis. Rede uitgesproken bij zijn Ambtsaanvaarding als Buitengewoon Hoogleraar aan de Rijksuniversiteit te Leiden, op 3 december 1919* (Leiden: Brill, 1919), 3.
33. For the history of these two statues and their nineteenth-century depictions, see the schedule in Brandes, Leydie Melville, and Knebel, *Beschrijving van Tjandi,* 27.

34. Rudolf A. H. R. Effert, *Volkenkundig verzamelen: het Koninklijk Kabinet van Zeldzaamheden en het Rijks Ethnografisch Museum 1816–1883*, PhD diss., University of Leiden, 2003, 185; Groot, *Van Batavia naar Weltevreden*, 192–93.
35. Krom, *De Sumatraansche*, 4–5; cf. Gerret P. Rouffaer, "Monumentale kunst op Java," *De Gids* 4, 9 (1901): deel II, 234.
36. Krom, *Inleiding tot de Hindoe-Javaansche Kunst I*, 6.
37. Ibid., 336; Nicolaas J. Krom, *Baraboedoer: Het heiligdom van het Boeddhisme op Java* (Amsterdam: H. J. Paris, 1930), 31.
38. Krom, *Inleiding tot de Hindoe-Javaansche Kunst I*, 335–36.
39. Martijn Eickhoff, "Archeologisch erfgoed: een onbeheersbaar concept," in *Erfgoed: De geschiedenis van een begrip*, ed. F. Grijzenhout (Amsterdam: AUP, 2007), 233–39; cf. Martijn Eickhoff, "C. J. C. Reuvens als erflater: Twee eeuwen 'genealogieën' van de Nederlandse archeologie," in *'Loffelijke verdiensten van de archeologie': C.J.C. Reuvens als grondlegger van de moderne Nederlandse archeologie*, ed. Erich H. P. Cordfunke et al. (Hilversum: Verloren, 2007), 139–41.
40. Caspar J. C. Reuvens, *Verhandeling over drie groote steenen beelden in den jare 1819 uit Java naar den Nederlanden overgezonden (1824)* Gedenkschriften in de hedendaagsche talen van der derde klasse van het Koninklijk Nederlandsch Instituut van Wetenschappen, Letterkunde en Schone Kunsten: Deel III (Amsterdam; Pieper & Ipenbuur, 1826), i–vii, 1–223.
41. Raffles, *The History of Java*, vol. 2, 41.
42. Reuvens, *Verhandeling over drie groote steenen*, 11.
43. Ibid., 13.
44. Engelhard aan Reuvens, Buitenzorg, 28–2–1827, BPL 885, Bijzondere collecties, Universiteit Leiden.
45. Krom, "Engelhard over de Javaansche Oudheden," 436. See also Kommers, *Besturen in een onbekende wereld*, 112–15.
46. 29–8–1832, Memorie ter bevordering der Javaansche Oudheid-Kunde, KBG DIR 0093, Arsip Nasional Republik Indonesia, Jakarta.
47. Ibid. For an English translation of parts of the version of the memorandum kept in the Rijksmuseum van Oudheden in Leiden, see Halberstma, *Scholars, Travellers and Trade*, 37–39. Whereas Halbertsma uses "misconduct," we have chosen "conduct" as it comes closer to the word "handelswijze" used by Reuvens in the version of the memorandum kept in the National Archive in Jakarta.
48. Ibid. See also Groot, *Van Batavia naar Weltevreden*, 261–63.
49. Eickhoff, "Archeologisch erfgoed," 237–38.
50. Bt 29 December 1840, no. 13, Algemene Secretarie, Arsip Nasional Republik Indonesia, Jakarta. See also Groot, *Van Batavia naar Weltevreden*, 308.
51. Bt 3 December 1842, no. 18, kept in: KBG Dir 1509, Arsip Nasional Republik Indonesia, Jakarta. See also Groot, *Van Batavia naar Weltevreden*, 310.

52. There are at least three copies of the drawing, kept in the Raffles collection of the British Museum and in the Horsfield collection of the India Office library, see Bastin and Rohatgi, *Prints of Southeast Asia*, 168.
53. Cf. "Java's Ruined Chandi's and the British Picturesque Ideal," *Bulletin of SOAS* 72, 3 (2009): 553.
54. Anderson, *Imagined Communities*, 155–85.

Bibliography

Aljunied, Syed M. K."Sir Thomas Stamford Raffles' Discourse on the Malay World: A Revisionist Perspective." *Sojourn* 20, 1 (2005): 1–22.

Anderson, Benedict. *Imagined Communities: Reflections on the Origin and Spread of Nationalism*. Revised edition. London, New York: Verso, 1991.

Barbanera, Marcello. "Metamorfosi delle rovine e identià cultural." In *Relitti rilitti. Metamorfosi delle rovine e identià culturale*, edited by Marcello Barbanera, 15–88. Torino: Bollati Boringhieri, 2009.

Bastin, John S. *The Development of Raffles' Ideas on the Land Rent System In Java and the Work of the Mackenzie Land Tenure Commission*. PhD diss., University of Leiden, 1954.

———. *The Native Policies of Sir Stamford Raffles in Java and Sumatra: An Economic Interpretation*. Oxford: Clarendon Press, 1957.

Bastin, John S. and Pauline Rohatgi. *Prints of Southeast Asia in the India Office Library: The East India Company in Malaysia and Indonesia 1786–1824*. London: Her Majesty's Stationery Office, 1979.

Bloembergen, Marieke, and Martijn Eickhoff. "Conserving the Past, Mobilising the Indonesian Future: Archaeological Sites, Regime Change and Heritage Politics in Indonesia in the 1950s." *Bijdragen tot de Taal-, Land- en Volkenkunde* 167, 4 (2011): 405–36.

———. "A Wind of Change on Java's Ruined Temples: Archaeological Activities, Imperial Circuits and Heritage Awareness in Java and the Netherlands (1800–1850)," *BMGN* (2013): 81–104.

Blix, Goran. *From Paris to Pompeii: French Romanticism and the Cultural Politics of Archaeology*. Philadelphia: University of Pennsylvania Press, 2009.

Blom, Jessy. *The Antiquities of Singasari*. Leiden: Burgersdijk & Niermans, 1939.

Boulger, Demetrius C. *The Life of Sir Stamford Raffles*. London: Knight, 1897.

Brandes, Jan L. A., Hendrik L. Leydie Melville, and J. Knebel, eds. *Beschrijving van Tjandi Singasari: en De Wolkentoneelen van Panataram*. Archeologisch Onderzoek op Java en Madura II. The Hague: Nijhoff, 1909.

Brongers, Johannes A. *Een vroeg begin van de moderne archeologie: Leven en werken van Cas Reuvens (1793–1835): Documentatie van een geleerden leven*. Nederlandse archeologische rapporten 23. Amersfoort: ROB, 2002.

Chutiwongs, Nandana. "Çandi Singasari—A Recent Study." In *Interpreting Southeast Asia's Past. Monument, Image and Text*, edited by Elisabeth A. Bacus, Ian C. Glover, and Peter D. Sharrock, 100–21. Singapore: NUS Press, 2008.

Cordfunke, Erich H. P. et al., eds. *'Loffelijke verdiensten van de archeologie': C. J. C. Reuvens als grondlegger van de moderne Nederlandse archeologie.* Hilversum: Verloren, 2007.

Deen Schildgen, Brenda. *Heritage or Heresy: Preservation and Destruction of Religious Art and Architecture in Europe.*New York: Palgrave Macmillan, 2008.

Eickhoff, Martijn. "Archeologisch erfgoed: een onbeheersbaar concept." In *Erfgoed: De geschiedenis van een begrip*, edited by F. Grijzenhout, 231–36. Amsterdam: AUP, 2007.

———. "C. J. C. Reuvens als erflater: Twee eeuwen 'genealogieën' van de Nederlandse archeologie." In *'Loffelijke verdiensten van de archeologie': C. J. C. Reuvens als grondlegger van de moderne Nederlandse archeologie,* edited by Erich H. P. Cordfunke et al., 135–47. Hilversum: Verloren, 2007.

Effert, Rudolf A. H. R. *Volkenkundig verzamelen: het Koninklijk Kabinet van Zeldzaamheden en het Rijks Ethnografisch Museum 1816–1883.* PhD diss., University of Leiden, 2003.

Groot, Hans. *Van Batavia naar Weltevreden: Het Bataviaasch Genootschap van Kunsten en Wetenschappen, 1778–1867.* Leiden: KITLV Uitgevers, 2009.

De Haan, Frederik. "De Historie van een Oudgast." *Tijdschrift voor Indische Taal-, Land- en Volkenkunde* 18 (1901): 195–225.

Halbertsma, Ruurd B. *Scholars, Travellers and Trade: The Pioneer Years of the National Museum of Antiquities in Leiden, 1818–1840.* London: Routledge, 2003.

Hoijtink, Mirjam. *Caspar Reuvens and the Museums of Antiquities in Europe, 1800–1840.* Turnhout: Brepols, 2012.

Hoock, Holger. *Empires of the Imagination: Politics, War and the Arts in the British World, 1750–1850.* London: Profile Books, 2010.

IJzerman, Jan W. *Beschrijving der Oudheden nabij de grens der residenties Soerakarta en Djogdajkarta.* Batavia: Landsdrukkerij, 1891.

Kommers, Jean H. M. *Besturen in een onbekende wereld: Het Europese binnenlandse bestuur in Nederlands-Indië: 1800–1830: Een Antropologische studie.* Vol. I. PhD diss., Katholieke Universiteit Nijmegen, 1979.

Kouznetsova, Seda. "Colin Mackenzie as Collector of Javanese Manuscripts and Manuscript BL MSS JAV. 29," *Indonesia and the Malay World* 36–106 (2008): 375–394.

Krom, Nicolaas J. *De Sumatraansche Periode der Javaansche Geschiedenis: Rede uitgesproken bij zijn Ambtsaanvaarding als Buitengewoon Hoogleraar aan de Rijksuniversiteit te Leiden, op 3 december 1919.* Leiden: Brill, 1919.

———. *Inleiding tot de Hindoe-Javaansche Kunst.* The Hague, 1923.

———. "Engelhard over de Javaansche Oudheden." *Bijdragen tot de Taal-, Land- en Volkenkunde van Nederlandsch Indië* 76 (1929): 435–48.

———. *Baraboedoer: Het heiligdom van het Boeddhisme op Java*. Amsterdam: H. J. Paris, 1930.

Kuitenbrouwer, Maarten. *Tussen oriëntalisme en wetenschap: Het Koninklijk Instituut voor Taal-, Land- en Volkenkunde in historisch verband 1851–2002*. Leiden: KITLV Uitgeverij, 2001.

Lunsingh Scheurleer, Pauline. "Collecting Javanese Antiquities: The Appropriation of a Newly Discovered Hindu-Buddhist Civilisation." In *Colonial Collections Revisited*, edited by Pieter ter Keurs, 71–114. Leiden: Research School CNWS, 2007.

Mackenzie, Colin. "Narrative of a journey to examine the remains of an ancient city and temples at Brambana in Java." *Verhandelingen van het Bataviaasch Genootschap der Kunsten en Wetenschappen* 7 (1814): ix, 1–53.

Raffles, Thomas S. "A Discourse delivered at a Meeting of the Society of Arts and Sciences in Batavia, on the Twenty-fourth day of April 1813, being the Anniversary of the Institute." *Verhandelingen van het Bataviaasch Genootschap der Kunsten en Wetenschappen* 7 (1814): 1–34.

———. *The History of Java*. 2 vols. London: Black, Parbury and Allen, 1817.

Reuvens, Caspar J. C. *Verhandeling over drie groote steenen beelden in den jare 1819 uit Java naar den Nederlanden overgezonden (1824)*. Gedenkschriften in de hedendaagsche talen van der derde klasse van het Koninklijk Nederlandsch Instituut van Wetenschappen, Letterkunde en Schone Kunsten. Deel III. Amsterdam; Pieper & Ipenbuur, 1826), i–vii, 1–223.

Rietbergen, Peter. "Oriëntalisme: een theorie van ficties—de fictie van een theorie? Een poging tot contextualisering en herinterpretatie." *Tijdschrift voor Geschiedenis* 111 (1998): 545–75.

Rouffaer, Gerret P. "Monumentale kunst op Java." *De Gids* 4, 9 (1901): deel II, 225–52.

Simmen, Jeannot. *Ruinen Faszination in der Graphik vom 16. Jahrhunderd bis in die Gegenwart*. Dortmund: Harenberg, 1980.

Somers, Jan A. *Nederlandsch-Indië: Staatkundige ontwikkelingen binnen een koloniale relatie*. Zutphen: Walburg Pers, 2005.

Springer, Carolyn. *The Marble Wilderness: Ruins and Representation in Italian Romanticism*. Cambridge: Cambridge University Press, 1987.

Tiffin, Sarah. "Raffles and the Barometer of Civilisation: Images and Descriptions of Ruined Candis in The History of Java," *JRAS* 3–18–3 (2008): 241–360.

———. "Java's Ruined Chandi's and the British Picturesque Ideal." *Bulletin of SOAS* 72–3 (2009): 553.

Trigger, Bruce C. "Alternative Archaeologies: Nationalist, Colonialist, Imperialist *Man*." *The Journal of the Royal Anthropological Institute* 19, 3 (1984): 355–70.

Zucker, Paul. *Fascination of Decay. Ruins: Relic—Symbol—Ornament*. Ridgewood: Gregg Press, 1968.

Chapter 8

Meeting Point Deshima: Scholarly Communication between Japan and Europe up till around 1800

Peter Rietbergen

Introduction

Try to imagine their first, spectacular, view: the blue bay of Nagasaki, the green mountains enclosing it, and then, Deshima—the artificial "island De." Try to imagine the thoughts of a Dutchman arriving there around the turn of the eighteenth century, or, for that matter, the thoughts of a Frenchman, a German, or a Swede: from the early seventeenth to the early nineteenth centuries, all Europe served the Dutch East India Company in Japan, either for profit or for knowledge.[1]

However, if we consider the situation around the turn of the century, we should first of all realize that few ships from Holland arrived at Deshima in those years, since communications with Europe were erratic, due to political problems in the Dutch Republic since 1795 and, a few years later, due to the imposition of Napoleon's continental system on Dutch overseas trade. Moreover, even in the late 1790s, the VOC authorities in Batavia had no ships to annually bring in goods and people from the Indies to Japan, as had been Company practice since the early 1600s. To continue to do so, they had to hire American vessels.

Inevitably, the British occupation of Java from 1811 to 1816 brought contacts to a complete standstill: the Company's "opperhoofden" in Japan, the directors of the Deshima trade, refused to acknowledge London's pretended political authority over all the Company's Asian possessions, the more so since, of course, they had feared and resisted British commercial competition for centuries already. For the same reason, and at the same time, these men were warding off Russian efforts to gain a foothold in the island empire, the same way as they had in the early seventeenth century worked to reduce and finally eliminate the influence of the Portuguese.

In all this, the "opperhoofden" were helped by the Japanese themselves. Admittedly, the strict seclusion policy—termed *Sakoku* by Tadao Shizuku in his famous 1801-treatise—which had been already in practice in the 1630s and was strictly imposed by the government at Edo from the 1670s onwards, had been moderated since the mid-eighteenth century, but in the 1790s it was intensified again.

In short, while for nearly two centuries a Dutch ship had been arriving at Deshima almost every year, precisely the years around 1800 represented an all-time low in direct and even indirect contacts between Europe and Japan. Only from 1817 onward, the commercial and cultural relationship that had, for two hundred years, been at the basis of a unique geography of knowledge in the widest sense of the word,[2] once more linked the world of the West to its farthest East through the tiny VOC-settlement on Deshima.

Even so, my first question remains a valid one. What would have been the thoughts of a VOC-servant—who, since the Company's bankruptcy in 1799, had become an official of the Dutch state? Well, I feel he would have wondered if the financial gain he expected from being posted to Japan did after all outweigh the dreary prospect of staying—for an entire year or more—on the tiny island that now came into view: a very small, almost prison-like place, really, since opportunities to actually leave Deshima were few, and traveling through Japan with any degree of freedom was completely impossible.[3]

However, if my VOC-servant cherished any expectations of his stay beyond stuffing his purse and his pocket, his dearest wish might well be to join the annual embassy to the shogunal court in Edo.

This journey, the so-called *Hof-Reis*, and the subsequent ceremonial reception of the Company's ambassador at the hugely impressive Edo castle was, of course, the way in which the Company paid its (extremely costly) tribute to the Tokugawa-shogun, Japan's de facto ruler, in order to get permission to continue its trade. Recently, I have stressed that we tend to forget that the *Hof-Reis* served a very central need in Japanese political representation: in forcing the Dutch to make this long, cumbersome, and expensive overland trip—rather than coming to Edo by ship—the *bakufu*, the shogunal government, showed its subjects along the country's much-frequented main roads its power to dominate and manipulate these red-haired barbarians. In doing so, it propagandistically enhanced its authority, which, one has to remember, remained shaky during the first decades of the seventeenth century—viz. the Shimabara revolt in the 1630s—and even in later times was not unchallenged either.[4]

Within the context of questions about the transfer and transformation of knowledge between the metropolitan centers in Europe and their commercial/colonial peripheries around the turn of the eighteenth century, it has to be stressed that the *Hof-Reis* was the one way in which the Dutch—and, thus, the world of Europe—were able to gain at least a modicum of knowledge about the mysterious country that saw itself as the "Abode of Amaterasu" and named itself *Ni-hon*, the "Foundation of the Sun."

I would like to add that as far as knowledge about Japan's mighty neighbor, China, was concerned, the situation really was no better. Since 1773, when the authorities of the Church of Rome had decided to dissolve the Society of Jesus—in part to counter the Jesuits' policy of religious-cultural accomodation or, even, syncretism—Europe had had to do without its major informants about life in China though, of course, some information did travel westward via the European factories in Canton. Nor were there any annual embassies to the imperial court in Beijing, enabling scholars to at least traverse the Middle Kingdom from south to north, in the manner of the Japanese *Hof-Reis*. Indeed, the first British embassy, the Macartney mission of 1793, was a failure, and the one undertaken by the VOC in 1795—the first since their abortive mid-seventeenth-century efforts to secure a permanent foothold in China—fared little better, though one of the ambassadors, the scholarly inclined Isaac Titsingh, a former envoy to Japan, was enthusiastic about the reception he received in Beijing and the insights he gained there.[5]

A Mutual Relationship?

It has to be stressed that the VOC's position in Japan was in no way a colonial or imperial one—in whatever way one wants to define these two concepts. Nor, however, did Japanese-Dutch communication on Deshima, and in Nagasaki and Edo, constitute an open exchange. Indeed, the situation was in many ways unique.

We should realize that till the mid-nineteenth-century, Europe would have had no knowledge about Japan if Nihon, after having expelled all other Europeans, had not decided to allow the *Oranda*, the Dutch, to settle on Deshima, and, moreover, had allowed them to remain there long after the economic imperative of the relationship had become negligible. Notwithstanding periods of increased mistrust, for the better part of two centuries, Nihon's rulers retained their policy of using the VOC to gain some information about the outside world, as well as acquire "exotic" objects that provided status because they were produced by that same, strange world. They desired to do so despite the fact that, traditionally, their knowledge of the world came via China, via the Chinese community in Nagasaki, where, it has to be added, imported books were screened as thoroughly as, or perhaps even more so, than the books that now started to arrive via Deshima. Whatever the Tokugawa's reasons—which, it has to be added, were not always shared by all Edo policy makers—it was precisely this decision that enabled the Company to serve as a channel through which Europe gained a vision of Japan.

Consequently, the case of Deshima provides us with the one situation in which the Company acted as an active agent in the field of scholarly or learned communication, albeit only because, in this specific instance, it could not well do otherwise if it wanted to realize the gains offered by the trade with Japan: silver, silk, and other precious wares. Even so, the VOC by and large invested only in the transfer of European knowledge to Japan rather than the other way round. Mostly, it did leave its officials to decide what, if anything, they were to make of the wondrous worlds they were allowed to visit in the Company's service. However, there were exceptions to this rule, as shown by the *nihil obstat* obviously given by the Company's Dutch directorate to the publication of the Japan-texts of François Caron and Arnoldus Montanus, in the late 1640s and late 1660s; these texts evidently served to improve the image of the VOC in European eyes, which tended to see the Company only as a money-grabbing entreprise.

By the end of the seventeenth century, however, the Deshima operation probably cost more than its material profits warranted. Yet, though they did no longer actively stimulate the dissemination of knowledge about Japan, during the eighteenth century the Gentlemen XVII seem to have felt that their continued cultural prestige in Europe still necessitated retaining their expensive establishment in Japan.

Intermediaries

Of the three groups who were the main agents in the information and knowledge transfer between the VOC-servants and Japan, arguably the first were the 30-something families/clans of the so-called Dutch interpreters of Nagasaki, the *Oranda tsuji*. They originated in, respectively, those Japanese who had assisted the Company while it still resided in Hirado in the 1620s, and in the men who, after first having served the Portuguese, retrained in Dutch after the Lusitanians had been forced to leave, and the VOC was, with almost equal force, urged to move to Nagasaki. The interpreters' hold over the contact situation was almost complete since, at least formally, the Dutch were forbidden to learn Japanese. Moreover, they had a hereditary and therefore vested interest in manipulating the relationship, both for reasons of material gain and of sociopolitical and cultural status. Partly in consequence of this situation, the Company considered the group to be greedy and untrustworthy, often calling its members incompetent in the very language they were supposed to translate. Precisely the few exceptions, however, to a large extent determined what knowledge of things Japanese those VOC-men who aspired thereafter were able to acquire.

The second group—in part an offspring of the first—were, of course, the *rangakusha*, the "Dutch Specialists," who operated in Nagasaki as well as, from the 1720s onward, in Edo. Other Japanese intellectuals—both from the daimyo- and samurai-class and, increasingly, from the bourgeoisie—used *rangaku*, "Dutch knowledge," as far as it suited their own curiosity or other interests. It is fascinating to note that from the late eighteenth century onward, the Edo-based "Dutch scholars," now posing as an expert professional body, systematically tried to denigrate the contribution to Dutch studies made by the Nagasaki men—mere, unenlightened, unscholarly interpreters.[6]

On the most basic level, the often forgotten third group consisted of the women with whom the VOC-men had sex, either occasionally or, as happened quite often, in common law marriages that, over the years, produced many Euro-Japanese children. Though the Company men were forced to take financial care of their offspring, given the cultural and moral views on both sides the life of these Eurasians was a complicated one although in some cases they seem to have made reasonable careers for themselves. However, whether they, and their mothers, came to constitute a structural link between successive generations of Company servants and Japanese society, thus facilitating and enhancing the transfer of culture, remains to be studied.

Japan in Europe: Knowledge Cumulative or Repetitive?

In analyzing, as I have done extensively some years ago,[7] what kind of information, or "knowledge," about Japan was transmitted to Europe via the VOC during the two hundred years of its existence there is one thing that I find striking: the lack of any cumulative effect, of scholarly progress, so to say.

It appears that since 1645 when François Caron published what was actually the first survey of Japan to reach Europe after the Portuguese travelogues of the late sixteenth century,[8] each successive generation did produce one or more authors who tried to repeat that feat—either by publishing works based on their own, first-hand knowledge, or by compiling texts from second-hand sources. None, however, used a new conceptual format or, indeed, added many new insights—the one exception being the Dutch polygraph Arnoldus Montanus. While he himself never visited Japan, he was given access to the Company's archives. His book, written in 1679, was, as I have shown elsewhere, a remarkable exercise in comparative cultural history.[9]

Things changed with the publication, in 1727, in London, of the *History of Japan*, by Johann Caspar Scheuchzer, the Swiss librarian of the famous Scottish collector Sir Hans Sloane. He edited a manuscript-text acquired by his master, which had been written in the 1690s by the late Engelbert Kaempfer, a one-time VOC-surgeon on Deshima—of Westphalian descent—and, moreover, a keen observer, analyst, and interpreter of the variegated cultures of South and East Asia. Scheuchzer, however, considerably altered the notes Kaempfer had made in Japan and, subsequently, during his retirement years in Germany. He also added other Japanese material from his master's extensive collection to Kaempfer's corpus.[10] Nor did he have any scruples in venting his own opinions about what he considered to be strange about Japan.

The Scheuchzer-Kaempfer book—of which the French and the Dutch versions appeared in 1729—not only was far more voluminous than any of its predecessors, it also was far more succesful, scholarly speaking. Indeed, in view of its systematic thoroughness and the many original sources it used and presented—including maps and other visuals collected by Kaempfer in Japan—it deservedly became the lodestar of Europe's views of and opinions on Japan for the next century. Perhaps it also succeeded in doing so because the ways in which Scheuchzer had rewritten Kaempfer's text did, actually, foreshadow the mood of the future—England's early eighteenth century, the so-called Augustan Age, enlightened and critical, to be sure, but also moralizing and judgmental—rather than reflect that of the past, Germany's late seventeenth century, the Baroque period characterized, perhaps, by a rather more wide-ranging but also less critical, yet more accepting, view of the world.

Meanwhile, it is astonishing to note that, during the entire seventeenth century, information about Japan that was transmitted to Europe reached ever-larger readerships. It is all the more astonishing when we realize that the number of first-hand accounts in which this information was contained was very limited—in a span of hundred years, no more than a dozen travelogues were published, and the really authoritative among them were very few indeed. Notwithstanding the scanty information—or, perhaps, precisely because of its scantiness—a wide variety of European texts now concerned itself with Japan.

Increasingly, the travel tales that Europeans wrote were (re-)issued in texts covering the entire known world, and these one- or, often, multivolume works now always included sections about Japan as well. Moreover, in the world of scholarship, learned monographs about such specific topics as geography, history, cultural customs, and so on often tended to adopt a comparative approach as well, in which the peculiar case of Japan figured with increasing prominence.

Inevitably, the new genre of the encyclopedia, too—created precisely to counter the unsettling effects of the proliferation of monographic scholarly literature—began to devote entries either to the country in its entirety or, even, to the various aspects of its history, religion, et cetera.

By and large, however, Japan remained a mystery, a world even stranger than, for example, China. Consequently, Japan also entered the realm of European literature. It figured in poems, in novels, and in childrens' stories. It became a setting for operas, plays, and other staged fantasies that catered to Europe's need for the exotic.[11]

In view of the many textual forms that now (re-)presented "Japan" to and in Europe, the intermediaries who used and, of course, transformed, not to say wilfully distorted, the images of Japan imparted by the publications that resulted from the Company's presence in Nihon, were a variegated lot, ranging from serious scholars to money-hungry hacks.

Meanwhile, the visual arts, too, showed a veritable "Japonaiserie." Indeed, the main halls of the sumptuous summer palace that August the Strong built on the banks of the Elbe at Pillnitz were decorated in a style people at the time thought to be "Japanese," though we now can see it was composed of elements of a fantasized, picture-book China. Actually, the paintings mostly derived from the illustrations of another highly popular VOC-text, the report in on the Company's 1655-embassy to Beijing written by Johan Nieuhof.[12]

Given the dearth of reliable texts on and images of Japan, one would have expected European readers to be eager to add new, updated information to what, by the 1730s, had become rather dated knowledge. However, despite, or perhaps because of, the success of Scheuchzer's *Kaempfer*-edition—which, of course, was already dated when it appeared, since Kaempfer had described Japan as he had seen it in the late 1680s—no new compendium of knowledge about Japan was published for many decades to come.

Another reason might well be that the Dutch East India Company was, after all, interested in commerce, primarily. It was not an "imperial agency" that had to gather as much useful data as possible to create the knowledge foundation necessary for the success of any imperial project. This, of course, only was the perspective of the Company's heirs, the nineteenth-century Dutch kings. Moreover, the few learned or cultured men who joined the VOC-staff on Deshima—mostly medical practitioners—may not necessarily have been widely read intellectuals who could not wait to gain fame with a new book on the forbidden empire.

Fact is that Carl Thunberg (1743–1828), the one person who, in the 1780s, undertook to write an up-to-date survey, while adding new material to the corpus amassed by Kaempfer, did not adopt a changed perspective. This, of course, is not to say that his book was not greeted with enthusiasm. Obviously, Kaempfer's text had been out of print for many years, and if only for that reason, a new study of Japan was welcome indeed.[13]

Thunberg was a young Swede who first studied botany with Linnaeus and then trained as a medical doctor. Upon his arrival in the Dutch Republic, where he was to complete his studies, he was introduced to members of the Amsterdam patriciate. Soon, a group of rich regents decided they wanted to sponsor a trip to the East Indies, where Thunberg should gather plants, curiosities, and information of all kinds imaginable. It was a situation that mirrored the one in which, for example, Alfred Russel Wallace traveled to Asia in the 1840s and 1850s, to work for wealthy British collectors. However, while he, in doing so, came to formulate the fundamentals of what became "the theory of evolution," Thunberg's accomplishments did not range that far. However, after a stay at the Cape, and on the island of Java, he joined the Deshima-factory in 1775 and remained in Japan for nearly one-and-a-half year, till December 1776. Both his contacts with the Nagasaki community and his trip to Edo provided him with the material for the two works that were to establish his scholarly fame.

Studying his return trip, it becomes clear to us not only who his patrons were, but also what kind of wider audience he was expected to appeal to. Arriving in London in October 1777, he was fêted by the members of the Royal Society, always keen on greeting scholars who returned from abroad with new information about exotic natures and cultures. On reaching Amsterdam in 1778, he was given a warm welcome by his sponsors. They now could show him the hundreds of curiosities he himself had shipped to the Netherlands during the previous years. Finally, in March 1779, Thunberg returned to Sweden again, with his scholarly luggage, containing coins, Chinese and Japanese books, and a large amount of seeds, dried plants, stones, and other specimens of natural history. The better part of this material ended up in the collections of the University of Uppsala, of which Thunberg was to be one of the most prominent professors during the following decades.

Beyond his reputation as a botanist, which he gained first and foremost through his *Flora Japonica*—published, with subsequent volumes, in Uppsala between 1784 and 1805—Thunberg's fame, certainly with the larger public, rested on his travelogue: the *Resa uti Europa, Afrika, Asia* (four volumes, Uppsala 1788–1794), and, more particularly, on the third and fourth volume of the book, in which he described his Japanese experiences. For precisely those 600 pages were widely read all over Europe. Although the German, English, and French editions of Thunberg's travels—published, respectively, in 1792 and 1794—contained material about his stay on the Cape and in the East Indies as well, for obvious commercial reasons he concentrated on the secrets of Japan.

Meanwhile, to cash in on the new interest in Japan occasioned by Thunberg's return, in 1780 a Dutch publisher hastened to translate his earlier 1779 and 1780 papers—written for the Swedish and the British Royal Society—on Japanese history and society,[14] resulting in a 40-something page booklet. In the following years, a Dutch translation of the full text of the *Resa* was apparently deemed unprofitable, probably because the English, French, and German versions that had appeared satisfied the Dutch market as well.

The entire Thunberg-episode shows, once again, to what extent all over Europe information gained through that one, privileged channel, the Dutch East India Company, was sought by an increasingly inquisitive, cosmopolitan society. At the same time, the very fact that it was through the VOC, only, that Europe could acquire knowledge about Japan inevitably was bemoaned by those dependent on it. Such late eighteenth-century "enlightened" critics as, for example, the Abbé Raynal did suggest that, perhaps, the time of monopoly companies was, or rather should be, over. Though the demise of the VOC was, unbeknownst to him, very near indeed, he did not realize that the other party involved—the Japanese authorities—as yet was not willing to "open" their country in an equally enlightened way.

As I indicated above, Thunberg's book followed the traditional ways of representing Japan. Indeed, his text, both as to its structure and its content, very much resembles the way Scheuchzer, 60 years before, had presented the diary and other fragments left by Kaempfer. To judge by Thunberg's notes to his own tale, he had found use for Kaempfer's information dozens of times, so much so that it is not always easy to determine where the German doctor's descriptions end and Thunberg's own observations begin.

Does this mean Thunberg plagiarized his predecessor(s)? Certainly not more so than most travel writers did—and indeed still do.[15] Has this diminished the impact of his text? Not really. The eighteenth century, as much as our own age, was a time wherein knowledge aged fast. Even in scholarly circles, questions long before answered were asked anew and, even more amazing, the same answers were produced with a sometimes baffling claim to originality. Moreover, the general public then reacted as it does now: a new title, especially one that was cleverly marketed, often was received as a new book—the fact that even a short visit to a well-stocked library might produce a number of other, earlier relevant titles on a given topic based on the same or even more extensive and authoritative information seems to have escaped many innocent or simply lazy readers.

What did Thunberg offer those who turned to him for information about Japan? First of all it is interesting to note that he allowed his own persona as author to enter the scene—a thing that men like Kaempfer or Montanus would not have thought appropriate.[16] This is, perhaps, indicative of the more individualistic turn European culture and even scholarship took in the Romantic period, the latter decades of the eighteenth century. For example, both in his autobiography and in his travelogue, Thunberg stressed the often dangerous nature of his scholarly endeavors, posing as a real adventurer-traveler—an image, of course, meant to enhance the status of his book in the eyes of his readers. And whether childlike naive, more than normally arrogant, or simply sensibly self-conscious, Thunberg tells his readers he was a huge success with the Japanese because of his knowledge and his limitless disposition to inform them, as well as [his] sound, kind, and friendly behavior.

Of course, his public was meant to understand this to be the very reason why Thunberg had been able to probe into Japanese society and culture as deeply as he suggested he had done. That, again, served to stress the reliability of his information and the value of his analyses.

Following the precepts of the "ars apodemica," the set of rules devised in sixteenth-century Europe for the presentation of information about countries and peoples, Thunberg opens with a survey of the geography and topography of Japan, giving, of course, due attention to the island empire's flora and fauna. He continues with chapters on the economy, the people and their language, the history of the Japanese empire, the structure of society, and the role of the shogunal government. In all this, he strives to avoid the judgmental stance so often taken by European travelers when confronted with other cultures—or so he says. To stress his objectivity, he explicitly tells his readers that he has tried to look both for the good and the bad, the useful and the harmful, in Japan. That any comparison, or even the very categories that serve as the instruments of comparison, could be nothing but subjective, European ones, Thunberg was unable, or perhaps unwilling, to perceive.

Titsingh's Japan: A New View?

Arriving on Deshima the year after Thunberg left, Isaac Titsingh (1745–1812) is the one person who stands out in the field of Europe's Japanese studies around the turn of the century.

Although Lequin has done—and, in fact, still is doing—a magnificent job in restoring Titsingh's scholarly reputation,[17] presenting his hero as a complete original, as Europe's first japanologist ever, is, I feel, entirely unwarranted. Though working a century later, Titsingh, like Kaempfer, used traditional European concepts and formats for presenting his findings as, for example, the age-old genre of the encyclopedic "Land and People"-description and, in a number of shorter texts he meant to publish, the by now traditional marriage- and funeral-rites approach to non-European cultures. Indeed, despite Lequin's claims, one must conclude that Titsingh very much followed in the footsteps of his predecessors. For example, his report on what we now call acupuncture and moxibustion does, in a way, repeat a seventeenth-century tract on the subject written by Willem ten Rhyne, a VOC-doctor on Deshima, published in Latin and in Dutch in 1683 and 1684. Indeed, it was Ten Rhyne who coined the word "acupuncture" in the first place. A few years later, Kaempfer even had added an original Japanese treatise on the topic as an appendix to his own work—as Titsingh planned to do, too. Likewise, in his survey of Japan's de facto rulers, the shoguns, Titsingh used Japanese coins, as had Thunberg before him.

Three times a director of the Dutch affairs on Deshima and two times a visitor to Edo, Titsingh also learned to read, speak, and write Japanese—no mean accomplishment. Like Kaempfer, Titsingh, too, relied heavily on the secret assistance of a Nagasaki interpreter he managed to befriend—his relationship with Namura Naosaburo mirrors Kaempfer's friendship with Imamura Geneimon.[18] With this man's help Titsingh systematically gathered the scholarly material which, as he himself wrote, he craved more than money.

Yet, much of Titsingh's research seems to have been conducted after he left Deshima, starting, really, when he had returned to Batavia—a town that, I should add, he felt was totally lacking in anything resembling an intellectual climate. From then on, till his death, the letters Titsingh wrote to his Japanese friends are, basically, long lists of questions about everything he wanted to include in the magnum opus he slowly started writing or, rather, composing. For while retaining the categories traditionally used in Europe to systematize data, he did plan to fill them with translations of Japanese primary material, only, rather than offer his own analysis of these texts. If he had succeeded in realizing his program, he would have presented Europe with, among other things, a translation of one of the great feats of the Sino-Japanese scholarly tradition—the *Wakan Sansai Zue*, the 1712 Japanese edition of a 1609 Chinese encyclopedia.

Alas, the great work Titsingh envisaged, a new "Description of Japan"—significantly he had decided to use the very title given to the Dutch translation of Kaempfer in 1729—never materialized. Nor, we should add, was it at all likely that he ever would have brought his own publication program to completion; it simply was far too ambitious. In the year of his death, 1812, when the fate of Europe hung in the balance, negotiations failed for printing the three-volume, 1000-page-long Dutch text that Titsingh had been able to finish; the times were not friendly to such costly, large-scale scholarly enterprises. Likewise, the English and French publishers decided not to take the risk.

Meanwhile, the reports written by a few Dutch scholars on the (non-)desirability of a publication are revealing. In 1812, one of the pundits of the newly founded Dutch Royal Academy declared:

> I feel I have to say I totally reject the literal translation of original works [. . .] if these works belong to the East of Asia and, therefore, contain

> superstitious nonsense and miserable tales whose continuous repetition makes reading unbearable.

The writer Anton Falck—not, perhaps, an authority in the field of Oriental Studies—also stated that Titsingh's opinion about the Japanese

> as a people whose civilisation equalled that of the peoples of Europe was highly debatable.[19]

Consequently, Titsingh's major work was not given to the Dutch public, despite efforts in the late 1820s yet to do so, after the favorable opinion voiced in 1824 by the former "opperhoofd" of Deshima, Hendrik Doeff.[20] Nor did anything happen when in 1831 Professor Sibout Klinkhamer again argued in favor of its publication, not only because by now British and French Orientalists already were editing parts of Titsingh's texts but also because, precisely, "the authenticity of the pieces he has contributed"[21] constituted their major scholarly value.

The inability of the Dutch scholarly establishment to recognize the importance of Titsingh's work is all the more surprising since, in 1816, King Willam I had acquired the huge collection of, mostly, Chinese antiquities and art amassed by Jean Theodore Royer, which he destined to be, among other things, a tool for the study of history.[22] To house it, the king now decided to give the Dutch public a "Royal Cabinet of Rarities." Moreover, in 1817, he ordered the new director of Deshima, Hendrik Cock Blomhoff, to assemble as many Japanese objects as possible, to further enrich this collection. In this, Blomhoff succeeded, creating one of the most extensive visualizations of Japanese daily life outside Japan.[23] Indeed, it is obvious that these and other collections did in some ways result in a definite "Japan hype" in, at least, artistic circles at The Hague.[24]

One must feel sorry for Titsingh, for precisely during these years the fate of his own collection of Japanese original material had been disappointing, to say the least.[25] Here, too, I must stress, he had acted as Kaempfer had before him, gathering not only Japanese manuscripts and books, maps, and woodblock prints, but also coins and other curios. In 1809, Titsingh decided to bequeath his so-called cabinet to the British Museum—where, a century earlier, due to the efforts of Sir Hans Sloane, but perhaps unbeknownst to Titsingh, Kaempfer's collection also had ended up. Titsingh hoped that, eventually, his "cabinet" would serve as the stimulus and the focus of wide-ranging research of everything connected with Japan. However, when he died, in Paris, the collection was sequestered by the French government and, later, bought by a Parisian book- and curio dealer. Finally, it was dispersed all over the Western world.

In short, the impact of Titsingh's undeniably extensive knowledge of Japan on the learned community was disappointingly small. Two friends of his, Louis Langlès and Joseph de Guignes, both sinologists, made a detailed inventory of his cabinet, but it was not printed. Another friend, the young, promising Oriental scholar Abel Remusat, in 1820 published Titsingh's own French text of the *Memoires et Anecdotes sur la dynastie regnante des Djogoun, souverains du Japon*, to which he added two other short pieces, one on *harakiri* and one on Japanese poetry—together with Titsingh's translations of *haiku*, probably the first European ones ever. Two years later, an English version appeared.

Finally, in 1834, the sinologist Julius Klaproth took care of the publication of *Nipon o Dai itsi ran, ou, Annales des empereurs du Japon, traduit par M. Isaac Titsingh avec l'aide de plusieurs interprètes attachés au comptoir hollandais de Nangasaki. Ouvrage revisé, complété et corrigé sur l'original japonais-chinois, accompagné de notes et précédé d'un Aperçu d'histoire mythologique du Japon.*[26] However, here again one may wonder whether this survey of Japan's emperors amounted to something completely different from the lists Kaempfer had compiled from Japanese sources, a hundred years before.

It is interesting to note that, in the course of the seventeenth and eighteenth centuries, two ways of "studying" Japan had developed. Mostly, European scholars would use the material available—always, of course, as presented and translated by their Japanese helpmates who, thus, in a way controlled the transfer—as the basis for their own analysis and interpretation. Only a few authors, like Kaempfer—even as he "spoke" through the Scheuchzer-edition—and Titsingh felt that their readers should also be able to judge the original Japanese texts by themselves, albeit in translated editions.

Given the limited relevance of Titsingh's works at the time, one must assume that around the turn of the eighteenth century, most Europeans still culled their knowledge about Japan either from Kaempfer or from Thunberg.[27] This only changed with the publication of Philipp von Siebold's *Nippon: Archiv zur Beschreibung Japans*-series, from the 1830s onward. But though Von Siebold's contributions were not negligible, one must admit that the way he systematically tried to play down or even outright denigrate the work of his predecessors was positively disgraceful, the more so since he used their findings liberally.

Among his forerunners Hendrik Doeff had, perhaps, been the most important. Due to the political circumstances in Europe, this former "opperhoofd" of Deshima had resided on the island longer than any other director before him. Having arrived in 1799, he only left in 1817. His great contribution to Japan's Dutch studies and, indeed, to knowledge of the world was his Dutch-Japanese dictionary; it combined phrases in turn-of-the-century Dutch with, quite revolutionary, translations in everyday, contemporary Japanese. During his return journey, in a shipwreck he tragically lost the Japanese wife he had wed and whom, quite exceptionally, he had decided to take home to "Patria." In the shipwreck he also lost his own copy of the dictionary. Luckily, the manuscript version he had left in Japan was completed and became Japan's one most important translation tools during the early nineteenth century. Doeff's pride of this achievement still rings through his *Herinneringen uit Japan*, which, however, only appeared in 1833. Significantly, he stated that, even then, more than a century after its publication, the Scheuchzer-Kaempfer *Beschrijving* was the best general survey available.[28]

Europe in Japan: Between Scholarship and Exoticism

As to Japan's knowledge of Europe, and, indeed, of the entire world outside its own Sino-Confucian universe, it is, of course, a known fact that the shogunal court as well as the Edo and Nagasaki intellectuals relied on the VOC as much as the handful of Company intellectuals relied on them. Specifically, much has been said, and written, about the ways Japanese scholars—medical practitioners prominently among them—tried to understand what Europe had produced especially in the field of the sciences. Of course, Beukers has done groundbreaking work precisely in this field.[29] However, it would be wrong to assume the Japanese had many ways to gain their Dutch knowledge, their Western wisdom. Basically, we can distinguish but two.

There were, of course, conversations—necessarily limited in scope and depth, if only because of the linguistic and other limitations inherent to the mostly supervised, restricted, and restrictive contacts between the Dutch and the Japanese.

And then there were books. In providing Japan with texts for translation by the Japanese interpreters or the *rangakusha*, the Company mostly acted on express orders from the shogunal authorities who, it has to be added, sometimes were prompted by specific requests from Japanese intellectuals. The books were either Dutch originals, or translations from other European languages. In both cases, the

Nagasaki interpreters, and the subgroup specifically charged with the translation of scholarly material, provided Japan with a mass of new information, whether it remained in manuscript form or was subsequently printed. Inevitably, the entire process was not one of translation alone. In many ways, not always well-understood, yet, the knowledge contained in these European texts was transformed, too, to suit a Japanese need for and understanding of the matters contained therein.

As I indicated earlier, during the reign of Tokugawa Yoshimune—that is, from the 1720s onward—the *Sakoku* restrictions were eased, and the shogun himself, recognizing the advantages of a controlled import of certain knowledge, ordered a few bright young men to start studying Dutch. Consequently, the *rangakusha* and others interested in European science and scholarship could order more books, and, also, translate them and disseminate the information more freely than before.

Moreover, in Japan, as in Europe, during the second part of the eighteenth century the ways of gaining and interpreting knowledge and of looking at the world underwent considerable change. As a few authors have stressed—recently Winkel among them[30]—in Japan, too, an economically and socially burgeoining bourgeoisie, in Edo as well as in the bigger commercial towns,[31] was intellectually exploring its world: both the outer world inside Japan itself, and the even wider world beyond Nihon's closed borders. It is fascinating to note that, mirroring developments in Europe, precisely in this period, well-to-do Japanese began to collect Western rarities, sometimes even decorating an entire room in the Dutch style—as did, in 1789, the Orange family when redecorating a room in their palace "Huis ten Bosch" in the (Sino-)Japanese fashion.

Indeed, just as Japan was a synonym for all things exotic in the West, so was *Oranda* a catchword for everything exotic in Japan,[32] such as the electrification machine that, according to a print from 1798, was for sale in an Osaka curio shop—much like the Japanese rarities that could be purchased in the "Japans Magazijn" at The Hague.

Even in visual culture, new information resulted in new perspectives.[33] Shiba Kokan, reading Gerard Lairesse's works on painting from the early eighteenth century—the 1701 *Grondlegginge ter Teekenkonst* among them—in 1783 had mastered the art of copper engraving, and also introduced Japan to the illusion of three-dimensionality. In the late 1780s, he left Edo for Nagasaki, desirous to learn more. Secretly visiting Deshima, he became friends with the then director Hendrik Romberg. He avidly studied *Het Menselyk Bedrijf*, the collection of emblematic prints published in 1694 by Johan and Caspar Luiken, and in his 1799 "Treatise on Western Painting," held a plea for, among other things, more realism in the depiction of man and nature.[34] Soon, the new genre of the *uki-e*, perspective prints, showed there was a market for (t)his style. Inevitably, this new point of view cannot but have influenced Japan's ways of looking at, and, consequently, judging things in other fields as well.

However, the situation in which Dutch, or, in a wider sense, foreign studies operated was a complex one. Often, Japanese scholars were stimulated by their government to either translate or write texts as a reaction to, precisely, the real or percieved threat of foreign danger. Thus, when the Russians started to send their ships to Japan's shores as part of czarist imperial expansion in Eastern Asia, both the Nagaski interpreters and a few *rangakusha* started translating one of the many Dutch versions of Johann Hübners highly succesful, early eighteenth-century geographical, originally German-language textbooks to know more about Eurasia.

Nevertheless when, in the late 1780s, Hayashi Shihei published his treatise on national security, the *Kaikoku heidan*, stressing the fact that the ocean was one unrestricted waterway between—significantly—Japan, China, and Holland, the Edo authorities were appalled. Acknowledging the truth of his observation, and the threat it posed for their hold over their secluded world, they quickly seized the printing blocks, and exiled the author to Japan's inhospitable far north.

By the 1790s, the relative openness that had characterized the previous decades came to an end. As indicated above, in 1801 Shizuki Tadao completed his *Sakoku ron*, the "Discussion about the Closed Country."[35] As Boot has shown recently,[36] it was, actually, a solid translation of a text more than a 100 years old, viz. appendix nr 6 of Scheuchzer's version of Kaempfer's *History*, in which the Swiss editor/translator had used an earlier Kaempfer piece,[37] an essay in which the German doctor had analyzed and endorsed the reasons behind Japan's seclusion policy. Now, Tadao's comments suggested that these reasons—based on a European, empirical view—should be used to restate the Tokugawa's traditional perspective.

However, since it impinged on the *bakufu*'s views of Japan's foreign relations, Shizuki's text could not be printed.[38] Yet, it was widely read in manuscript copies and may well have strengthened the prevalent—political and, also, cultural—view that Japan should refrain from opening up to the outside world. This did not spell the end of Japan's *rangaku,* though. On the contrary.

Admittedly, by the late 1790s, most Japanese interested in the knowledge the *Oranda* could provide, must have noticed the lack of any new information coming from Europe. However, they may not have known the reason. One of the main sources through which Nihon was informed about changes in the wider world were the so-called Dutch Newsletters, the *Oranda fusetsugaki*, annual reports on "global politics." Obviously, these could be easily manipulated to suit the policies of the Company. Indeed, the VOC-authorities did so precisely at the turn of the eighteenth century: the occupation of the Dutch Republic by France could not be explained to Edo, since it would greatly diminish the status of the Company and its superiors, and, moreover, it had brought Holland under the sway of a Roman Catholic power, the one thing the shogunal government, remembering the imperialist policy of the Portuguese in the late sixteenth century, feared most.

But despite the dearth of new information from Europe, precisely around 1800 the Japanese did intensify their translation efforts and make the most of the limited data that were available. For example, in the 1790s, the Otsuki family, who operated a private academy of Dutch Studies in Edo,[39] started the custom of throwing a New Year's party, inviting all *rangakusha* who might be staying in the capital. The first party, in 1795, even was organized around the theme of the translation of books imported from Europe and, specifically, the work on a Dutch-Japanese dictionary undertaken by the Otsuki. The tradition of these gatherings continued into the late nineteenth century.

In this cultural-political context, Otsuki Gentaku wrote his "Holland Explained," in 1799—taking to task those Japanese who, he felt, continued to spread all kinds of mistaken information about the Netherlands, particularly the "incompetent" Nagasaki interpreters. Not surprisingly, in 1811 he was one of the moving forces behind the establishment—within the wider context of the shogunal bureaucracy—of an official agency for "Dutch," meaning "foreign" studies. And in 1815, Sugita Genpaku even produced what one might call a first effort at writing the history of Japanese scholarship in this field, with his "The Beginning of Dutch Studies."

In short, the turn of the eighteenth century presents us with a contradictory situation regarding Dutch/European-Japanese cultural contacts. Both the Japanese authorities and the Dutch were concerned about the way changes in the world might influence and, indeed, threaten their respective positions, as well as about the larger effect of their traditional interaction within their own societies. Tightening control over the transfer of knowledge that was deemed politically and economically vital was one way of coping with the problems, real or imagined.

Meanwhile, in both cultures the thirst for information among the general public increased. Both in Japan and in Europe, the groups involved continued using the texts that were available, transforming the knowledge contained therein in ways that inevitably influenced their already changing societies.

Notes

1. Peter Rietbergen, "Wie verre reizen doet... Compagniesdienaren en andere schrijvende Reizigers," in *Kennis en Compagnie: De Verenigde Oost-Indische Compagnie en de moderne Wetenschap*, ed. Leonard Blussé, et al. (Amsterdam: Balans, 2002), 164–84. For a general survey of the relations between the VOC and Japan, see Leonard Blussé, et al., *Bewogen betrekkingen: 400 Jaar Nederland-Japan* (Hilversum: Teleac/Not, 2000). Also, Peter Rietbergen, *Japan verwoord: Nihon door Nederlandse Ogen* (Leiden: Hotei, 2003). Two older surveys, Donald B. Keene, *The Japanese Discovery of Europe, 1720–1798* (Stanford: Stanford University Press, 1969), and George B. Sansom, *The Western World and Japan* (Tokyo: Tuttle, 1977), now need to be revised, also in view of increased Japanese scholarship in this field.
2. David Livingstone, *Putting Science in Its Place: Geographies of Scientific Knowledge* (Chicago: The University of Chicago Press, 2003).
3. Of course, the so-called Deshima Diaries—the "Dagboeken" kept by the VOC-opperhoofd—constitute the most relevant source for the study of Dutch-Japanese relations. Presently, a complete edition is not available. For the eighteenth century, the Deshima Marginalia-project has provided relevant access: Paul van der Velde, et al., *The Deshima Diaries, Marginalia 1700–1740* (Tokyo: The Japan-Netherlands Institute, 1992); Leonard Blussé, et al., *The Deshima Diaries, Marginalia 1740–1800* (Tokyo: The Japan-Netherlands Institute, 2004).
4. Peter Rietbergen, "Ten hove gegaan, ten hove ontvangen," in *Hof en Handel: Aziatische vorsten en de VOC, (1620–1720)*, ed. Peter Rietbergen and Elsbeth Locher-Scholten (Leiden: KITLV Press 2004), 277–302.
5. Cf. Jan Duyvendak, "The Last Dutch Embassy to the Chinese Court (1794–1795)," *T'oung Pao* 33 (1937): 1–137; Frank Lequin, *Isaac Titsingh in China (1794–1796)* (Alphen aan de Rijn: Canaletto, 2005).
6. For Sugita Genpaku's rather prejudiced survey, see his 1815 *Rangaku kotohajime* [Dawn of Western Science], translated by Ryozo Matsumoto (Tokyo: Hokuseido Press, 1969).

7. The following section is mostly based on the research contained in Rietbergen, *Japan Verwoord.*
8. It was included in the collection of VOC-texts published by: Isaac Commelin, *Begin ende Voortgangh van de Vereenighde Nederlantsche Geoctroyeerde Oost-Indische Compagnie,* vol. 3 (Amsterdam: Joannes Jansonius, 1645/1646), 72–98.
9. Arnoldus Montanus, *Gedenkwaerdige Gesantschappen der Oost-Indische Maetschappy (...) aen de Kaisaren van Japan* (Amsterdam: Jacob van Meurs, 1669).
10. Kaempfer's orginal notes and fragments were only published recently: Beatrice M. Bodart-Bailey, ed. and trans., *Kaempfer's Japan: Tokugawa Culture Observed, by Engelbert Kaempfer* (Honolulu: University of Hawai Press, 1999). Whatever their merits over and above the Scheuchzer version, it has to be borne in mind that they had no influence whatsoever on Europe's views of (seventeenth-century) Japan during the eighteenthto the twentieth centuries.
11. Rietbergen, *Japan Verwoord,* especially Chapters 4 and 6.
12. Peter Rietbergen, "Zover de Aarde reikt: De werken van Johan Nieuhof (1619–1672) als illustratie van het probleem der cultuur- en mentaliteitsgeschiedenis tussen specialisatie en integratie," *De Zeventiende Eeuw*3, 1 (1986): 17–40.
13. Peter Rietbergen, "Becoming Famous in the Eighteenth Century: Carl-Peter Thunberg (1743–1828) between Sweden, the Netherlands and Japan," *De Achttiende Eeuw* 36, 1 (2004): 50–61. Only very recently, Thunberg's travel notes have become available to non-Swedish readers. They have been transcribed into English as volume six of: Leonard Hansen, ed., *The Linnaeus Apostles* (London: IK Foundation, 2007–2012).
14. Carl Thunberg, *Verhandeling over de Japansche natie, haare zeeden, gebruiken, en haare munten. Beschreven en bijeen verzameld op eene reize naar Japan* (Amsterdam: Jan Schneider, 1780).
15. Rietbergen, *Wie verre reizen doet,* passim.
16. Peter Rietbergen, "Japan: The 'Un-Knowable Other'? Two Seventeenth-Century European Models for 'Knowing' Japan," *LIAS* 29, 1 (2002): 63–80.
17. Frank Lequin, *The Private Correspondence of Isaac Titsingh,* vols 1–2 (Amsterdam: Gieben, 1990–1992); Lequin, *Isaac Titsingh (1745–1812): Een passie voor Japan* (Alphen aan de Rijn: Canaletto, 2002). Most of the following section is based on material gathered by Lequin.
18. Titsingh's dealings with the Japanese authorities and his use of the interpreters can be reconstructed from his "secret diary," which is part of the *Deshima Diaries, Marginalia*-edition: ut supra, note 2.
19. Falck's report has been added as appendix 1, no. 10 to: Lequin, *Private Correspondence,* vol.2, 810–15.

20. Doeff's report is appendix 1, no. 16 in: Lequin, *Private Correspondence*, vol. 2, 846–47.
21. Klinkhamer's report is appendix 1, no. 17 in: Lequin, *Private Correspondence*, vol. 2, 849–63.
22. Cf. Jan van Campen, *De Haagse jurist Jean Theodore Royer (1737–1807) en zijn verzameling Chinese voorwerpen* (Hilversum: Verloren, 2000).
23. See Susan Legène, *De bagage van Blomhoff en Van Breugel* (Leiden: KITLV Press 1998).
24. Marika Keblusek, *Japansch Magazijn: Japanse kunst en cultuur in 19e-eeuws Den Haag* (Leiden: Hotei, 2000).
25. Frank Lequin, *A la recherche du Cabinet Titsingh: Its History, Contents and Dispersal; Catalogue Raisonné of the Founder of European Japanology* (Alphen aan de Rijn, Canaletto, 2003).
26. See also Timon Screech, *Secret Memoirs of the Shoguns: Isaac Titsingh and Japan 1779–1822* (London: Routledge, 2006).
27. See Rietbergen, ut supra notes 1, 3, and 5.
28. Hendrik Doeff, *Herinneringen uit Japan* (Haarlem: Erven Bohn, 1833), 5.
29. For example, Harm Beukers, *Red-Hair Medicine: Dutch-Japanese Medical Relations* (Amsterdam: Rodopi, 1991).
30. Margarita Winkel, *Discovering Different Dimensions: Explorations of Culture and History in Early Modern Japan* (PhD diss., University of Leiden, 2004).
31. Cf. C. Nakane, et al., *Tokugawa Japan: The Social and Economic Antecedents of Modern Japan* (Tokyo: University of Tokyo Press, 1990).
32. See for the context: Martha Chaiklin, *Cultural Commerce and Dutch Commercial Culture: The Influence of European Material Culture on Japan, 1700–1850* (Leiden: CNWS, 2003).
33. For a general survey: Calvin French, et al., *Through Closed Doors: Western Influence on Japanese Art, 1639–1853* (Rochester: Oakland University, 1977). Also, Willem van Gulik, "Verbeelding van de werkelijkheid," *Kennis en Compagnie*, 149–63.
34. On Shiba Kokan: Robert Parthesius, *Japanse verwondering: Shiba Kokan (1747–1818), kunstenaar in de ban van het Westen* (Amsterdam: Amsterdams Historisch Museum, 2000); Calvin French, *Shiba Kokan: Artist, Innovator, and Pioneer in the Westernization of Japan* (New York: Weatherhill, 1974).
35. On Shizuki: Yoshida Tadashi, "La science newtonienne selon Shizuki Tadao (1760–1806)," in *Repenser l'ordre, repenser l'heritage: paysage intellectuel du Japon (XVIIe–XIXe siècles)* ed. Francis Gerard, et al. (Geneva: Droz, 2002), 377 sqq. Also, Torii Yumiko, "Shizuki Tadao's Awareness of International Politics," *Journal of the Japan-Netherlands Institute* 9 (2008): 107–24. In fact, the entire instal-

ment of the 2008 *JJNI* is devoted to Shizuki and his intellectual world.

36. Wim Boot, "Shizuki Tadao's *Sakoku-ron*," *Journal of the Japan-Netherlands Institute* 9 (2008): 88–106.
37. It had actually been published (also posthumously) as part of: Engelbert Kaempfer, *Amoenitatum exoticarum...fasciculi V* (Lemgo: Henricus Meyerus, 1712).
38. It was only printed in 1850 and again, in 1891 and 1914.
39. Cf. Conrad Krieger, *The Infiltration of European Civilization in Japan during the Eighteenth Century* (Leiden: Brill, 1940).

Bibliography

Beukers, Harm. *Red-Hair Medicine: Dutch-Japanese Medical Relations.* Amsterdam: Rodopi, 1991.

Blussé, Leonard, et al. *Bewogen betrekkingen: 400 Jaar Nederland-Japan.* Hilversum: Teleac/NOT, 2000.

Blussé, Leonard, et al. *The Deshima Diaries, Marginalia 1740–1800.* Tokyo: The Japan-Netherlands Institute, 2004.

Bodart-Bailey, Beatrice M., ed. and trans. *Kaempfer's Japan: Tokugawa Culture Observed by Engelbert Kaempfer.* Honolulu: University of Hawai Press, 1999.

Boot, Wim. "Shizuki Tadao's *Sakoku-ron*." *Journal of the Japan-Netherlands Institute 9* (2008): 88–106.

Chaiklin, Martha. *Cultural Commerce and Dutch Commercial Culture: The Influence of European Material Culture on Japan, 1700–1850.* Leiden: CNWS, 2003.

Commelin, Isaac. *Begin ende Voortgangh van de Vereenighde Nederlantsche Geoctroyeerde Oost-Indische Compagnie.* Vol. 3. Amsterdam: Joannes Jansonius, 1645/1646, 72–98.

Doeff, Hendrik. *Herinneringen uit Japan.* Haarlem: Erven Bohn, 1833.

Duyvendak, Jan. "The Last Dutch Embassy to the Chinese Court (1794–1795)." *T'oung Pao* 33 (1937):1–137.

French, Calvin. *Shiba Kokan: Artist, Innovator, and Pioneer in the Westernization of Japan.* New York: Weatherhill, 1974.

French, Calvin, et al. *Through Closed Doors: Western Influence on Japanese Art, 1639–1853.* Rochester: Oakland University, 1977.

Genpaku, Sugita. *Rangaku kotohajime* [Dawn of Western Science]. Translated by Ryozo Matsumoto. Tokyo: Hokuseido Press, 1969.

Hansen, Leonard, ed. *The Linnaeus Apostles.* 11 vols. London: IK Foundation, 2007–2012.

Kaempfer, Engelbert. *Amoenitatum exoticarum...fasciculi V.* Lemgo: Henricus Meyerus, 1712.

Keblusek, Marika. *Japansch Magazijn: Japanse kunst en cultuur in 19e-eeuws Den Haag.* Leiden: Hotei, 2000.

Keene, Donald B. *The Japanese Discovery of Europe, 1720–1798.* Stanford: Stanford University Press, 1969.

Krieger, Conrad. *The Infiltration of European Civilization in Japan during the Eighteenth Century*. Leiden: Brill, 1940.
Legène, Susan. *De bagage van Blomhoff en Van Breugel*. Leiden: KITLV Press, 1998.
Lequin, Frank. *The Private Correspondence of Isaac Titsingh*. 2 vols. Amsterdam: Gieben, 1990–1992.
———. *Isaac Titsingh (1745–1812): Een passie voor Japan*. Alphen aan de Rijn: Canaletto, 2002.
———. *A la recherche du Cabinet Titsingh: Its History, Contents and Dispersal; Catalogue Raisonné of the Founder of European Japanology*. Alphen aan de Rijn: Canaletto, 2003.
———. *Isaac Titsingh in China (1794–1796)*. Alphen aan de Rijn: Canaletto, 2005.
Livingstone, David. *Putting Science in Its Place: Geographies of Scientific Knowledge*. Chicago: The University of Chicago Press, 2003.
Montanus, Arnoldus. *Gedenkwaerdige Gesantschappen der Oost-Indische Maetschappy (....) aen de Kaisaren van Japan*. Amsterdam: Jacob van Meurs, 1669.
Nakane, C., et al. *Tokugawa Japan: The Social and Economic Antecedents of Modern Japan*. Tokyo: University of Tokyo Press, 1990.
Parthesius, Robert. *Japanse verwondering: Shiba Kokan (1747–1818), kunstenaar in de ban van het Westen*. Amsterdam: Amsterdams Historisch Museum, 2000.
Rietbergen, Peter. "Zover de Aarde reikt: De werken van Johan Nieuhof (1619–1672) als illustratie van het probleem der cultuur- en mentaliteitsgeschiedenis tussen specialisatie en integratie." *De Zeventiende Eeuw* 2, 1 (1986):17–40.
———. "Wie verre reizen doet...Compagniesdienaren en andere schrijvende Reizigers." In *Kennis en Compagnie: De Verenigde Oost-Indische Compagnie en de moderne Wetenschap*, edited by L. Blussé et al., 164–84. Amsterdam: Balans, 2002.
———. *Japan verwoord: Nihon door Nederlandse Ogen*. Leiden: Hotei, 2003.
———. "Becoming Famous in the Eighteenth Century: Carl-Peter Thunberg (1743–1828) between Sweden, the Netherlands and Japan." *De Achttiende Eeuw* 36, 1 (2004): 50–61.
———. "Ten hove gegaan, ten hove ontvangen." In *Hof en Handel: Aziatische vorsten en de VOC, (1620–1720)*, edited by Peter Rietbergen and Elsbeth Locher-Scholten, 277–302. Leiden: KITLV Press 2004.
Sansom, George B. *The Western World and Japan*. Tokyo: Tuttle, 1977.
Screech, Timon. *Secret Memoirs of the Shoguns: Isaac Titsingh and Japan 1779–1822*. London: Routledge, 2006.
Tadashi, Yoshida. "La science newtonienne selon Shizuki Tadao (1760–1806)." In *Repenser l'ordre, repenser l'heritage: paysage intellectuel du Japon (XVIIe–XIXe siècles)*, edited by Francis Gerard, et al., 377sqq. Geneva: Droz, 2002.

Thunberg, Carl. *Verhandeling over de Japansche natie*. Amsterdam: Jan Schneider, 1780.
Van Campen, J. *De Haagse jurist Jean Theodore Royer (1737–1807) en zijn verzameling Chinese voorwerpen*. Hilversum: Verloren, 2000.
Van der Velde, Paul, et al. *The Deshima Diaries, Marginalia 1700–1740*. Tokyo: The Japan-Netherlands Institute, 1992.
Van Gulik, Willem. "Verbeelding van de werkelijkheid." In *Kennis en Compagnie De Verenigde Oost-Indische Compagnie en de moderne Wetenschap*, edited by Leonard Blussé, et al., 149–63. Amsterdam: Balans, 2002.
Winkel, Margarita. *Discovering Different Dimensions: Explorations of Culture and History in Early Modern Japan*. PhD diss., University of Leiden, 2004.

Chapter 9

Ethnographic Fieldwork in the Dutch Cape Colony at the Beginning of the Nineteenth Century: *De Kaffers aan de Zuidkust van Afrika* (1810) by Ludwig Alberti*

Siegfried Huigen

In eighteenth-century travel accounts, ethnography was a set part of a more comprehensive, encyclopedic interest in the foreign world. People were described, as well as landscapes, plants, and animals. But between 1740 and 1790 the first signs of specialization can be detected in the German-speaking countries and in Russia; ethnography started developing into a discipline.[1] Following on a comparable, somewhat later development in France, the first Dutch language ethnographic monograph, *De Kaffers aan de Zuidkust van Afrika, Natuur- en Geschiedkundig beschreven* (The Kafirs on the South Coast of Africa, described naturalistically and historically) by Ludwig (or Lodewyk) Alberti (1768–1812) was published in 1810 in the Netherlands.[2] Alberti's book offers a description of the western Xhosas, which, at the time he wrote about them, occupied the country on both sides of the Fish River and part of the south coast of South Africa. Alberti's book was accompanied by a large picture atlas in oblong format, *Zuid-Afrikaansche Gezichten* (South African Views) with three large ethnographic pictures out of a total of four. In 1811 the French edition of both books was published. After Alberti's death in 1812 the original German text of *De Kaffers aan de Zuidkust van Afrika* was published as *Die Kaffern auf der Südküste von Afrika nach ihren Sitten und Gebräuchen aus eigener Ansicht beschrieben* (The Kaffirs on the South Coast of Africa, described from personal experience according to their manners and customs) on the initiative of his family in 1815. The modern English edition is a free translation of this text.[3]

Alberti's book is unique in more than one respect. It is the oldest Dutch ethnographic description of the "Kaffirs," as the Xhosas were generally known at the time, the first Dutch ethnographic monograph and also the only practical application of the ethnographic questionnaire of Joseph-Marie Degérando (or De Gérando; 1772–1842), the *Considérations sur les diverses méthodes a suivre dans l'observation des peuples sauvages* (A consideration of the various methods to be followed in the observation of savage peoples; 1800).[4] This questionnaire occupies an important position in the history of anthropology as the earliest attempt to lay a theoretical basis for ethnographic descriptions. Lastly, *De Kaffers aan de Zuidkust van Afrika* is a very early example of applied anthropology in the sense of ethnography intended to serve administrative purposes within a colonial context.[5]

Although in South African historiography Alberti's book is an important source for the history of the Xhosas (for instance, in the ethnohistorical study of the Xhosas, Peires's *The House of Phalo*),[6] the text as such has never been analyzed. In this essay I set out to explore the structure of *De Kaffers aan de Zuidkust van Afrika* and study the way in which Alberti followed the questionnaire and the ethnographic paradigm in Degérando's *Considérations*. I will use the Dutch version of Alberti's book as the basis of my investigation, as it is the oldest published version and the only one authorized by Alberti.[7] This version appeared at the time when Dutch claims to the Cape Colony were still valid, which is important to a third aspect I wish to investigate: the way in which Alberti attempted, by means of his book, to realize his ambition to be (re-)appointed as *landdrost* (district administrator) of the district of Uitenhage in the eastern parts of the Cape Colony.

Alberti and the Dutch Colonial World

Ludwig Alberti was born in the German county of Waldeck in 1768. He came to the Netherlands in 1784 as lieutenant of a regiment of Waldecker mercenaries and left for the Cape in 1802 as commander of a company of Waldeckers. In 1803 he and a small number of troops were posted far from Cape Town, at Fort Frederick, on the site of the present-day Port Elizabeth. In October of that year, Alberti became commander of the fort and in February 1804 *landdrost* of the newly proclaimed district of Uitenhage as well. Here he stayed until the Cape Colony was reoccupied by the British in 1806, which was the last time that the Cape Colony changed hands. The British had captured the (then) Dutch Cape Colony in 1795 after the revolutionary, so-called Batavian Republic had been established in the Netherlands, which was an ally of France. The interim British administration remained until 1803. This was because in 1802 it was agreed at Amiens that the Cape Colony would be returned to the Batavian Republic. The Batavian administration lasted from 1803 to 1806, when the British reoccupied the Cape. The short Batavian period in the history of the Cape Colony was characterized by a flurry of activity aimed at gathering as much information on South Africa as possible, especially about the hinterland of the colony. Alberti's book was the result of this more comprehensive Dutch effort at intelligence gathering with the purpose of drafting new policies for the Cape Colony.[8]

As *landdrost* of Uitenhage, Alberti had a particularly difficult task. He had to keep the peace in a volatile border district with a detachment that was far too small. Since the last years of the rule of the Dutch East India Company, the eastern border of the Cape Colony had been the site of confrontations between Xhosas, Khoisan, and European colonists. In this conflict the Khoisan (the modern collective name for "Hottentots" and "Bushmen") were the weakest party. By the end of the eighteenth century a large number of them had been assimilated, either voluntarily or by coercion, into the Xhosa community or worked for the European farmers as cowherds. The Xhosas and the European farmers expanded in opposite directions: the farmers eastward and the Xhosas westward. In the Zuurveld, an area between the Zondags River and the Fish River that had been regarded as the eastern border of the Cape Colony by the colonial authorities since 1778, these migrations collided, which gave rise to devastating border wars that continued far into the nineteenth century and eventually led to the subjugation of the Xhosas in the mid-nineteenth century.[9]

The European farmers on the eastern frontier caused problems in other respects as well. Between 1795 and 1801 they rose up against colonial authority three times. Despite the conflict with the Xhosas about grazing and cattle, the colonists sometimes concluded brief alliances with Xhosa tribes against the colonial government. After 1803 the government of the so-called Batavian Republic tried to stabilize this explosive border area. The main measure taken to this end was the proclamation of the new border district of Uitenhage of which Alberti was appointed *landdrost*. Although he had an insufficient number of troops at his disposal, nevertheless, he carried out his task successfully. The main achievement of the short-lived Batavian government was that peace reigned on the eastern border between 1803 and 1806. For this Alberti deserves most of the credit. By regularly negotiating with the Xhosas and adopting a clever divide-and-rule policy he was able to keep the warring parties apart.[10] After his departure in 1806 the border conflicts soon flared up again.

After the surrender of the Cape, Alberti returned to the Netherlands. He subsequently fought in Spain as a major in a Dutch detachment. After the publication of the Dutch edition of his book, he accompanied the former Cape governor Janssens to the Netherlands Indies, where Napoleon—the Netherlands having become part of the French Empire in July 1810—had appointed Janssens governor-general. In May 1811 Janssens arrived in Batavia with 500 men. His rule was short-lived, because on August 3, 1811, a British force landed on Java. After Batavia had fallen, Janssens retreated to the fort in Meester Cornelis, a suburb of Batavia. During the British attack on the fort (August 10) Alberti was hit by a bullet. Almost a year later, on June 2, 1812, Alberti died of "fever" and was buried in the cemetery of Weltevreden.[11]

Ethnographic System

Alberti wanted to realize several objectives with his book. In the first place the book is a letter of application, dedicated to Louis Napoleon (1778–1846), Napoleon's brother and king of Holland (1806–1810) at the time the book was printed. The Kingdom of Holland ("Koninkrijk Holland") was created in 1806 by Napoleon Bonaparte as a vassal state of France, with Napoleon's third brother, Louis Napoleon Bonaparte, as head of state. Because Louis Napoleon did not live up to his brother's expectations—he tried to serve Dutch interests instead of his brother's—the kingdom was dissolved in 1810, after which the Netherlands were annexed by France until 1813.

Alberti presents himself to the king in his "Voorberigt" (preface) as the future administrator of his old district of Uitenhage. Nowhere had he been happier in all his life than as *landdrost* of this tumultuous border district, he says. In 1810 such an open application was still possible, because the Cape Colony had been occupied only temporarily by England in 1806. The Cape Colony and other Dutch possessions officially became British territories only with the Treaty of London in 1814. In view of what Alberti says in the preface, the book was intended to prove his competence as a district administrator. His competence consisted of ethnographic and administrative knowledge, with his knowledge as an ethnographer strengthening his capacity as an administrator. Nevertheless, the book is in the first place an ethnographic monograph.

The first twenty and the last two chapters of *De Kaffers aan de Zuidkust van Afrika* contain a description of the Xhosas. Most of this part is in the form of what is now known in anthropology as a realist tale, a description in which the authoritative author hides behind a list of facts about the daily life of the indigenous population.[12] Alberti emphasizes the novelty of his information. According to him, the only publications on this subject that appeared before 1810 were Jacob van Reenen's report on a Dutch expedition to the Eastern Cape in search of survivors of the wreck of the English ship Grosvenor and a passage in John Barrow's *An Account of Travels into the Interior of Southern Africa* of 1801.[13] In Alberti's view, these texts were of no great value, because his predecessors had paid "Kafferland," the country of the Xhosas, only brief visits and had not penetrated very far into it either. He himself, on the other hand, had traveled to "Kafferland" repeatedly, and he had also noticed how his initial observations differed from those he made later.[14]

Alberti's list of predecessors is not entirely complete, because he fails to mention the description the Dutch missionary Johannes van der Kemp (1747–1811) had first published in the *Transactions of the London Missionary Society* in 1801. This description was reprinted in 1804 together with other reports by missionaries under the title "An Account of the religion, customs, population, government, language, history, and natural productions of Caffraria."[15] During his term as *landdrost* Alberti had had official contact with Van der Kemp, although he was probably not aware of the existence of this article, because I have not been able to identify borrowings from Van der Kemp's report.[16] The ethnographic part of Van der Kemp's article was in any case a good deal shorter than Alberti's book; it consists mainly of loose observations. On the other hand, Van der Kemp did publish linguistic data, a subject Alberti hardly deals with.[17]

Alberti not only published the most comprehensive ethnography of the Xhosas of that time, but also went at it systematically. In his own words, he had attempted "to record that which everyone wanted to know about a nation that still seems to live almost in a natural state."[18] In practice he followed the guidelines drawn up by Baron Joseph-Marie de Gérando, or, during the period that is of interest here, "citoyen"

Degérando (1772–1842) in 1800 for ethnographic descriptions, the *Considérations sur les diverses méthodes a suivre dans l'observation des peuples sauvages.*[19] The most obvious connection between Degérando and Alberti is a quotation from the *Considérations* in the chapter in *De Kaffers aan de Zuidkust van Afrika* in which the question concerning the extent to which it is desirable to "civilize" primitive peoples is discussed.[20]

When Degérando's *Considérations* and *De Kaffers aan de Zuidkust van Afrika* are put side by side, it becomes clear that Alberti relied on Degérando for the selection of his material as well as for the structure of his book. Like Degérando, Alberti confines himself almost exclusively to the study of the culture of the foreign people. Although he mentions a "physical" description in the title of his book, besides a "historical" description, the former does not amount to much more in practice than a short description of the area in which the Xhosas live (Chapter 2), their external appearance (Chapter 3) and their physical strength (Chapter 5). The rest of the book is dedicated mainly to their culture. Here he follows Degérando's distinction between the description of the person as an individual and as part of the community. Specific issues that Degérando urged travelers to investigate turn up in Alberti—more or less in the same order as in Degérando. The list below illustrates these similarities (table 9.1). Alberti is very dependent upon Degérando. When he formulates his aim as being "to record that which everyone wanted to know about a nation that still seems to live almost in a natural state," then these are to a large extent the problems Degérando regarded as relevant in this regard in his *Considérations.*[21]

Table 9.1 Parallels between Alberti's *Kaffers* [. . .] and Degérando's *Considérations*

Chapters from Alberti's *Kaffers*		Degérando's *Considérations* (marginal notes)
Voorberigt		Premier défault (peu de séjour [. . .] au milieu d'eux) (133–124)
I	Origin of the Xhosas	Huitième défault (origine) (137), Traditions (166)
II	Location, extent, and nature of the region of the Xhosas in the vicinity of the Cape	Climat (145)
III	Physical form and external appearance of the Xhosas	Etat des sauvages et d'abord de l'individu. Son existence physique (145)
IV	Food of the Xhosas	Aliments (145)

continued

Table 9.1 Continued

Chapters from Alberti's *Kaffers*		Degérando's *Considérations* (marginal notes)
V	Physical strength of the Xhosas	Forces et actions physiques (146)
VI	Normal sleeping and resting of the Xhosas	Repos (146)
VII	Clothing and ornaments of the Xhosas	Vêtements (146–7)
VIII	Physical and moral education of children	Education physique (148), Education morale des enfants (157)
IX	Diseases, medicines, and age of the Xhosas	Effets moraux de maladies (147) Longévité (148)
X	Language, drawing, writing, and counting skills and mental capacity	Facultés (152–154)
XI	God, religion, superstition, withcraft, and moral uncleanness of the Xhosas	Dieu; Esprits; Idées sur son existence; Immortalité (151–152)
XII	Domestic life and activities of the Xhosas	Société domestique (155)
XIII	Respect shown by children, family relations, and respect for elders among the Xhosas	Autorité des pères (155)
XIV	Influence and morals of Xhosa women	Femmes. Leur état (155), Pudeur (156)
XV	Love and marriage among the Xhosas	Amour;Mariage; Son caractère, ses effets; Divorce, polygamie (156–157)
XVI	Social life of the Xhosas	
XVII	Government, respect, tributes, and heredity of the Xhosa chiefs	Rapports politiques intérieurs; Magistrats. Leur titre. Leur autorité (158)
XVIII	Jurisprudence of the Xhosas	Délits (161)
XIX	War and peace of the Xhosas	Guerre; Causes; Circonstances; Art militaire, Armes; Courage; Effets de la guerre, Paix, Alliance (159–160)
XX	Funerals and mourning among the Xhosas	Tombeaux (166)
XXI	Relations between the Xhosas and the colonists	–
XXII	Thoughts about the treatment of the Xhosas as regards the peace in and prosperity of the colony	–
XXIII	Thoughts about civilizing the Xhosas	On p. 248–249 Alberti quotes Degérando to counter the latter's opinions about civilizing "savages"
XXIV	Characteristics and particulars of Gaika	–
XXV	Popular view of the origin of the Xhosas	Huitième défault (origine) (137), Traditions (166)

Degérando and the Société des Observateurs de l'Homme

Degérando had drawn up his *Considérations* at the request of the Société des Observateurs de l'Homme to ensure that scientific travelers brought back ethnographic information that could be used for comparative anthropological research. This was prompted by the French expedition led by Nicolas Baudin (1754–1803) to Australia and a prospective new expedition by the French-Surinamese traveler François le Vaillant (1753–1824) to Africa. However, Le Vaillant never left on a new expedition, and the man who was responsible for anthropological observations during Baudin's expedition, François Péron (1775–1810), did not follow Degérando's guidelines.[22] In the literature on Degérando and the Société des Observateurs, Alberti is not mentioned, although he was the only one who implemented Degérando's ideas, which occupy an important position within the history of French anthropology and the development of cultural anthropology.[23] Discussions of the implementation of Degérando's ideas to date have ended in a description of the failed application thereof by Péron. The successful implementation by Alberti has so far been overlooked.

Degérando and the Société des Observateurs de l'Homme play an important part in the history of anthropology, because they were among the first to attempt to put the study of man and human culture on a theoretical footing. The Société was founded in 1799 to promote the scientific study of the "physical, moral and intellectual existence" of man. The activities of the society were short-lived, however, because it was already disbanded in 1804 as a result of internal problems. Its members included the naturalists Cuvier, Lamarck, Jussieu, and Saint-Hilaire, the physicians Cabanis and Pinel, and the travelers Bougainville and Le Vaillant. The young linguist Degérando was at that time one of the less well known members of the society.

The composition of the society already shows that the approach to anthropology was encyclopedic. No discipline enjoyed priority. Besides the physical characteristics, the cultural qualities of man were to be studied. With a view to the former, Georges Cuvier (1769–1832) drew up the guidelines for the collection of physical data on foreign people, the *Note instructive sur les recherches à faire relativement aux differences anatomiques des diverses races d'hommes* (Instructions for investigations to be conducted with regard to the anatomical differences between human races), for the Le Vaillant and Baudin expeditions. Here one could learn, for example, how to prepare skeletons and skulls (you had to boil them). Degérando, on the other hand, gave guidelines for the collection of cultural data. The objective of the Observateurs was (useful) self-knowledge under the motto "Gnoti te auton"—know yourself. By studying the Other, one could discover what all people had in common. The point of departure for the research program was a dualistic view of man. Following in the tracks of the naturalist Buffon and in opposition to the prevailing materialistic views of man (such as those of the Idéologues),[24] the Observateurs held on to the paradigm of the *homo duplex*, man as a combination of moral and physical characteristics, where the one cannot be reduced to the other. The Observateurs defended human dignity against attacks by the materialists. In the Observateurs' view, man's moral capacities were separate from the body. Subject to the primacy of reason, man was perfectible and not determined by physical characteristics. Scientific racism as it became current in the nineteenth century was precluded in this paradigm. The Observateurs held that human differences were not caused by the body but were of climatological, cultural, social, or political origin. While they did draw an ethnocentric distinction between "civilized" and "savage" (on the whole Europeans were civilized, but some fellow Europeans could also be savage), the differences between these categories were gradual. All people formed one large family and shared one "Esprit." They differed from one another only in "intervals" of civilization. In principle all of mankind went through the same development, some groups having merely advanced further on the long road of civilization than others. This made the study of savages historically relevant. The savages were the contemporary ancestors of the more highly developed Europeans. One could see from the savages how one's own ancestors had lived.[25] That the savages had fallen behind the rest of humanity was, according to the Observateurs, a result of their geographic isolation. Besides studying savages, the Observers also saw it as their task to liberate savages from their isolation. The anthropology of the Observers was a philanthropic science with an explicit "mission civilatrice."[26]

The main outlines of the philosophy of the Observers were also present in Degérando's *Considérations*. The primary aim of his document, however, was to standardize anthropological research, to ensure that the scientific traveler—the *voyageur-philosophe*—took the necessary trouble and collected information that could be used for comparative anthropological research. Degérando wanted to construe the habitus of the observer.[27] His observer had to study the foreign society attentively, learn the language of his study objects, and integrate himself into the foreign society as far as possible. Through prolonged residence he had to become a *concitoyen*, a fellow citizen—an approach that brought Degérando close to modern ideas of ethnographic fieldwork.[28] The study of man consisted, according to Degérando, of three main parts, which are dealt with individually in the *Considérations*: the study of the physical nature and circumstances of the group observed (Degérando deals with this rather briefly), the study of man as an "intellectual and moral being" in his different life phases, and (via a description of family relationships) as a member of the community of which he formed part.[29] These main categories are elaborated in the *Considérations* into more specific questions and subquestions. As part of the psychological characteristics of the individual, for example, resting/sleeping ("Repos") had to be studied. This gave rise to the following more specific questions:

> We must be told how many hours he [the savage] sleeps; whether his sleep is deep; whether he is at peace or troubled by dreams; what the nature of his dreams may be; whether he has a set hour for going to sleep; whether he finds being awake inconvenient or a nuisance; in what position he sleeps or rests.[30]

In the sixth chapter of his book, "Ordinary sleep and rest of the Xhosas," Alberti provides some direct answers to these questions.[31] Here we learn that the Xhosas are not fond of sleeping and that when they sleep, their sleep is peaceful and light; it is therefore easy to wake them up. In the morning they are well rested. "Otherwise these people very much like to be busy and enjoy almost no other rest than is necessary to satisfy their real needs."[32] According to Degérando's guidelines, such a description of the sleeping pattern would provide information about the mental characteristics of the foreign individual.

Alberti's Ethnographic Innovations

In broad outlines, Alberti follows the main categories into which Degérando divided his topics. The physical living conditions and physical characteristics are discussed briefly in the beginning (Chapters 2–5), after which the rest of Alberti's book describes the culture of the Xhosas, their moral and intellectual existence, in the words of Degérando and the Observateurs. Within this section, Alberti also follows the pattern set by Degérando by first examining man as an individual—his sleeping pattern, clothes, education, in Chapters 6–10)—and then proceeding, via family relationships, to a description of the characteristics of the society in which he lives (Chapters 11–20).

Alberti is also aware of the quality requirements Degérando stipulated for a description. When he criticizes Barrow for not having stayed in "Kafferland" long enough, he knows that much depends on the frequency and duration of the contact. Alberti also notes that there is a huge difference between the impressions he gained after his first visit to the Xhosas and his subsequent opinions, after repeated visits.[33] This still falls far short of the requirement that Degérando set for the observer, namely that he had to become a "concitoyen" of the people he observed, but it does show that Alberti was aware of this problem area. At the same time he thereby admits that he went through a learning process from being an outsider to becoming fairly familiar with a foreign culture, a trope also found in twentieth-century ethnography.[34] Furthermore, in the preface Alberti describes limitations under which he had collected his information. He did not speak the language of the Xhosas (Degérando attached great value to the mastery of language), and so had to rely on interpreters.[35] Partly for this reason he was not able to determine much about the intellectual capacity of the Xhosas and the origin of their "superstitious acts."[36] Obtaining information on the latter was also impossible because they did not know anything about it themselves and because they had no priests, the group that normally would be able to provide information on matters of this kind.[37] Such an admission of one's own limitations also has a rhetorical function; it increases the credibility of other statements, which then have obviously passed the test of self-criticism.

Alberti had made use of indigenous informers. An important source was the Xhosa chief "Gaika" or Ngqika (1775–1829), to whom he devotes a short chapter, "Characteristics and particulars of Gaika," which intended to present a description of Ngqika's character by sketching situations and quoting statements showing Ngqika's prudence.[38] At the time when Alberti was in contact with him, Ngqika was the most influential leader of the western Xhosas. Earlier on Ngqika had been an important source of information for John Barrow.[39]

In broad outlines, Alberti follows the scientific paradigm of Degérando and the Observateurs. Alberti is interested mainly in the culture of the foreign "nation"; he sees this as a product of historical circumstances, which—as we saw earlier—he could not always identify.[40] In his view the Xhosas did not differ significantly from Europeans: they were not another "race," let alone an inferior one, but rather a "wild or rather half-wild nation."[41] They were on a different civilization interval, but formed part of the same humanity as the Europeans.

However, Alberti was skeptical of the *mission civilatrice* of the Observateurs. In one of the concluding chapters of his book he raises the question of whether it is desirable to civilize them, as Degérando and the Observateurs would like to do. Alberti devotes the Chapter 23: Thoughts on civilizing the Kaffirs ("Gedachten over de beschaving der Kaffers") to this issue.[42] Alberti reformulates this question as "whether the true happiness of a wild, or rather half-wild nation is indeed promoted by civilizing it or not."[43] The related question is whether the civilized or the uncivilized peoples should be regarded as happier. At first glance, "civilized man" is better off, but the savage is happier because he has fewer needs, which are therefore easier to meet. Civilization on the other hand creates new needs and reduces happiness.

Alberti then quotes (in Dutch translation) Degérando's exhortation to pass on the benefits of civilization, but not the drawbacks:

> [Give them] our arts, but not our decay; our moral code, but not the examples of our vice; our sciences, but not our dogmas, about which we argue fiercely; the advantages of civilization, but not its abuses: carefully hide from them how much people devour each other, even in the most enlightened countries, because of disputes and destroy themselves through their passions.[44]

On this Alberti comments as follows:

> If these instructions were faithfully carried out, the benefits of civilization would be less subject to doubt: however, the extremes juxtaposed by Mr Degerando [*sic*] are so closely interrelated that it will not be easy to communicate the benefits without the disadvantages becoming intermingled with them at the same time.[45]

In other words: Alberti says that the advantages and disadvantages of civilization are difficult to separate. It would therefore be cruel to change the happy and healthy situation of the Xhosas if in the process the disadvantages of civilization are passed on.

At the end of this chapter Alberti makes this question even more complex by involving considerations of colonial administration. Assuming that civilizing the Xhosas would be advantageous to them, it may only be done if it does not disadvantage the colony. In any case, missionaries must not be involved in this, because "such persons are usually profoundly ignorant, come from the lowest classes of the population and are religious fanatics throughout."[46] Alberti had a personal aversion to missionaries in any case.[47]

With this parting thought Alberti takes his leave of Degérando's themes. Colonial administration plays no part anywhere in Degérando because he focuses on the description of wild people outside a colonial context. Alberti, on the other hand, deals fairly extensively (37 pages in total) with the troubled relations between the Xhosas and the Cape Colony in Chapters 21 and 22: "Betrekkingen tussen de Kaffers en de Volkplanting" (Relations between the Kaffirs and the Colony) and "Gedachten over de behandeling der Kaffers ten opzigte van de rust en welvaart der Volkplanting" (Thoughts about the treatment of the Kaffirs in regard to peace and prosperity of the Colony).[48] The former chapter contains a historical overview of the border conflict up to the departure of the Dutch administrators in 1806. The next chapter outlines a policy for the border region. In these chapters Alberti is no longer the neutral ethnographer, but assumes the persona of an experienced colonial administrator with relevant ethnographic and political knowledge.

According to Alberti's historical overview, the conflict on the eastern border of the Cape Colony is not due to the Xhosas.[49] He follows the premise that the Xhosas as a savage people behave well when they are treated well. Unfortunately the "Christian neighbors" in the colony have not treated the Xhosas so well, which has had a corrupting effect on them. He had already expressed this opinion in a letter to Governor Janssens in 1805.[50] In his book Alberti wrote that the problems were aggravated by divisions among the Xhosas, an anarchist attitude among the "Jacobite" colonists toward the government, an alliance between the Xhosas and the Khoikhoi who had been maltreated by the colonists, and an unbalanced policy of the temporary British government (1795–1803). It was only with British force (before 1803) and much Dutch diplomacy (1803–1806) that it had been possible to temporarily restore peace. For a permanent arrangement, however, it would be necessary to evict the Xhosas who had settled within the borders of the colony. Owing to the threat of war, the Dutch administration was unfortunately unable to do this.

In the next chapter, "Gedachten over de behandeling der Kaffers ten opzigte van de rust en welvaart der Volkplanting," Alberti provides a blueprint for a permanent solution to the problems in the border area.[51] These proposals contained two key points: expelling the Xhosas from the colony and delegation of authority to the *landdrost* of the border district of Uitenhage, the position Alberti had occupied, and again aspired to. Once the Xhosas had been driven out, it would be the *landdrost's* duty to keep the warring parties apart and to maintain good relations with the Xhosas. "This landdrost should be an energetic person, diligent in serving the Colony, but also of an enlightened mind about uncivilized nations and fair in his treatment of them."[52] He subsequently explains this "enlightened mind about uncivilized nations." "Enlightened" meant that the *landdrost* of Uitenhage should possess ethnographic knowledge. He had to know how Xhosas think and thus be able to press the right buttons in order to achieve the desired effect. The example he uses to illustrate his thought is typical. He wanted this example to show how ethnographic knowledge helps to exercise power:

> Lastly, like knowledge of human nature (*menschenkennis*) in general provides the greatest certainty about all acts between one person and another, one should also know the mind of a wild or semi-wild nation well in order to influence it to advantage and to achieve the desired goal. It is with regard to this that the preceding considerations produce adequate conclusions to serve as the basis of subsequent acts. For example, the *Kaffers* hold the will of the supreme power in high esteem, which they are accustomed to respect. Very different, therefore, is the impression which the prohibition of the governor of the colony makes on them from that of the colonists when the latter put some request or other to them, and for this purpose use an imperative tone of voice.[53]

In other words: like "knowledge of human nature" or *menschenkennis* is indispensable in bourgeois society, you must also know the "mind of a wild or semi-wild nation" in order to follow an effective policy with regard to such a nation.[54] For example, if you know to whom the Xhosas have assigned authority, you also know as colonial administrator how you can exercise authority over them. Ethnographic knowledge is therefore indispensable to an effective administrator, and this knowledge is contained in Alberti's book. It is also striking in this pronouncement that Alberti does not in principle draw a distinction between ethnographic knowledge and bourgeois *savoir vivre*. Both are means of self-assertion; the European world is therefore not materially different from that of the "wild nations."

Conclusion

Alberti adroitly brought the threads together in the concluding chapters and proved his eligibility for the office of *landdrost* of Uitenhage. The first twenty chapters testified especially to his ethnographic knowledge, after which he stated in the twenty-second chapter that this knowledge was indispensable to someone who had to perform the duties of the *landdrost* of Uitenhage. In the conclusion, he presents himself as a pragmatic administrator to whom the interests of the colony come first. No matter how pure the character of the Xhosas as a "wild or semi-wild nation" may be, the contact with the hypocritical colonists has unfortunately spoiled them, and it is better to push them beyond the border. According to Alberti, they even penetrated far into the colony on their hunting trips, which caused unrest among the colonists living there. The paradoxical position of the colonial official as ethnographer already comes to the fore in this early ethnographic description. Although Alberti is sympathetic to the Xhosas and their leader Ngqika, in the end the *raison d'état* wins.

Alberti's book was no mean feat. Somewhere between 1800, when Degérando's *Considérations* appeared in Paris, and before 1810, the year in which *De Kaffers aan de Zuidkust van Afrika* was published, he came into possession of the latest and at that time the best ethnographic questionnaire that provided him with a guideline for the selection of his material and the structure of this book. Alberti was the only one who successfully applied this questionnaire. He used this questionnaire as a convenient descriptive model, without worrying about Degérando's underlying endeavor to calibrate ethnographic information for further processing by scholars in Europe. Moreover, Alberti deserves the credit for realizing—based on his practical experience—that ethnographic knowledge can be relevant to colonial administration. This made his work not only the first theoretically founded ethnography to be published in the Netherlands, but also the first to give an initial impetus to applied anthropology. It was no more than an impetus, because the application of ethnographic knowledge remains limited to one example, which must in the first place serve to illustrate Alberti's competence as an enlightened colonial administrator.[55] Alberti used it to show how suitable he was to be appointed *landdrost* of Uitenhage once again, so that he could return to the region where he had spent the happiest time of his life.[56]

Notes

*An earlier version of this essay appeared as a chapter in Siegfried Huigen, *Knowledge and Colonialism. Eighteenth-Century Travellers in South Africa* (Leiden/Boston: Brill, 2009). I would like to thank Jean Kommers and Han Vermeulen for their helpful comments and suggestions.

1. According to British and American historiography, cultural anthropology came into being only when anthropological museums (1836–37) and ethnological societies (1839–43) were established. However, Stagl and, particularly, Vermeulen have convincingly argued that

German-speaking scholars already developed an ethnographic discourse and research program in Central and Eastern Europe during the eighteenth century. These scholars already used the terms *Ethnographie*, *Ethnologie*, *Völkerkunde*, and *Volkskunde*. See H. F. Vermeulen, "Origins and Institutionalization of Ethnography and Ethnology in Europe and the USA," in *Fieldwork and Footnotes: Studies in the History of European Anthropology*, ed. H. F. Vermeulen and A. A. Roldan (London: Routledge, 1995), 5–16; J. Stagl, *A History of Curiosity: The Theory of Travel, 1550–1800* (Chur: Harwood Academic Publishers, 1995); H. F. Vermeulen, "Early History of Ethnography and Ethnology in the German Enlightenment: Anthropological Discourse in Europe and Asia, 1710–1808" (PhD diss., University of Leiden, 2008). The history of German anthropology has, until recently, largely been ignored in British and American histories of anthropology (cf. G. Penny and M. Bunzl, "Introduction: Rethinking German Anthropology, Colonialism, and Race," in *Wordly Provincialism: German Anthropology in the Age of Empire*, ed. Penny, Glenn and Matti Bunzl (Ann Arbor: University of Michigan Press, 2003), 1–30.

2. L. Alberti, *De Kaffers aan de Zuidkust van Afrika. Natuur- en Geschiedkundig beschreven* (Amsterdam: Maaskamp, 1810). For an overview of the early Dutch ethnography see F. L. Ellen, "The Development of Anthropology and Colonial Policy in the Netherlands: 1800–1960,"*Journal of the History of Behavioral Sciences* 12 (1976): 303–24; H. F. Vermeulen and J. Kommers, "Introduction: Histories of Anthropology in the Netherlands," in *Tales from Academi: History of Anthropology in the Netherlands. Part I.*, ed. H. F. Vermeulen and J. Kommers (Saarbrücken: Verlag für Entwicklungspolitik, 2002), 1–63. The Dutch edition of *De Kaffers aan de Zuidkust van Afrika* [. . .] was translated from the German by the professor of theology Konijnenburg (Alberti, *De Kaffers*, "Voorberigt").
3. L. Alberti, *Zuid-Afrikaansche Gezichten* (Amsterdam: Maaskamp,1810); *Description physique et historique de Cafres sur la côte méridionale de l'Afrique* (Amsterdam: Chez E. Maaskamp, 1811;, *Die Kaffern auf der Südküste von Afrika nach ihren Sitten und Gebräuchen aus eigener Ansicht beschrieben* (Gotha: Becker, 1815); *Ludwig Alberti's Account of the Tribal Life and Customs of the Xhosa in 1807* (Cape Town: Balkema, 1968).
4. J. M. De Gérando, *Considérations sur les diverses méthodes a suivre dans l'observation des peuples sauvages*, in *Aux origines de l'anthropologie française. Les Mémoires de la Société des Observateurs de l'Homme en l'an VIII*, ed. J. Copans. and J. Jamin (Paris: Le Sycomore, 1978 [1800]), 126–69.
5. Gulliver defines the genre of applied anthropology as follows: "In my view [. . .] 'applied anthropology' involves both a particular orientation of research and a mode of presentation and use of its results. Research

and its findings are specifically directed to the concerns of governments and their officers and are intended to be applied to policy making and administration" (P. H. Gulliver, "An Applied Anthropologist in East Africa during the Colonial Era," in *Social Anthropology and Development Policy*, ed. R. Grillo and A. Rew (London/New York: Tavistock, 1985), 37. With regard to applied anthropology, Kommers refers to the "pretence" of an evident significance for colonial administration (J. Kommers, "Snouck Hurgronje als koloniaal etnograaf: De Atjèhers (1893–1894)," *Sharqiyyât* 8, 2 [1996]: 97). Alberti indeed does not progress beyond pretence.

6. J. B. Peires, *The House of Phalo: A History of the Xhosa People in the Days of Their Independence* (Johannesburg: Ravan, 1981).
7. Cf. Alberti, *Die Kaffern*, 202.
8. Huigen, *Knowledge and Colonialism*, 169–89.
9. Peires, *The House of Phalo*; H. Giliomee, "The Eastern Frontier, 1770–1812," in *The Shaping of South African Society, 1652–1840*, ed. R. Elphick and H. Giliomee (Cape Town: Maskew Miller/Longman, 1989), 421–71; N. Mostert, *Frontiers: The Epic of South Africa's Creation and the Tragedy of the Xhosa People* (London: Jonathan Cape, 1992).
10. W. M. Freund, "Society and Government in Dutch South Africa: The Cape and the Batavians, 1803–6," (PhD. diss., Yale University, 1971), 319, 324.
11. Alberti, *Die Kaffern*, "Vorbericht"; Alberti, *Ludwig Alberti's Account of the Tribal Life*, ix–xiv.
12. Cf. J. Van Maanen, *Tales of the Field: On Writing Ethnography* (Chicago: University of Chicago Press, 1988), 46–54.
13. Earliest publication in 1792, in the English translation by Riou (J. van Reenen, *A Journal of a Journey from the Cape of Good Hope, Undertaken in 1790 and 1791, by Jacob van Reenen, and Others of his Countrymen, in Search of the Wreck of the Hounourable East India Company's Ship, the Grosvenor* [London: Nicol, 1792]); J. Barrow, *An Account of Travels into the Interior of Southern Africa, in the Years 1797 and 1798*, vol. 1 (London: Cadell & Davies, 1801), 193–221.
14. Alberti, *De Kaffers*, ii.
15. J. T. van der Kemp, "An Account of the Religion, Customs, Population, Government, Language, History, and Natural Productions of Caffraria," *Transactions of the London Missionary Society. Vol. 1. From its Institutions in the Year 1795 to the End of the Year 1802* (London: Williams, 1804), 432–68.
16. I. H. Enklaar, *Life and Work of Dr. J. Th. van der Kemp, 1747–1811: Missionary Pioneer and Protagonist of Racial Equality in South Africa* (Cape Town: Balkema. 1988), 151, 155–56.
17. Van der Kemp, "An Account," 442–58.
18. Alberti, *De Kaffers*, iii.

19. In the text below I always refer to the modern edition of the *Considérations sur les diverses méthodes a suivre dans l'observation des peuples sauvages* published by Copans and Jamin.
20. Alberti, *De Kaffers*, 248–49.
21. Ibid., iii.
22. G. W. Stocking, *Race, Culture and Evolution: Essays in the History of Anthropology* (New York/London: The Free Press/Collier-Macmillan, 1968), 31–34.
23. Cf. Stocking, *Race, Culture and Evolution*, 13–41, F. C. T. Moore, "Translator's Note," in J. M. Degérando, *The Observations of Savage Peoples* (London: Routledge & Kegan Paul, 1969), 1–58; S. Moravia, *La scienza dell' huomo nel Settecento* (Bari: Laterza, 1970); De Gérando, *Considérations*; J. L. Chappey, *La Société des Observateurs de l'Homme (1799–1804):. Des anthropologues au temps de Bonaparte* (Paris: Société des études robespierristes, 2002); A. Robben, "Beginnings." Chappey's comprehensive and thorough study is the most complete study of Degérando and the Observateurs, but once again no reference is made to Alberti. For the description of the ideas of the Observateurs, I rely especially on Chappey.
24. The Idéologues, such as Antoine Destutt de Tracy (1754–1836), author of *Elements d'idéologie*, viewed man's psychological faculties (the human "ideology") as part of his physiology. According to Destut de Tracy, man's spiritual life could be traced back to sensory stimuli.
25. This remained a persistent topos in subsequent nineteenth-century evolutionary anthropology. See in this regard J. Fabian, *Time and the Other: How Anthropology Makes Its Object* (New York: Columbia University Press, 1983).
26. Chappey, *La Société des Observateurs de l'Homme.*
27. Ibid., 268.
28. De Gérando, *Considérations*, 131–32, 135, 138.
29. Ibid., 138, 140, 134.
30. Ibid., 146.
31. Alberti, *De Kaffers*, 48–51.
32. Ibid., 50.
33. Ibid., *De Kaffers*, ii.
34. J. Clifford, *The Predicament of Culture. Twentieth-Century Ethnography, Literature, and Art* (Cambridge: Harvard University Press, 1988), 40; Van Maanen, *Tales of the Field*, 75–78.
35. Alberti, *De Kaffers*, 86, 256.
36. Ibid., iii, iv.
37. Ibid., *De Kaffers*, iv; 94.
38. Ibid., *De Kaffers*, 253–57.
39. Huigen, *Knowledge and Colonialism*, 156–58.
40. Alberti, *De Kaffers*, ii, iii.
41. Ibid., 246.
42. Ibid., 246–52.

43. Ibid., 246.
44. Ibid., 248, De Gérando, *Considérations*, 132.
45. Alberti, *De Kaffers*, 249.
46. Ibid., 250.
47. There was a personal reason for this dislike of missionaries. Alberti and the Batavian government were already irritated with the missionaries in the Cape Colony in 1804, especially with Van der Kemp, because he advised the Khoikhoi not to become soldiers (the Batavian government had a regiment of indigenous troops) and was reluctant to make the Khoikhoi who were in his care available as labor for Fort Frederick, of which Alberti was the commander (cf. Freund, "Society and Government in Dutch South Africa," 381, 384; Enklaar, *Life and Work of Dr. J. Th. van der Kemp*, 151, 155–56, 161). Alberti had also in 1804 advised against allowing Van der Kemp to do missionary work among the Xhosas (Freund, "Society and Government in Dutch South Africa," 391).
48. Alberti, *De Kaffers*, 207–33; 234–45.
49. Ibid., 207–33.
50. "Nobody can be more convinced of the incompetence of Europeans to deprive peaceful peoples of other parts of the world of the possession of their original places of residence in order to gain some political or financial advantage from it than I am; this conviction cannot result in anything else than utter disgust at all the unnecessary murders or even only maltreatment of individuals or nations living in this outpost" (South Africa, Cape Archives, letter from Alberti to Janssens, June 12, 1805, CA, BR 377, fol. 67).
51. Alberti, *De Kaffers*, 234–45.
52. Ibid., 242.
53. Ibid., 244.
54. The historical dictionary of the Dutch language (*Woordenboek der Nederlandsche Taal*, last modified July 2, 2012, http://gtb.inl.nl/), defines "menschenkennis" (knowledge of human nature) as follows: "Duidelijk inzicht in het karakter der menschen in 't algemeen, of in dat van de menschen binnen een bepaalden kring" (having a clear understanding of the nature of people in general, or the nature of people in a specific social circle). In the German version of Alberti's book, a similar term, *Menschenkenntniß*, is used (Alberti, *Die Kaffern*, 192).
55. The impact of Alberti's book on anthropological discourse at the beginning of the nineteenth century remained limited. At the time of publication, Alberti's book was only reviewed as an ethnographic text in German-speaking countries. While *De Kaffers aan de Zuidkust van Afrika* in the Dutch journal *Vaderlandsche Letteroefeningen*, in 1810, was still called a work of history, the 1815 German edition was regarded as an example of *Völkerkunde* (ethnology) in Austria and Germany. None of the reviewers showed any interest in Alberti's opinions on the application of ethnography within an imperial context (*Algemeene*

Vaderlandsche Letteroefeningen; Eerste Stuk, 1810, 659–661; *Annales des Voyages de la géographie et de l'histoire* 17 [1812], 128–38, 234–41; *Allgemeine Literatur-Zeitung. Zweyter Band, Junius*, 1813, 350–51; *Ergänzungsblätter zur Jenaischen Allgemeine Literatur-Zeitung* 49 (1815): 110–12; *Wiener Allgemeine Literaturzeitung*, August 27, 1816, 1089–97).

56. Alberti, *De Kaffers*, viii.

Bibliography

Alberti, L. *De Kaffers aan de Zuidkust van Afrika. Natuur- en Geschiedkundig beschreven*. Amsterdam: Maaskamp, 1810.

———. *Zuid-Afrikaansche Gezichten*. Amsterdam: Maaskamp,1810.

———. *Description physique et historique de Cafres sur la côte méridionale de l'Afrique*. Amsterdam: Chez E. Maaskamp, 1811.

———. *Die Kaffern auf der Südküste von Afrika nach ihren Sitten und Gebräuchen aus eigener Ansicht beschrieben*. Gotha: Becker, 1815.

———. *Ludwig Alberti's Account of the Tribal Life and Customs of the Xhosa in 1807*. Cape Town: Balkema, 1968.

J. Barrow, J. *An Account of Travels into the Interior of Southern Africa, in the Years 1797 and 1798*. Vol. 1. London: Cadell & Davies, 1801.

Chappey, J. L. *La Société des Observateurs de l'Homme (1799–1804): Des anthropologues au temps de Bonaparte*. Paris: Société des études robespierristes, 2002.

Clifford, J. *The Predicament of Culture. Twentieth-Century Ethnography, Literature, and Art*. Cambridge: Harvard University Press, 1988.

Copans J. and J. Jamin. *Aux origines de l'anthropologie française: Les Mémoires de la Société des Observateurs de l'Homme en l'an VIII*. Paris: Le Sycomore, 1978.

De Gérando, J. M. "Considérations sur les diverses méthodes a suivre dans l'observation des peuples sauvages." In *Aux origines de l'anthropologie française. Les Mémoires de la Société des Observateurs de l'Homme en l'an VIII*, edited by J. Copans and J. Jamin, 126–69. Paris: Le Sycomore, 1978 [1800].

Ellen, F. L. "The Development of Anthropology and Colonial Policy in the Netherlands: 1800–1960." *Journal of the History of Behavioral Sciences* 12 (1976): 303–24.

Enklaar, I. H. *Life and Work of Dr. J. Th. van der Kemp, 1747–1811: Missionary, Pioneer and Protagonist of Racial Equality in South Africa*. Cape Town: Balkema. 1988.

Fabian, J. *Time and the Other: How Anthropology Makes Its Object*. New York: Columbia University Press, 1983.

Freund, W. M. "Society and Government in Dutch South Africa: The Cape and the Batavians, 1803–6." PhD. diss., Yale University, 1971.

Giliomee, H. "The Eastern Frontier, 1770–1812." In *The Shaping of South African Society, 1652–1840*, edited by R. Elphick and H. Giliomee, 421–71. Cape Town: Maskew Miller/Longman, 1989.

Gulliver, P. H. "An Applied Anthropologist in East Africa during the Colonial Era." In *Social Anthropology and Development Policy*, edited by R. Grillo and A. Rew, 37–57. London/New York: Tavistock, 1985.

Huigen, S. *Knowledge and Colonialism: Eighteenth-Century Travellers in South Africa*. Leiden/Boston: Brill, 2009.

Kemp, J. T. van der. "An Account of the Religion, Customs, Population, Government, Language, History, and Natural Productions of Caffraria." In *Transactions of the London Missionary Society. Vol. 1. From Iits Institutions in the Year 1795 to the End of the Year 1802*, 432–68. London: Williams, 1804.

Kommers, J. "Snouck Hurgronje als koloniaal etnograaf: *De Atjèhers* (1893–1894)." *Sharqiyyât* 8, 2 (1996): 87–115.

Maanen, J. van. *Tales of the Field: On Writing Ethnography*. Chicago: University of Chicago Press, 1988.

Moore, F. C. T. "Translator's Note," in J-M. Degérando, *The Observations of Savage Peoples*. London: Routledge & Kegan Paul, 1969. 1–58.

Moravia. S. *La scienza dell' huomo nel Settecento*. Bari: Laterza, 1970.

Mostert, N. *Frontiers: The Epic of South Africa's Creation and the Tragedy of the Xhosa People*. London: Jonathan Cape, 1992.

Penny, G. and M. Bunzl, "Introduction: Rethinking German Anthropology, Colonialism, and Race." In *Wordly Provincialism: German Anthropology in the Age of Empire*, edited by G. Penny and M. Bunzl, 1–30. Ann Arbor: University of Michigan Press, 2003.

Peires, J. B. *The House of Phalo: A History of the Xhosa People in the Days of Their Independence*. Johannesburg: Ravan, 1981.

Reenen, J. van. *A Journal of a Journey from the Cape of Good Hope, Undertaken in 1790 and 1791, by Jacob van Reenen, and Others of His Countrymen, in Search of the Wreck of the Hounourable East India Company's Ship, the Grosvenor*. London: Nicol, 1792.

Stagl, J. *A History of Curiosity: The Theory of Travel, 1550–1800*. Australia: Harwood Academic Publishers, 1995.

Stocking, G. W. *Race, Culture and Evolution: Essays in the History of Anthropology*. New York/London: The Free Press/Collier-Macmillan, 1968.

Vermeulen, H. F. "Origins and Institutionalization of Ethnography and Ethnology in Europe and the USA." In *Fieldwork and Footnotes: Studies in the History of European Anthropology*, edited by H. F. Vermeulen and A. A. Roldan, 39–59. London: Routledge, 1995.

Vermeulen, H. F. "Early History of Ethnography and Ethnology in the German Enlightenment: Anthropological Discourse in Europe and Asia, 1710–1808." PhD. diss., University of Leiden, 2008.

Vermeulen, H. F. and J. Kommers. "Introduction: Histories of Anthropology in the Netherlands." In *Tales from Academia. History of Anthropology in the Netherlands. Part I.*, edited by H. F. Vermeulen and J. Kommers, 1–63. Saarbrücken: Verlag für Entwicklungspolitik, 2002.

Woordenboek der Nederlandsche Taal. Last modified July, 2010. http://gtb.inl.nl/.

Chapter 10

Intellectual Wastelands? Scholarship in and for the Dutch West Indies up to ca. 1800

Gert Oostindie

Early modern science had been crucial to European expansion. A constant series of new accomplishments in cartography, maritime technology, the art of warfare, and the like preceded and accompanied Western colonialism. While much of this technological progress may have been one-directional, European exploration and settlement in faraway continents also stimulated the further development of metropolitan science. As will become evident from this chapter, the Dutch case is no exception to this rule.

Early modern Dutch overseas expansion was ambitious and multidirectional. By the mid-seventeenth century, the Dutch East India Company (VOC) and its West Indian counterpart (WIC) were in the process of establishing an extensive series of settlements throughout Asia, Africa, and the Americas. Recent scholarship has corrected the traditional wisdom that the Dutch Atlantic was of little significance both to the wider Atlantic world and to the metropolis.[1] Even so, the Republic's presence in West Africa and the Americas was of limited importance—in comparison to both the VOC domain and to the colonial empires of other European powers in the Americas.

The one permanent WIC stronghold on the West African coast, in Elmina, was of little demographic significance, housing less than 300 Europeans on average throughout the early modern period. After the loss of Brazil (1654) and New Netherland (1667), the Dutch Atlantic Empire became mainly an Caribbean affair. This modest emporium consisted of six small islands in the Caribbean Sea and several colonies in the Guianas, on the northeastern coast of South America. Curaçao, just off the coastline of present-day Venezuela, served mainly as a commercial hub. So did the northerly Caribbean islet of St. Eustatius ("Statia") in the second half of the eighteenth century. Suriname, the major colony on the "Wild Coast" of the Guianas, was developed into a typical Caribbean plantation economy entirely dependent on African slave labor. The neighboring Dutch colonies of Berbice, Demerara, and Essequibo were only fully exploited as such after the British takeover during the Napoleonic wars.

This chapter probes into the significance of metropolitan science for the major Dutch Caribbean colonies, in this case Curaçao and Suriname, and vice versa. An underlying assumption is that the circulation of knowledge and ideas between Europe and the colonies, and more specifically between the Republic and the Dutch West Indies, was more intensive than many historians have assumed. At the same time the argument of this chapter runs that up to the early nineteenth century, this circulation of ideas had little significance to the field of "pure" science, and only limited repercussions in the field of the "arts" or applied sciences.

Exciting Beginnings: Johan Maurits in Brazil

From a science perspective, Dutch colonialism in the Atlantic had exciting beginnings in Dutch Brazil (1630–54).[2] The famous governor Johan Maurits van Nassau-Siegen (1637–44) invested heavily in the field of arts and sciences, an approach not shared by most VOC and WIC patrons. The contemporary Brazilian remembrance of this "humanist prince" emphasizes his patronage of painting, both ethnographic and naturalistic (Albert Eckhout and Frans Post), as well as his active support for the hard sciences.

In Maurits's entourage were the German cartographer, mathematician, and astronomist Georg Markgraf and the medical doctor Willem Piso, a Dutchman educated in Leiden and Caen, France. The two studied the local flora and fauna and also engaged in medical, meteorological, and astronomic research, resulting in their massive joint publication *Historia Naturalis Brasiliae* (1648). The volumes written by Piso, *De Medicina Brasiliensi*, are still considered pioneering works on tropical medicine. Markgraf's volumes have a similar standing in naturalist scholarship. The *Historia Naturalis Brasiliae* was concluded by a historical and ethnographic account written by WIC official Johannes de Laet.

Brazil was fully recaptured by the Portuguese in 1654. The "Dutch"—among these many from other Northern European countries as well as a considerable number of Sephardic Jews and *cristãos novos*—departed for the Netherlands, the Caribbean, and North America. Future Dutch colonialism in the Caribbean would expand on lessons learned in Brazil on sugar plantations, Atlantic slave trading, and African slavery and would also be characterized by an unusual importance of Jewish settlers. But the scholarly and artistic fervor characteristic of the Brazilian adventure would not be replicated.

Why was that? The exceptional story of Dutch Brazil may be explained by the fact that the Republic itself was in its heyday at the time, also from an intellectual point of view. Moreover, this was a time of great optimism about the WIC and its prospects in the Americas, and investments apparently followed suit. Numbers will have mattered too. By 1645 the colony counted some 6,000 Europeans, a significant number indeed in comparison to the numbers of whites in the later Caribbean colonies. Finally, Johan Maurits himself made a crucial difference, as a visionary Maecenas able to recruit the necessary means in Europe and as an autocratic ruler able to have his way.

Natural Sciences

With possibly some 8 million inhabitants in Java alone in the early nineteenth century, the Dutch East Indies dwarfed the metropolitan population of 2.3 million. The contrast with the number of inhabitants of the Caribbean colonies is even starker. The total population of Suriname was some 55,000 around 1800, while that of Curaçao was less than a mere 20,000. The overwhelming majority in the two Caribbean colonies were descendants of enslaved Africans. In the late eighteenth century, slaves made up almost 90 percent of the population in Suriname and 60 percent in Curaçao.

The geographical expanse of the Dutch orbit again offers remarkable contrasts. The surface of Java alone easily tripled that of the Republic. Suriname had over four times the metropolitan surface, but much of the colony was "useless" tropical rainforest. Curaçao, with a mere 444 square kilometres, is in an altogether different category. Tiny, arid, and barren, the island had been categorized among the *islas inútiles* (useless islands) by the Spanish soon after its discovery.

While it therefore may not come as a surprise that European interest gravitated toward the Dutch colonies in the Cape and Asia, it is equally understandable that interest in the Dutch West Indies focused mainly on Suriname. This colony on the Wild Coast housed a thriving plantation sector along the Atlantic coast and spectacular biodiversity in the vast interior. In addition Suriname harbored a population of some significance, divided among the multiethnic capital of Paramaribo, the overwhelmingly African plantation zone, and the interior inhabited by Amerindians and Maroons, runaway slaves and their descendants.

From the early sixteenth century, illusions of Eldorado and the existence of rich gold mines had inspired European explorers to venture into the interior of the Guianas. The fate of most of these expeditions was dismal, as gold was seldom mined, and many explorers found a miserable death in the tropical forest. Systematic geological research of Suriname only started in the late nineteenth century, long after the Humboldtian revolution had gathered momentum elsewhere in the Americas and beyond.

The development of cartography was limited. While the earliest map of the Wild Coast dates from 1529 and the development of Suriname's plantation sector was accompanied by a successive series of adequate maps of the coastal region, serious reconnaissance of the interior only started in the second half of the nineteenth century. The same applies to archaeology and the earth sciences.[3] Barring some preliminary searches around 1820, scholarly cartography and geological research to the West only took off with the first "Dutch West Indian Scientific Expedition" of 1884–1885. By then intrinsic scholarly interest, the lure of mineral resources, and better facilities stimulated the organization of expeditions. Founded in 1873, the KNAG/Royal Dutch Geographical Society played a crucial role to this end, privileging however the Dutch East Indies over Suriname, let alone the Dutch Caribbean islands.[4]

Local academic interest and scholarly expertise in physical geography must have been virtually nonexistent. A contemporary medical doctor in Suriname alluded to daily meteorological observations in the 1770s, but systematic registration did not start before the late 1840s.[5] Practical knowledge however did matter. Planters' manuals and the like did discuss issues such as types of soil and seasonal climatic variations in great detail, as these directly influenced productivity in the plantation sector.

While most of the published histories, descriptions, travelogues, planters' manuals, and so on from the colonial period date from the nineteenth century, we do have a good number of such works from the preceding period. The more elaborate of these include fascinating though not always reliable descriptions of the colony's natural habitat, both its flora and fauna. The authors of such works, mainly visitors, would often add drawings to their evocations of majestic nature. Few of these works however had serious scholarly merits, nor any pretence to this end.

One major and certainly the earliest exception to this observation is Maria Sybilla Merian's work on the flora and fauna of Suriname.[6] While her personal history as a talented and enterprising female researcher and artist in a male-dominated world has attracted much interest, her lasting scholarly contribution as one of the "Great Naturalists" clearly stands by itself. While Merian concentrated on insects and butterflies, she also drew flora and reptiles. Published as early as 1705, her *Metamorphosis insectorum Surinamensium* was to become a source of inspiration for scholars as well as a first introduction of Surinamese nature to a wider European audience.

Merian's drawings have been praised for their accuracy, but her ambitions were not that of a scholarly entomologist, as she repeatedly affirmed. Her work did however stimulate European scholarship. Much of her drawings was collected by Hans Sloane and hence included in the treasures of the British Museum. Carl Linnaeus would extensively use the *Metamorphosis* for his systematic classification of nature. The tenth edition of the Linnaean *Systema naturae* (1758) has over 100 references to her work.

Other than this exceptional story, there is only scattered evidence of European scholars taking the study of Surinamese nature seriously before the nineteenth century. Little was heard of the botanist Isaak Eliazer Augar arriving in Suriname in 1734. First attempts to found botanical gardens in Paramaribo date from 1743 and 1787, but apparently did not last long.[7] One amazing history has only recently been rediscovered. In 1754 Linnaeus sent a dedicated Swedish student to Suriname to collect specimens of the local flora. Upon his return to Sweden, this Daniel Rolander fell into disrepute with Linnaeus, possibly because of the issue of intellectual ownership of his findings. His *Diarium Surinamicum, quod sub itinere exotico conscripsit Daniel Rolander, tomus I & II, 1754–1756* was not published, and Linnaeus thwarted Rolander's academic career. His manuscript, archived in the Danish Botanical Library, is only now being published, in a bilingual Latin-English edition.[8]

We do have another remarkable instance of the Surinamese contribution to European botanical and hence pharmaceutical research. In 1730, the enslaved African Quassie demonstrated the medical uses of a local type of bitter wood. Word spread around quickly. Through Carel Gustaf Dahlberg, a Swedish plantation holder residing in the colony, a specimen was sent to Sweden and would eventually be classified in the Linnaean system as *Quassia amara L.* (1761). Dahlberg's interest was probably strengthened by none other than Daniel Rolander, the tutor to his children. A few years later, C. M. Blom published a scholarly treatise, perhaps a doctoral thesis, on this *kwasibita*. In 1751 Quassie was manumitted, and in 1776 he was even invited to travel to the Netherlands to visit with the Stadholder. The high distinction awarded to him by Willem V probably honored his role as a spy for the colonial regime no less than his herbal expertise.[9]

The Antilles attracted little naturalist attention. Floral data collected by the English botanist Leonard Plukenet were included in Linnaeus's *Species Plantarum* (1753) and added to by Nicolaus Joseph Jacquin (1763). The latter was born and educated in the Netherlands but had moved via Paris to Vienna, whence the Holy Roman emperor Francis I sent him on a scientific expedition to the West Indies and Central America (1755–59). Jacquin also provided the first descriptions of the flora of the Windwards Antilles. Upon their return, both Plukenet and Jacquin became leading botanists in Europe. Nonetheless it seems that systematic scholarly research on the flora and fauna of the islands had to await the early twentieth century.[10]

While contemporary authors were deeply impressed by the fauna of Suriname and presented spectacular descriptions and at times depictions, they offered no systematic analyses (and at times some very confusing information at that). The scholarly study in this field again would have to wait until the later nineteenth century. Specimens of Surinamese wildlife, particularly birds, did find their way though to eighteenth-century European collections. George Edwards's 1743 *A Natural History of Birds* included ten birds from Suriname. In the 1760s, the above-mentioned Dahlberg apparently sent over some 700 zoological specimens to Linnaeus. Many of these specimens were eventually preserved in the St. Petersburg Museum.

The Dutch medical doctor Philippe Armand Fermin, practicing in Suriname and in the meantime a productive author of various books on both nature and society in Suriname, equally built up his own collection, parts of which ended up in Hans Sloane's collections. His *Histoire naturelle de l'Hollande équinoxiale* (1765) and particularly his *Description générale, historique, géographique et physique de la Colonie de Surinam* (1769) have long been considered reliable and pioneering works.[11]

Science for Development: Agro-Industrial Technology

All over the Caribbean, the advent of the Europeans implied the decimation or at best marginalization of the native Amerindian populations and a virtual restart of history. Whether under Spanish, British, French, Dutch, or Scandinavian rule, the West Indies were reconfigured for plantation purposes. Old World populations came to the West Indies, mainly voluntarily from Europe, exclusively as enslaved labor from Africa. Old World animals were shipped to the Caribbean, from horses and cattle through sheep and goats to dogs and chickens. Even the major crops of sugarcane and coffee that would form the mainstay of the new plantation sector and the raison d'être for the West Indian colonial project as such were imported from the Old World.[12]

While for ecological reasons Curaçao was not suitable to serve as such, Suriname was the quintessential Caribbean plantation colony. And it is precisely in this field that one finds a strong interaction between metropolitan science, particularly the "art" of applied science, and colonial agro-industrial technology. Following Smith and Marx as well as the abolitionist ideology, slavery has long been misinterpreted as anachronistic from an economical (in contrast to an ethical) point of view. In fact, throughout its history the Caribbean plantation achieved remarkable technological innovations by the standards of its time, while stubbornly clinging to enslaved labor. This applies particularly to the dominant institution of the sugar plantation, one of the world's first agro-industrial enterprises, essentially "factories in the field."

The Surinamese plantation was no exception to this rule. Relevant technological innovations developed either in Europe or in competing Caribbean colonies were quickly implemented. Well into the nineteenth century, Suriname even belonged to the technologically most advanced and hence most productive category of New World plantations. This applies in some extent to the agrarian part of production, but particularly to the subsequent processing technology. Surinamese planters' manuals testify to a vivid interest in issues such as finding the optimal match between type of soil and crop, crop selection, manure and fallow, irrigation and drainage, and so on.

In the processing phase of the milled sugarcane, the Surinamese planters clearly participated in a wider international circulation of technological expertise. The one unique dimension of the Surinamese plantation was the adaptation of the Dutch polder system to plantation agriculture. The metropolitan technology of water management was adapted to create sophisticated polder plantations in Suriname. This remarkable innovation was crucial to irrigation and drainage as well as to transportation for all types of plantations, and in addition provided water-powered energy for the sugar mills. The ensuing increase of productivity was astonishing and enviable to competitors. Abbé Raynal bestowed upon the Dutch the honor of "having domesticated the ocean in the New World, just like they did in the Old."[13]

Of course, we are looking here not at "pure science" but rather at the adaption of metropolitan technology to local needs. Perhaps we would think of this as "science for development," in contemporary usage. Certainly all this technological innovation made for spectacular increase of production and productivity. But little in the character of this plantation complex corresponds to the notion of genuine sustainable and equitable development.

Medical Science

The demographic performance of the populations of the Dutch West Indian colonies was dismal well into the nineteenth century. High morbidity, high mortality, and low fertility characterized Suriname too. The figures for Curaçao were better, but even there natural reproduction was probably not accomplished before the nineteenth century. This grim reality affected the European and African inhabitants and particularly first arrivals alike. While the constant threat of deadly disease haunted the minority of European colonists themselves as well, they also had to worry about the overwhelming majority of African origins. To their owners, these slaves represented indispensable and costly capital.

Over time, there was a slow but unmistakable improvement in the demographic performance of all segments of the population. The explanations for this progress probably lie mainly with the diminishing importance of the African slave trade and hence to the slow creolization of the local population, resulting in better resistance to endemic disease and lesser importation of foreign epidemics. It is not altogether clear at what point this logic became clear to the local elites. Anyway, out of evident self-interest the local authorities and planters were always interested in implementing specific measures that would increase natural reproduction. Such measures, which would be termed *lotsverbetering* (amelioration policies) after the formal abolition of the slave trade (1807/1814), had a longer history. Planters' wisdom as expressed in manuals and the like departed from the maxim that "wise" rule combined discipline—enforced if necessary by the whip and worse—with some balance between high work demands and "reasonable" treatment.

Next to factors such as nutrition, housing, and room for cultural expression, health issues were constantly discussed among the local elites and in all planters' manuals. What to do about high morbidity, high mortality, and low fertility? What were the explanations for these problems and hence what would be the logical preventive or curative measures?

Small wonder, there was a vivid interest in medical expertise. It seems that up to the later nineteenth century, this interest was not met by significant medical research or valorization of European medical scholarship—but on the other hand, some local advance was made, particularly in prevention. This conclusion need not surprise us. The major breakthroughs in medical knowledge date from the later nineteenth century and beyond. As late as 1800, European medical science had not that much to offer either to its own populations or to colonial subjects in the tropics. European doctors working from a miasmatic paradigm and practicing "heroic" measures such as bleeding and blistering often jeopardized rather than cured their ailing patients.[14]

This sobering fact need not blind us to some modest advances made and particularly to the genuine efforts to gain a better understanding of illnesses and to improve preventive and curative practices.[15] There was something of an institutional tradition in medical care, ranging from government hospitals, pharmacies, and quarantine establishments to lay provisions on individual plantations. Contemporaries bitterly complained about the low levels of medical expertise on these plantations as well as on ships. Only the governmental institutions were considered as more or less state of the art, increasingly so with more state supervision after 1800.[16]

European medical science being as modestly developed as it was, "state-of-the-art" may have been a sorry disclaimer. It would take much longer before widespread deadly or devastating illnesses such as cholera, yellow fever, and lepra would be fully understood and become curable.[17] Yet what we do observe is that resident physicians such as Carel Doerffel (1758), Philippe Fermin (1764), G. W. Schilling (1769), and Jacob Voegen van Engelen (1786–1788) did make valiant efforts to describe and categorize the many illnesses they encountered. So, incidentally, did medical specialists serving the maritime trade with the West Indies (Bontius, 1694, Titsingh, 1752 and others), including the Atlantic slave trade (Gallandat, 1769). This tradition would be continued on increasingly firmer ground in the nineteenth century by authors such as medical doctor F. A. Kuhn (1828).

The few relevant medical innovations that were made in the eighteenth century spread rather quickly to the colonies. The most telling example is the preventive treatment of smallpox. The technique of inoculation (with human material) first introduced in Europe from Turkey was implemented in the British West Indies in the 1770s. There are claims that the same innovation had already reached Curaçao, through Boston, by the 1740s.[18] The safer technique of vaccination with animal material was discovered shortly before 1800 and quickly adapted in the British West Indies. Possibly British pressure secured the early-nineteenth-century adoption in the Dutch West Indies of this second preventive innovation as well, later however than would have been advisable.[19]

In sum, then, the contribution of the Caribbean colonies to the development of the medical sciences in the metropolis was meager at best, while metropolitan science had only limited significance to the health situation in the colonies. At the interface of the Netherlands and the West Indies were physicians and pharmacists trained in the metropolis, who deployed their certified—some of these had even completed dissertations—but imperfect expertise in the Atlantic colonies. An additional note on the circulation of knowledge is the fact that between 1713 and 1830, some 20 students from the *British* West Indies took a degree in medicine at Leiden University, many of these probably returning to the colonies.[20] In all, these transatlantic exchanges added up to some incremental growth in medical expertise by the late eighteenth century.

Ethnography

As Benjamin Schmidt has demonstrated in *Innocence Abroad* (2001), early Dutch colonization in the Americas resulted in an avalanche of publications linking New World settings with ideas about national identity in the young Republic at home. By the time of the narrowing of the Dutch Empire to the Caribbean, this stream of narratives on the Americas was wearing thin and became more down to earth. The Republic would remain the major European publisher of books on faraway European colonies and newly discovered tropical territories, but its own West Indian colonies were a sideline at best.

While several dozens of contemporary travelogues, planter's manuals, historical essays, and the like were published on Suriname prior to 1800, the number of publications on the islands dating from this period is negligible. Small wonder, the available publications offer no consistent body of early ethnography. Contemporary accounts do offer invaluable if usually heavily biased information on the Amerindians of Suriname and the enslaved Africans and the emerging Creole populations of both colonies. This proto-ethnography is used and discussed critically in modern historiography, particularly regarding the issue of race and enslavement.

Suffice it to briefly address one crucial debate here, the influence of the European Enlightenment and subsequently abolitionism on how contemporaries in the Dutch colonies wrote about enslaved Africans and thought about the future of slavery.[21] Going through the body of eighteenth-century writings, we do observe—alongside the predictable bigotry—that later authors tend to peruse the work of their predecessors and at the same time add new observations based on their own first-hand experiences. Over time, we witness an increasing acceptation that nonwhites are fully human beings, but a near omnipresent refusal to accept them as equals. As far as slavery is concerned, there were very few authors in the Dutch orbit advocating the abolition of the Atlantic slave trade prior to its actual forced ending by the British (1807).[22] This is a telling reminder that Enlightenment does not necessarily produce abolitionism; it did in some places, it failed to do so elsewhere.

The next question then becomes, why did European abolitionism have so little an impact in the Dutch West Indies? The immediate and obvious answer is that most authors were stubborn advocates of slavery simply because their way of living was intricately linked to the system. In this, they did not differ much from their counterparts in other slave colonies throughout the Americas. The major difference has rather to do with the puzzling absence of an abolitionist movement in the Netherlands up to the 1840s. Whereas abolitionist *philosophes,* religious leaders, political scientists, and other public intellectuals were vociferous and crucially influential in the United Kingdom and France, this type of enlightened discourse was virtually absent among the Dutch. Much has been said about this perhaps counterintuitive fact. For the present purposes, we can only conclude that both the metropolis and the colonies were way behind the intellectual currents of Western Europe.[23]

As an afterthought to this conclusion, it may be useful to point out that this backwardness cannot be explained by an absence of circulation of ideas. Even if their numbers were modest, people traveling between the West Indies and the Republic must have brought with them observations and ideas about the colonies and their governance.[24] Moreover, the Netherlands continued to be open to publications from all major European countries, and the rare descriptions we have of late-eighteenth-century libraries in the Dutch West Indies likewise include works of prominent French *philosophes* and the like. Even more directly, just as news from the French Revolution (1789) had immediately reached the Caribbean and soon sparked its Haitian equivalent (1791), so too news of the Haitian revolution quickly spread to Curaçao and inspired several ultimately suppressed revolts there (1795, 1800).[25]

Colonial Intellectuals and Learned Societies

Bearing in mind the modest numbers of the Dutch West Indian populations, particularly of the free segment, coupled with the low level of local educational facilities and the crassly exploitative character of their societies, one might not expect much of an intellectual life in either Suriname or Curaçao. And so it was, even in the more developed colony of Suriname. David Nassy, the principal author of the *Essai historique,* affirmed that his country "lacked whatever one needs to train and develop a mediocre intellect."[26]

An educational sojourn in the metropolis therefore was requisite for the children of the local elites. Over one hundred Dutch West Indians, mainly from Suriname, indeed studied in the Netherlands in the eighteenth century. Law was the favorite discipline, followed by medicine—both studies of immediate practical use.[27] The greater numbers of university-trained professionals in the colonies however were recent immigrants from Europe, often with no intention to make their residence permanent.

Up to the early nineteenth century, public intellectuals were white and part of the establishment. As a group they were divided along lines of birth (European or colonial) and religion (Christian or Jewish). Probably more important also from their own perspective, they shared the determination to distance themselves from the nonwhite population. Most works published in the eighteenth century were written by individual authors, and it seems difficult to reconstruct any type of intellectual network here other than the obvious observation that among the small colonial elites, virtually all relevant men—female participation in intellectual, political, or technology debates was extremely rare—must have known one another.

Departing from this premise, it is worthwhile to search for institutions built to stimulate intellectual exchange within the colony and with the outside world. In the period under study, there were no serious educational facilities to this end. In the second half of the eighteenth century, however, we do find some organizations aiming at the advancement of an intellectual debate beyond strict utilitarianism.

In the first place, there was a remarkable Masonic activity. The first Masonic lodge of Curaçao was established in 1757, one of the first throughout the Americas. In the next years several more lodges were founded. Apparently there was no separate Jewish lodge on the island. Inspired by the Masonic ideology and the metropolitan example, a department of the Maatschappij tot Nut van 't Algemeen was founded only in 1817 and lasted a mere ten years.[28]

The first lodge in Suriname followed in 1761. In the late 1770s, there were no less than six lodges with over two hundred members; by 1800 the colony still boasted four. A remarkable fact about the Masonry in Suriname is that there were separate Jewish lodges, but that most lodges were of a mixed religious character. Anecdotic evidence suggests that the greater part of the male colonial elites, including high officials and even governors, participated in one or more Masonic lodges.[29]

There was nothing revolutionary about the Masonic movement in the colonies; a mid-nineteenth-century pamphlet would even comment that the lodge of Suriname counted "the most brutal slave executioners" among its leadership. True to the ideals of international Masonry, the lodges did debate on ways to advance the development of the colonies. For this reason, Masons were instrumental in establishing local departments of the metropolitan Maatschappij tot Nut van 't Algemeen. The department for Suriname was established in 1794—much earlier therefore than its Curaçaoan counterpart—lasted until 1800, and was revived in 1816.[30]

For Suriname, one loose Jewish association of self-appointed intellectuals deserves special attention. The five-member Association of Learned Jewish Men headed by David Nassy first served as a debating club, but ended up publishing one of the most noteworthy studies in the colony's historiography, the above-mentioned *Essai historique*, published in French (1788) and next in Dutch (1791). This book, actually authored by Nassy alone, is the fruit of a circle of men who defined themselves firmly as both Jews and *criollos*. The double message of the book was, first, that Jews are honorable and productive members of their societies, wherever; and second, that they have made this particular colony their home.[31]

Another type of institution, extant only in Suriname, had no other pretence than to provide a forum for debating ways to optimize plantation production. Several such debating clubs were listed at one time or another. The one book actually published as a result of such deliberations discloses a truly utilitarian agenda—and testifies to the commonsensical approach of its membership to plantation agriculture, ruling slaves, and medicine.[32] Mention may also be made of short-lived associations such as the Collegium Medicum and the naturalist Kollegie van Natuur-Onderzoekingen (both ca. 1780), the Suriname branch of the Oeconomische Tak van de Hollandsche Maatschappij der Wetenschappen, and journals dedicated to medical issues (*De Surinaamsche Artz*, 1786–1788) or plantation culture (*De Surinaamse landman*, 1801–1805).[33]

The last quarter of the eighteenth century also saw the founding, in Suriname, of associations with literary and theatrical ambitions. Mention may be made of De Surinaamsche lettervrinden (1786), the *Letterkundige Uitspanningen* (1785–1787), and the Jewish literary association Docendo Docemur (1783) in which, again, Nassy was a leading member. We find no scholarly pretense here either.[34] Scattered evidence regarding membership does suggest that a prosopographic study of the local elites would disclose high levels of overlap in the membership of the various institutions discussed above, no matter the discrepancy of their official ambitions, intellectual orientations, and religious belonging.

What literature was available to the educated class? For Curaçao, we have no inventories of libraries dating from the eighteenth century, probably because these were extremely rare. There were no public libraries. The first relevant data are from the 1830s and from the library of the prominent Jewish citizen Haim Abinun DeLima. In his multilingual library, travel accounts, literature and political, geographical, historical, and reference works dominated. In contrast to Suriname, planters' manuals were rare on the island.[35]

Again, and typically, we have more indications for libraries in Suriname. In the *Essai historique*, Nassy boasted that the combined private libraries of a series of prominent residents of the colony—including the governor and other metropolitan officials, medical doctors, and many Christian and Jew "laymen"—were second to none to libraries throughout the Americas and could rival with large libraries in Europe. David Nassy's private library alone was an impressive multilingual collection of four to five hundred books, two-thirds of it nonfiction. With many volumes on physics, chemistry, physiology, the hard sciences were well represented. So were the medical and pharmaceutical sciences, certainly because Nassy practiced these trades.[36]

The figure of David Isaac Cohen de Nassy merits some additional attention. Nassy was an extraordinary individual on many accounts. While he failed as a coffee planter, he was more successful as a mainly self-taught medical doctor, an expert in pharmaceutical knowledge, a dignitary in the Jewish community, a sworn multilingual interpreter, and a public intellectual standing tall in the Surinamese elite. His remarkable career testifies to his individual capacities and drive, but moreover gives a remarkable insight into the possibilities for circulation of ideas in the Atlantic world.

David Nassy was born in Suriname in 1747, an isolated outpost of the colonial world. Nevertheless, as we have seen, he would be able to link up to much of the intellectual currents of Europe. Later in life, he also ventured abroad. Partly for health and family reasons, he moved to Philadelphia in 1792. He published a study on yellow fever the next year and was elected as a member of the American Philosophical Society in 1795. He practiced as medical doctor both in Philadelphia and next in the Danish West Indian island of St. Thomas. Upon his return to Suriname in 1795, he was not granted permission to pursue this profession. In vain he urged for the establishment of higher education modeled after, no less, Harvard and Princeton. He died in Suriname in 1806.[37]

Whether Nassy left a lasting intellectual mark is a moot point. Expat official François Lammens, who had arrived in Paramaribo ten years after Nassy's passing away, wrote that among the local elites the arts and sciences were "never" a subject of discussion and that local cultural and intellectual associations were ephemeral in this materialist society.[38]

Sciences in the Dutch Colonial Orbit: Some Comparisons

So what was the balance sheet of the interaction between Dutch colonialism in the Americas and science around 1800? It seems that, after the loss of Brazil, the significance of European science to the Dutch West Indies was modest beyond the spheres of agro-industrial technology and, to a much lesser degree, medicine. In the opposite direction, findings collected by expatriates and some local amateurs in the Dutch West Indies, particularly Suriname, did find their way to the centers of European science. This applies mainly to the work of naturalists.

In his chapter on Indonesia in this book, Peter Boomgaard suggests that there is little reason to overestimate the scientific dimension of Dutch colonial expansion in Asia prior to the nineteenth century. Other historians have also remarked that neither the Dutch East India Company nor the West India Company displayed much interest in scholarly research—individual researchers, even those employed to other ends by the companies, mostly worked at their own initiative and with their own means. Even Stadtholder Willem IV's collections of tropical exotica—mainly naturalia, much of this bequeathed by officials from the VOC, and in great part not acquired in Dutch colonies but rather in China and other faraway places—did not reflect anything like a concerted metropolitan effort to add a scholarly dimension to colonialism.[39] This conclusion equally applies to the West Indies. Bearing in mind the lesser importance of the West Indies, the lower number of European settlers there, and the fact that at an early stage the WIC lost the desire and means to become a serious Maecenas for the arts and sciences, we may actually be surprised that at least Suriname did produce some intellectual and even scholarly activity in the first place.

The role of pre-nineteenth-century science in the colonies was mainly of a utilitarian character. This conclusion certainly holds for the Dutch West Indies. There was some room for the circulation of scholarly knowledge, intellectual currents, and the like. But utilitarianism dictated learning and only applied science mattered, as in technology and medicine. Interest in fundamental science was nonexistent. As for the humanities, the one crucial topos of enlightened thinking with direct relevance to the West Indies—the rethinking of race and enslavement—was consistently ignored.

In *Main Currents in Caribbean Thought*, Gordon K. Lewis characterized the intellectual debates in the Caribbean colonies as "crassly materialist and spiritually empty" and expressions of "cultural philistinism." Likewise, in *Slavery and Human Progress*, David Brion Davis observed that by the eighteenth century, slave societies throughout the Americas "were acquiring the image of social and cultural wastelands."[40] The above offers little to contradict such bleak assessments—hence the title of this chapter.

Yet, we should allow for the question mark, as well as some nuance. After all, what expectations may we nurture in the first place about the intellectual climate in such small societies characterized by enslavement and racism? Perhaps it is not the low level of intellectual activity that should surprise us, but rather the exceptional position of individuals such as David Nassy.

Meanwhile, back in the Netherlands, throughout the eighteenth century, scholarly interest in the West was even less than it was for the East. Major changes in all of this would have to await the mid-nineteenth century, well after the Humboldtian revolution. By then the significance of the Dutch East Indies to the metropolis had increased spectacularly, while the West Indies had become peripheral in the Dutch orbit and mind. All major scholarly institutions established in the metropolis with a full or partial colonial mission focused on the East Indies. The Koninklijk Instituut voor Taal-, Land- en Volkenkunde (KITLV, 1851) is a telling point in case. Likewise, virtually all prominent "colonial" scholars admitted as members by the Koninklijke Akademie van Wetenschappen (founded in 1808) worked on the East Indies rather than the West Indies.[41]

Are the findings on the Dutch West Indies remarkable in a wider Caribbean perspective? Following Lewis, we might not be surprised.

But again, there is need for nuance. The Spanish, French, and British Caribbean colonies all had larger populations and (consequently) more of an educational and scholarly tradition. The first university of the Americas was founded in 1538 in Santo Domingo, while Cuba had a university since 1728. With the Bourbon reforms of the 1760s, the Spanish Crown began to seriously revive academia in Spanish America. In the 1780s, French Saint-Domingue boasted an impressive academic infrastructure directly linked to the then internationally leading Royal French Academy of Sciences. The first institution for higher education in the British West Indies was established in 1743 (Codrington College, Barbados). Several local learned societies were established in the later eighteenth century all over the British West Indies, even if subsequent academic development in the Anglophone Americas was concentrated in North America.[42]

In comparison, educational and certainly scholarly conditions in the small and relatively isolated Dutch West Indies were scarce. Metropolitan support for colonial science, as Cook indicates a Dutch forte in its Golden Age, had become weak in the eighteenth century, much in contrast to the French and to a lesser degree the British and Spanish cases.[43] We do find a remarkably international set of scholars, professionals, and aficionado amateurs working in the Dutch West Indies, from Germans such as Markgraf, Doerffel, and Kuhn through Swedes such as Rolander and Dahlberg to Englishmen such as Plukenet. This however is not specific to the sphere of scholarship in the Dutch colonies, but rather reflects the fact that throughout the colonial period and in all Dutch colonies, a good number of the Europeans were not of Dutch origins, but rather German, Scandinavian, and so on—not to mention the great number of Dutch of Huguenot and Jewish descent.[44]

There were parallels as well. One of these is the significance of Masonic lodges. Another parallel is the apparently fluid circulation of ideas and practical technology, not just within each of the national subsystems of the Caribbean, but equally between these. We may well assume that in all colonial empires, there was a keen interest in the scholarly and technological advances made elsewhere. This required some sophistication in retrieval, as the greater part of scholarly publications were made in the metropolis and usually in the metropolitan language. Scholarship in the Dutch orbit possibly presented a wider linguistic range, but ironically transfer of knowledge within the Dutch West Indies seems to have been weak. Suriname's major outside contacts were with the Netherlands, North America, and the British West Indies. Curaçao was primarily oriented toward the Spanish Main,

Santo Domingo, Saint-Domingue, and North America. Direct links between Suriname and Curaçao were few.

The extant historiography emphasizes the utilitarian priorities of science's involvement all over the colonial world, and certainly in the Caribbean. James McClellan summarized this for pioneering Saint-Domingue: "Utility and the promise of utility constituted the reasons why science and medicine became enlisted in the service of colonial development."[45] Subsequent historical research has not come up with fundamentally different conclusions. Neither does this chapter on the role of science in and for the Dutch West Indies.

Notes

1. Gert Oostindie and Jessica Vance Roitman, "Repositioning the Dutch in the Atlantic, 1680–1800," *Itinerario* 36 (2012): 129–60.
2. The scholarly literature on Dutch Brazil is abundant and has found its way to numerous well-kept websites as well, including full-text facsimiles of the *Historia Naturalis Brasiliae.* On scholarship in Dutch Brazil, see Harold J. Cook, *Matters of Exchange: Commerce, Medicine, and Science in the Dutch Golden Age* (New Haven/London: Yale University Press, 2007), 210–25.
3. Th.E. Wong et al., ed., *The History of Earth Sciences in Suriname* (Amsterdam: Royal Netherlands Academy of Arts and Sciences/ Netherlands Institute of Applied Geoscience, TNO, 1998).
4. J. H. Westermann, "Geologisch onderzoek van de Nederlandse Antillen," *Nieuwe West-Indische Gids* 46 (1968): 65–72; Arnold Wentholt, ed., *In kaart gebracht met kapmes en kompas: Met het Koninklijk Nederlands Aardrijkskundig Genootschap op expeditie tussen 1873 en 1960* (Heerlen/Utrecht: APB/KNAG, 2003).
5. J. Voegen van Engelen, *De Surinaamsche artz 1786–1788: Facsimile met een inleiding van Prof. Dr. G.A. Lindeboom* (Utrecht: Natuurwetenschappelijke Studiekring voor Suriname en de Nederlandse Antillen, 1981 [1788]), 24; E. van der Kuyp, *Surinaamse medische en paramedische kroniek, tijdvak 1494–1899* (Paramaribo: Universiteit van Suriname, Faculteit der Medische Wetenschappen, 1985), 33.
6. Cook, *Matters of Exchange,* 332–38; Walter H. Lack, "Maria Sibylla Merian: The Metamorphosis of Insects," in *The Great Naturalists,* ed. Robert Huxley (London: Thames & Hudson, 2007), 118–23; Ella Reitsma, *Maria Sibylla Merian & dochters: Vrouwenlevens tussen kunst en wetenschap* (Zwolle: Waanders, 2008).
7. Van der Kuyp, *Surinaamse medische en paramedische kroniek,* 12; C. F. A. Bruijning and J. Voorhoeve, eds., *Encyclopedie van Suriname* (Amsterdam: Elsevier, 1977), 366.
8. Jeremy Hance, "Carl Linnaeus's Forgotten Apostle Rediscovered: An Ecological Account of 18th Century Suriname," last revised August

11, 2008. http://news.mongabay.com/2008/0811-hance_rolander.html. Stephanie Pain, "The Forgotten Apostle," *New Scientist* 195 (August 4, 2007): 41–45; Tinde van Andel, Paul Maas, and James Dobreff, "Ethnobotanical Notes from Daniel Rolander's *Diarium Surinamicum* (1754–1756): Are These Plants Still Used in Suriname Today?" *Taxon* 61 (2012): 852–63.

9. H. D. Benjamins and Joh. F. Snelleman, eds., *Encyclopaedie van Nederlandsch West-Indië* (The Hague: Nijhoff, 1914), 595; Van der Kuyp, *Surinaamse medische en paramedische kroniek*, 15; Gert Oostindie and Emy Maduro, *In het land van de overheerser II: Antillianen en Surinamers in Nederland 1634/1667–1954* (Dordrecht: Foris, 1986), 109–10; Frank Dragtenstein, *"Trouw aan de blanken": Quassie van Nieuw Timotibo, twist en strijd in de 18de eeuw in Suriname* (Amsterdam: KIT Publishers, 2004); Stephen Snelders, *Vrijbuiters van de heelkunde: Op zoek naar medische kennis in de tropen 1600–1800* (Amsterdam: Atlas, 2012), 193–212.
10. A. L. Stoffers, "Botanisch onderzoek van de Nederlandse Antillen," *Nieuwe West-Indische Gids* 46 (1968): 73–89.
11. Thomas E. Penard, "Historical Sketch of the Ornithology of Surinam," *West-Indische Gids* 6–7 (1925): 145–68.
12. Alfred W. Crosby Jr., *The Columbian Exchange: Biological and Cultural Consequences of 1492* (Westport, CT: Greenwood Press, 1972); *Ecological Imperialism: The Biological Expansion of Europe, 900–1900* (Cambridge: Cambridge University Press, 1986).
13. My translation from G. F. Raynal, *Histoire philosophique et politique des établissements et du commerce des Européens dans les deux Indes* (Amsterdam, 1774), IV, 336. On the sugar and slavery nexus, see Peter Boomgaard and Gert Oostindie, "Changing Sugar Technology and the Labour Nexus: The Caribbean, 1750–1900," *Nieuwe West-Indische Gids/New West Indian Guide* 63 (1989): 3–22. On Surinamese plantations and technological innovation, including polders, see Karel Davids, "Sources of Technological Change in the Dutch Guiana, c. 1670–1860," in A. *Mundialización de la Ciencia y Cultura Nacional*, ed. Lafuente, A. Elena, and M. L. Ortega (Madrid: Universidad Autónoma de Madrid, 1993), 659–71; Gert Oostindie, *Roosenburg en Mon Bijou: Twee Surinaamse plantages, 1720–1870* (Dordrecht: Foris, 1989); Gert Oostindie and Alex van Stipriaan, "Slavery and Slave Cultures in a Hydraulic Society: Suriname," in *Slavery and Slave Cultures in the Americas*, ed. Stephan Palmié (Knoxville: University of Tennessee Press, 1995), 78–99; and Alex van Stipriaan, *Surinaams contrast: Roofbouw en overleven in een Caraïbische plantagekolonie 1750–1863* (Leiden: KITLV Uitgeverij, 1993).
14. B. W. Higman, *Slave Populations of the British Caribbean 1807–1834* (Baltimore: The John Hopkins University Press, 1984), 272; see also Michael Craton, *Searching for the Invisible Man: Slaves and Plantation Life in Jamaica* (Cambridge: Harvard University Press,

1978), 128–31; Oostindie, *Roosenburg*, 139–49; A. M. G. Rutten, *Apothekers en chirurgijns: Gezondheidszorg op de Benedenwindse eilanden van de Nederlandse Antillen in de negentiende eeuw* (Assen: Van Gorcum, 1989), 59.

15. Snelders, *Vrijbuiters*, 213–16.
16. D. J. P. Arrias and E. van der Kuyp, "De historie van de quarantainedienst in Suriname," *Vox Guyanae* 3 (1958): 79–86; Van der Kuyp, *Surinaamse medische en paramedische kroniek;* J.Ph. de Palm, ed., *Encyclopedie van de Nederlandse Antillen* (Zutphen: Walburg Pers, 1985), 169; Rutten, *Apotheker;* L. W. Statius van Eps and E. Luckman-Maduro, eds., *Van scheepschirurgijn tot specialist: 333 jaar Nederlands-Antilliaanse geneeskunde* (Assen: Van Gorcum, 1973); J. Voegen van Engelen, *De Surinaamsche artz 1786–1788: Facsimile met een inleiding van Prof. Dr. G.A. Lindeboom* (Utrecht: Natuurwetenschappelijke Studiekring voor Suriname en de Nederlandse Antillen, 1981), 77–84.
17. Research in Suriname would play a significant role in the search for the explanation of lepra. See H. E. Menke, R. J. B. Wille, W. R. Faber, and T. Pieters, "Bijdragen van Nederland en zijn koloniën aan de kennis over de oorzaak van lepra in de 19e eeuw," *Nederlands Tijdschrift voor Geneeskunde: Tevens orgaan der Koninklijke Nederlandse Maatschappij ter Bevordering der Geneeskunst* 151 (2007): 825–30.
18. Statius van Eps and Luckman-Maduro, *Van scheepschirurgijn tot specialist*, 10, 41–42. Rutten, *Apothekers en chirurgijns*, 117.
19. Higman, *Slave Populations*, 27; Van der Kuyp, *Surinaamse medische en paramedische kroniek*, 24–28; Rutten 1989, 117–20.
20. A. M. G. Rutten, "Van meesterwinkel tot botanica: Drie eeuwen farmacie op de Nederlandse Antillen," *Pharmaceutisch Weekblad* 126 (1991): 345, and "Nederlandse invloed in de historie van de Curaçaose gezondheidszorg 1634–1900," *Pharmaceutisch Weekblad* 26 (1991): 392; Benjamins and Snelleman, *Encyclopaedie*, 451; Van der Kuyp, *Surinaamse medische en paramedische kroniek*, 17.
21. For an extensive discussion on images of enslaved Africans, slavery, race, and Christianization, see Gert Oostindie, "Same Old Song? Perspectives on Slavery and Slaves in Suriname and Curaçao," in Gert Oostindie, ed., *Fifty Years Later: Antislavery, Capitalism and Modernity in the Dutch Orbit* (Pittsburgh: University of Pittsburgh Press, 1996), 143–78.
22. Likewise, we will see that well into the 1830s and even 1840s, even after the British and French Emancipation acts, a majority of the authors in the Dutch West Indian colonies refuse to consider the ending of slavery or, facing the inevitable, will come up with all kinds of proposals aiming at postponement.
23. Gert Oostindie, ed., *Fifty Years Later*; A. N. Paasman, *Reinhart: Nederlandse literatuur en slavernij ten tijde van de verlichting* (Leiden: Nijhoff, 1984); Angelie Sens, "*Mensaap, heiden, slaaf*": *Nederlandse visies op de wereld rond 1800* (Den Haag: SDU Uitgevers, 2001). The

early physical-anthropological work by Petrus Camper, soon to be used in European racist pseudo-science, apparently went unheeded in the Dutch West Indies; see Allison Blakely, *Blacks in the Dutch World: The Evolution of Racial Imagery in a Modern Society* (Bloomington: Indiana University Press, 1993), 185–87.

24. For example, Michiel van Groesen, "Officers of the West India Company, Their Networks, and Their Personal Memories of Dutch Brazil," in *The Dutch Trading Companies as Knowledge Networks*, ed. Siegfried Huigen, Jan de Jong, and Elmer Kolfin (Leiden: Brill, 2010), 50–51, 55.
25. Wim Klooster and Gert Oostindie, ed., *Curaçao in the Age of Revolutions, 1795–1800* (Leiden: KITLV Press, 2011).
26. David Nassy, et al., *Geschiedenis der kolonie van Suriname: Behelzende derzelver opkomst, voortgang, burgerlyke en staatkundige gesteldheid, tegenwoordigen staat van koophandel, en eene volledige en naauwkeurige beschryving van het land, de zeden en gebruiken der ingezetenen* (Amsterdam: Allard en van der Plaat, 1791), vol. 1, 3, my translation.
27. Oostindie and Maduro, *Land van de overheerser*, 29–30, 171–72.
28. The first lodge on Statia was founded as early as 1747 through British intervention. De Palm, *Encyclopedie*, 497–99; Wim Rutgers, "Schrijven is zilver, spreken is goud: Oratuur, auratuur en literatuur van de Nederlandse Antillen en Aruba." PhD diss., Utrecht University, 1994,74–75.
29. Bruijning and Voorhoeve, *Encyclopedie*, 656–58; Lila Gobardhan-Rambocus, *Onderwijs als sleutel tot maatschappelijke vooruitgang: Een taal- en onderwijsgeschiedenis van Suriname, 1651–1975* (Zutphen: Walburg Pers, 2001), 45–46; Van Kempen, *Een geschiedenis* (2003), vol.1, 255, 268–69 and vol. 2, 338–39; De Palm, *Encyclopedie*, 498.
30. Quotation from Michiel van Kempen, *Een geschiedenis van de Surinaamse literatuur*, 4 vols. (Breda: De Geus, 2003), vol. 2, 339, my translation. The Dutch mother institution of the Maatschappij was founded in 1784 and soon had departments throughout the country. See Benjamins and Snelleman, *Encyclopaedie*, 455–56; Michiel van Kempen, *Een geschiedenis van de Surinaamse literatuur. V: Bijlagen, lijst van geraadpleegde archieven, bibliografie, summary, résumé* (Paramaribo: Okopipi, 2002), 39–43, and *Geschiedenis*, vol. 2, 339–41.
31. Robert Cohen, *Jews in Another Environment: Surinam in the Second Half of the Eighteenth Century* (Leiden: Brill, 1991), 73, 175; Gordon K Lewis, *Main Currents in Caribbean Thought: The Historical Evolution of Caribbean Society in Its Ideological Aspects 1492–1900* (Baltimore: The John Hopkins University Press, 1983), 97; Nassy, *Geschiedenis*, 1791.
32. Eensgezindheid, *Verzameling van uitgezochte verhandelingen, betreffende den landbouw in de kolonie Suriname; Opgesteld door het Landbouwkundig Genootschap: De Eensgezindheid, gevestigd in de devisie*

Matappika, binnen dezelve kolonie (Amsterdam: Gartman & Uylenbroek, 1804).

33. Bruijning and Voorhoeve, *Encyclopedie,* 67, 235, 473; Davids, "Sources,", 665; Gobardhan-Rambocus, *Onderwijs,* 45; Van Kempen, *Geschiedenis,* vol. 1, 267, 269, 274 and vol. 2, 331; Voegen van Engelen, *De Surinaamsche artz.*
34. Gobardhan-Rambocus, *Onderwijs,* 38, 41, 43; Van Kempen, *Geschiedenis,* vol. 5, 35–38.
35. Monique Alberts-Luijdjens, "Historical Development of Libraries in the Netherlands Antilles and Aruba," in *Caribbean Libraries in the 21st Century: Changes, Challenges, and Choices,* ed. Cheryl Peltier-Davis and Shamin Renwick (Medford, NJ: Information Today, 2007), 31–40; Rutgers, "Schrijven is zilver," 78–81, 84. In 1792, Edward Luther Low, the first owner of a printing press and editor of *The Sint-Eustatia Press,* offered 36 titles in 159 volumes for sale. These works were all in English, mainly history, literature, and travel accounts. A later book offered by another printer (1812) included encyclopedias and the *Dictionaire* of the Académie française (Rutgers, "Schrijven is zilver," 66–67).
36. Benjamins and Snelleman, *Encyclopaedie,* 139; Cohen, *Jews in Another Environment,* 97, 106–23, 181–251; Gobardhan-Rambocus, *Onderwijs,* 44, 47; Van Kempen, *Geschiedenis,* vol. 1, 255, 262–64 and vol. 2, 325–29; Nassy, *Geschiedenis,* vol. 2, 69. Public libraries were only established after 1800.
37. Cohen, *Jews in Another Environment,* 102, 119; Nathalie Zemon Davis, "David Nassy's "Furlough" and the Slave Matthaeus," in *New Essays in American Jewish History Commemorating the Sixtieth Anniversary of the Founding of the American Jewish Archives,* ed. Pamela S. Nadell et al. (Cincinnati: The American Jewish Archives, 2010), 79–94; Gobardhan-Rambocus, *Onderwijs,* 42–43; Van Kempen, *Geschiedenis,* vol. 1, 261–63; Nassy, *Geschiedenis;* Ineke Phaf-Rheinberger, *The "Air of Liberty": Narratives of the South Atlantic Past* (Amsterdam: Rodopi, 2008), 66–67.
38. Adriaan François Lammens, *Bijdragen tot de kennis van de Kolonie Suriname [...] tijdvak 1816 tot 1822,* ed. G. A. de Bruijne (Amsterdam/Leiden: VU/KITLV, 1982), 72, 94–95.
39. Siegfried Huigen, "Introduction," in *The Dutch Trading Companies as Knowledge Networks,* ed. Siegfried Huigen, Jan de Jong, and Elmer Kolfin (Leiden: Brill, 2010), 8, 11; Edwin van Meerkerk, "Colonial Objects and the Display of Power: The Curious Case of the Cabinet of William and the Dutch India Companies," in *The Dutch Trading Companies as Knowledge Networks,* ed. Siegfried Huigen, Jan de Jong, and Elmer Kolfin, (Leiden: Brill, 2010), 415–35.
40. Lewis, *Main Currents,* 109, 327; David Brion Davis, *Slavery and Human Progress* (New York: Oxford University Press, 1984), 80.

41. On the history of the KNAW, see Klaas van Berkel, *De stem van de wetenschap: Geschiedenis van de Koninklijke Nederlandse Akademie van Wetenschappen. Deel 1, 1808–1914* (Amsterdam: Bert Bakker, 2008). On the history of the KITLV, now an institute of the KNAW, see Maarten Kuitenbrouwer, *Tussen oriëntalisme en wetenschap: Het Koninklijk Instituut voor Taal-, Land- en Volkenkunde in historisch verband, 1851–2001* (Leiden: KITLV Uitgeverij, 2001). One illustration may suffice. The KITLV was founded in 1851 by the prominent former governor-general and minister of colonies J. C. Baud and two scholars working on Indonesia, Taco Roorda and Gerrit Simons; the latter were both appointed members of the KNAW. Several more nineteenth-century scholars working on Asia are mentioned by Van Berkel, *De stem* (e.g., J. H. C. Kern, C. van Vollenhoven). Only two members with some West Indian interest are mentioned, not because of their West Indian connection. One is G. J. Mulder, a chemist with one short publication on the nutrition of slaves in Suriname, the other only a corresponding member, J. C. Rijk, once governor to Suriname and afterward minister of the Marine. Baud's successor Johannes van den Bosch was also a corresponding member of the academy; again, his main interest was in the East Indies.
42. On the Spanish Americas, see Iris H. W. Engstrand, *Spanish Scientists in the New World: The Eighteenth-Century Expeditions* (Seattle: University of Washington Press, 1981) and, particularly on the Spanish Caribbean, Stuart McCook, *States of Nature: Science, Agriculture, and Environment in the Spanish Caribbean, 1760–1940* (Austin: University of Texas Press, 2002). On Saint-Domingue, see James E. McClellan, *Colonialism and Science: Saint Domingue in the Old Regime* (Baltimore: The John Hopkins University Press, 1992). On the British Empire and particularly the West Indies, see Alan G. Cobley, "The Historical Development of Higher Education in the Anglophone Caribbean," in *Higher Education in the Caribbean: Past, Present and Future Directions*, ed. Glenford D. Howe (Barbados etc.: The University of the West Indies Press, 2000), 1–23; Richard Drayton, *Nature's Government: Science, Imperial Britain, and the "Improvement" of the World* (New Haven: Yale University Press, 2000); and Richard H. Grove, *Green Imperialism: Colonial Expansion, Tropical Island Edens and the Origins of Environmentalism, 1600–1860* (Cambridge: Cambridge University Press, 1995). Codrington College was the only such institution in the British West Indies until the founding, in 1921, of the Imperial College of Tropical Agriculture in Trinidad (Cobley, "The Historical Development," 2).
43. Cook, *Matters of Exchange*.
44. Kruijtzer, "European Migration."
45. McClellan, *Colonialism and Science*, 289.

Bibliography

Alberts-Luijdjens, Monique. "Historical Development of Libraries in the Netherlands Antilles and Aruba." In *Caribbean Libraries in the 21st Century: Changes, Challenges, and Choices*, edited by Cheryl Peltier-Davis and Shamin Renwick, 31–40. Medford, NJ: Information Today, 2007.

Andel, Tinde van, Paul Maas, and James Dobreff. "Ethnobotanical Notes from Daniel Rolander's *Diarium Surinamicum* (1754–1756): Are These Plants Still Used in Suriname Today?" *Taxon* 61 (2012): 852–63.

Arrias, D. J. P. and E. van der Kuyp. "De historie van de quarantainedienst in Suriname." *Vox Guyanae* 3 (1958): 79–86.

Benjamins, H. D. and Joh. F. Snelleman, eds. *Encyclopaedie van Nederlandsch West-Indië*. The Hague: Nijhoff, 1914.

Berkel, Klaas van. *De stem van de wetenschap: Geschiedenis van de Koninklijke Nederlandse Akademie van Wetenschappen. Deel 1, 1808–1914*. Amsterdam: Bert Bakker, 2008.

Blakely, Allison. *Blacks in the Dutch World: The Evolution of Racial Imagery in a Modern Society*. Bloomington: Indiana University Press, 1993.

Boomgaard, Peter and Gert Oostindie. "Changing Sugar Technology and the Labour Nexus: The Caribbean, 1750–1900." *Nieuwe West-Indische Gids/New West Indian Guide* 63 (1989): 3–22.

Bontius, J. *Oost- en West-Indische warande, vervattende aldaar de leef- en geneeskonst, enz: Hier nevens is bijgevoegt de nieuw verbeterde chirurgijns scheepskist*. Amsterdam: Jan ten Horn, 1694.

Bruijning, C. F. A. and J. Voorhoeve, eds. *Encyclopedie van Suriname*. Amsterdam: Elsevier, 1977.

Cobley, Alan G. "The Historical Development of Higher Education in the Anglophone Caribbean." In *Higher Education in the Caribbean: Past, Present and Future Directions*, edited by Glenford D. Howe, 1–23. Barbados, etc.: The University of the West Indies Press, 2000.

Cohen, Robert. *Jews in Another Environment: Surinam in the Second Half of the Eighteenth Century*. Leiden: Brill, 1991.

Cook, Harold J. *Matters of Exchange: Commerce, Medicine, and Science in the Dutch Golden Age*. New Haven/London: Yale University Press, 2007.

Craton, Michael. *Searching for the Invisible Man: Slaves and Plantation Life in Jamaica*. Cambridge: Harvard University Press, 1978.

Crosby Jr., Alfred W. *The Columbian Exchange: Biological and Cultural Consequences of 1492*. Westport, CT: Greenwood Press, 1972.

———. *Ecological Imperialism: The Biological Expansion of Europe, 900–1900*. Cambridge: Cambridge University Press, 1986.

Davids, Karel. "Sources of Technological Change in the Dutch Guiana, c. 1670–1860." In *Mundialización de la Ciencia y Cultura Nacional*, edited by A. Lafuente, A. Elena, and M. L. Ortega, 659–71. Madrid: Universidad Autónoma de Madrid, 1993.

Davis, David Brion. *Slavery and Human Progress*. New York: Oxford University Press, 1984.

Doerffel, Carel. *Aanmerkingen over de in Curaçao van tyd tot tyd in zwang gaande ziekten*. Haarlem: J. Bosch, 1758.

Dragtenstein, Frank. *"Trouw aan de blanken": Quassie van Nieuw Timotibo, twist en strijd in de 18de eeuw in Suriname*. Amsterdam: KIT Publishers, 2004.

Drayton, Richard. *Nature's Government: Science, Imperial Britain, and the "Improvement" of the World*. New Haven: Yale University Press, 2000.

Eensgezindheid. *Verzameling van uitgezochte verhandelingen, betreffende den landbouw in de kolonie Suriname; Opgesteld door het Landbouwkundig Genootschap: De Eensgezindheid, gevestigd in de devisie Matappika, binnen dezelve kolonie*. Amsterdam: Gartman & Uylenbroek, 1804.

Engstrand, Iris H. W. *Spanish Scientists in the New World: The Eighteenth-Century Expeditions*. Seattle: University of Washington Press, 1981.

Fermin, Philippe. *Traité des maladies les plus fréquentes à Surinam, et des remèdes les plus propres à les guerir*. Maastricht, 1764.

Gallandat, D. H. *Noodige onderrigtingen voor de slaafhandelaaren*. Middelburg: Gillissen, 1769.

Gobardhan-Rambocus, Lila. *Onderwijs als sleutel tot maatschappelijke vooruitgang: Een taal- en onderwijsgeschiedenis van Suriname, 1651–1975*. Zutphen: Walburg Pers, 2001.

Goodman, David C. *Power and Penury: Government, Technology and Science in Philip II's Spain*. Cambridge: Cambridge University Press, 1988.

Groesen, Michiel van. "Officers of the West India Company, Their Networks, and Their Personal Memories of Dutch Brazil." In *The Dutch Trading Companies as Knowledge Networks*, edited by Siegfried Huigen, Jan de Jong, and Elmer Kolfin, 39–58. Leiden: Brill, 2010.

Grove, Richard H. *Green Imperialism: Colonial Expansion, Tropical Island Edens and the Origins of Environmentalism, 1600–1860*. Cambridge: Cambridge University Press, 1995.

Higman, B. W. *Slave Populations of the British Caribbean 1807–1834*. Baltimore: The John Hopkins University Press, 1984.

Huigen, Siegfried. "Introduction." In *The Dutch Trading Companies as Knowledge Networks*, edited by Siegfried Huigen, Jan de Jong, and Elmer Kolfin, 1–16. Leiden: Brill, 2010.

Kempen, Michiel van. *Een geschiedenis van de Surinaamse literatuur. V: Bijlagen, lijst van geraadpleegde archieven, bibliografie, summary, résumé*. Paramaribo: Okopipi, 2002.

———. *Een geschiedenis van de Surinaamse literatuur*. 4 vols. Breda: De Geus, 2003.

Klooster, Wim and Gert Oostindie, eds. *Curaçao in the Age of Revolutions, 1795–1800*. Leiden: KITLV Press, 2011.

Kruijtzer, Gijs. "European Migration in the Dutch Sphere." In *Dutch Colonialism, Migration and Cultural Heritage*, edited by Gert Oostindie, 97–154. Leiden: KITLV Press, 2008.

Kuhn, F.A. *Beschouwing van den toestand der Surinaamsche plantagie-slaven: Eene oeconomisch-geneeskundige bijdrage tot verbetering deszelven*. Amsterdam: Sulpke, 1828.

Kuitenbrouwer, Maarten. *Tussen oriëntalisme en wetenschap: Het Koninklijk Instituut voor Taal-, Land- en Volkenkunde in historisch verband, 1851–2001*. Leiden: KITLV Uitgeverij, 2001.

Kuyp, E. van der. *Surinaamse medische en paramedische kroniek, tijdvak 1494–1899*. Paramaribo: Universiteit van Suriname, Faculteit der Medische Wetenschappen, 1985. [Special issue of *Surinaams Medisch Bulletin* 9, 2 (1985).]

Lack, Walter H. "Maria Sibylla Merian: The Metamorphosis of Insects." In *The Great Naturalists*, edited by Robert Huxley, 118–23. London: Thames & Hudson, 2007.

Lammens, Adriaan François. *Bijdragen tot de kennis van de Kolonie Suriname [...] tijdvak 1816 tot 1822*. Edited by G. A. de Bruijne. Amsterdam/Leiden: VU/KITLV, 1982.

Lewis, Gordon K. *Main Currents in Caribbean Thought: The Historical Evolution of Caribbean Society in Its Ideological Aspects 1492–1900*. Baltimore: The John Hopkins University Press, 1983.

McCook, Stuart. *States of Nature: Science, Agriculture, and Environment in the Spanish Caribbean, 1760–1940*. Austin: University of Texas Press, 2002.

McClellan, James E. *Colonialism and Science: Saint Domingue in the Old Regime*. Baltimore: The John Hopkins University Press, 1992.

Meerkerk, Edwin van. "Colonial Objects and the Display of Power: The Curious Case of the Cabinet of William and the Dutch India Companies." In *The Dutch Trading Companies as Knowledge Networks*, edited by Siegfried Huigen, Jan de Jong, and Elmer Kolfin, 415–35. Leiden: Brill, 2010.

Menke, H. E., R. J. B. Wille, W. R. Faber, and T. Pieters, "Bijdragen van Nederland en zijn koloniën aan de kennis over de oorzaak van lepra in de 19e eeuw." *Nederlands Tijdschrift voor Geneeskunde: Tevens orgaan der Koninklijke Nederlandse Maatschappij ter Bevordering der Geneeskunst* 151 (2007): 825–30.

Nassy, David et al. *Geschiedenis der kolonie van Suriname: Behelzende derzelver opkomst, voortgang, burgerlyke en staatkundige gesteldheid, tegenwoordigen staat van koophandel, en eene volledige en naauwkeurige beschryving van het land, de zeden en gebruiken der ingezetenen*. Amsterdam: Allard en van der Plaat, 1791.

———. *Observations on the Cause, Nature, and Treatment of the Epidemic Disorder, Prevalent in Philadelphia*. Philadelphia, 1793.

Oostindie, Gert. *Roosenburg en Mon Bijou: Twee Surinaamse plantages, 1720–1870*. Dordrecht: Foris, 1989.

———, ed. *Fifty Years Later: Antislavery, Capitalism and Modernity in the Dutch Orbit*. Pittsburgh: University of Pittsburgh Press, 1996.

———. "Same Old Song? Perspectives on Slavery and Slaves in Suriname and Curaçao." In *Fifty Years Later: Antislavery, Capitalism and Modernity in the Dutch Orbit*, edited by Gert Oostindie, 147–78. Pittsburgh: University of Pittsburgh Press, 1996.

Oostindie, Gert and Alex van Stipriaan. "Slavery and Slave Cultures in a Hydraulic Society: Suriname." In *Slavery and Slave Cultures in the Americas*, edited by Stephan Palmié, 78–99. Knoxville: University of Tennessee Press, 1995.

Oostindie, Gert and Emy Maduro. *In het land van de overheerser II: Antillianen en Surinamers in Nederland 1634/1667—1954*. Dordrecht: Foris, 1986.

Oostindie, Gert and Jessica Vance Roitman. "Repositioning the Dutch in the Atlantic, 1680–1800." *Itinerario* 36 (2012): 129–60.

Paasman, A. N. *Reinhart: Nederlandse literatuur en slavernij ten tijde van de verlichting*. Leiden: Nijhoff, 1984.

Pain, Stephanie. "The Forgotten Apostle." *New Scientist* 195 (August 4, 2007): 41–45.

Palm, J. Ph. de, ed. *Encyclopedie van de Nederlandse Antillen*. Zutphen: Walburg Pers, 1985.

Penard, Thomas E. "Historical Sketch of the Ornithology of Surinam." *West-Indische Gids* 6–7 (1925): 145–68.

Phaf-Rheinberger, Ineke. *The "Air of Liberty": Narratives of the South Atlantic Past*. Amsterdam: Rodopi, 2008.

Raynal, G. F. *Histoire philosophique et politique des établissements et du commerce des Européens dans les deux Indes*. Amsterdam, 1774.

Reitsma, Ella. *Maria Sibylla Merian & dochters: Vrouwenlevens tussen kunst en wetenschap*. Zwolle: Waanders, 2008.

Rutgers, Wim. "Schrijven is zilver, spreken is goud: Oratuur, auratuur en literatuur van de Nederlandse Antillen en Aruba." PhD diss., Utrecht University, 1994.

Rutten, A. M. G. *Apothekers en chirurgijns: Gezondheidszorg op de Benedenwindse eilanden van de Nederlandse Antillen in de negentiende eeuw*. Assen: Van Gorcum, 1989.

———. "Nederlandse invloed in de historie van de Curaçaose gezondheidszorg 1634–1900." *Pharmaceutisch Weekblad* 26 (1991): 390–400.

———. "Van meesterwinkel tot botanica: Drie eeuwen farmacie op de Nederlandse Antillen." *Pharmaceutisch Weekblad* 126 (1991): 344–47.

Schilling, G. W. "Verhandeling over de melaatschheid. PhD diss., Utrecht University, 1769.

Schmidt, Benjamin. *Innocence Abroad: The Dutch Imagination and the New World, 1570–1670*. Cambridge: Cambridge University Press, 2001.

Sens, Angelie. "*Mensaap, heiden, slaaf*": *Nederlandse visies op de wereld rond 1800*. Den Haag: SDU Uitgevers, 2001.

Snelders, Stephen. *Vrijbuiters van de heelkunde: Op zoek naar medische kennis in de tropen 1600–1800*. Amsterdam: Atlas, 2012.

Statius van Eps, L. W. and E. Luckman-Maduro, eds. *Van scheepschirurgijn tot specialist: 333 jaar Nederlands-Antilliaanse geneeskunde*. Assen: Van Gorcum, 1973.

Stipriaan, Alex van. *Surinaams contrast: Roofbouw en overleven in een Caraïbische plantagekolonie 1750–1863*. Leiden: KITLV Uitgeverij, 1993.

Stoffers, A. L. "Botanisch onderzoek van de Nederlandse Antillen." *Nieuwe West-Indische Gids* 46 (1968): 73–89.

Titsingh, Abraham. *Geneeskonst der heelmeesters ten dienste van de zeevaart.* Amsterdam: Adriaan Wor, 1752.

Voegen van Engelen, J. *De Surinaamsche artz 1786–1788: Facsimile met een inleiding van Prof. Dr. G.A. Lindeboom.* Utrecht: Natuurwetenschappelijke Studiekring voor Suriname en de Nederlandse Antillen, 1981 [1788].

Wentholt, Arnold, ed. *In kaart gebracht met kapmes en kompas: Met het Koninklijk Nederlands Aardrijkskundig Genootschap op expeditie tussen 1873 en 1960.* Heerlen/Utrecht: APB/KNAG, 2003.

Westermann, J. H. "Geologisch onderzoek van de Nederlandse Antillen." *Nieuwe West-Indische Gids* 46 (1968): 65–72.

Wong Th.E. et al., ed. *The History of Earth Sciences in Suriname.* Amsterdam: Royal Netherlands Academy of Arts and Sciences/Netherlands Institute of Applied Geoscience, TNO, 1998.

Zemon Davis, Nathalie. "David Nassy's "Furlough" and the Slave Matthaeus." In *New Essays in American Jewish History Commemorating the Sixtieth Anniversary of the Founding of the American Jewish Archives,* edited by Pamela S. Nadell et al., 79–94. Cincinnati: The American Jewish Archives, 2010.

Index

Note: Locators followed by 't' denotes tables.

The manufacturer's authorised representative in the EU is Springer Nature Customer Service Centre GmbH, Europaplatz 3, 69115 Heidelberg, Germany. If you have any concerns regarding our products, please contact ProductSafety@springernature.com

Printed and bound by CPI Group (UK) Ltd, Croydon, CR0 4YY
13/10/2026
02232375-0010